Chemistry of the Amazon

ACS SYMPOSIUM SERIES **588**

Chemistry of the Amazon

Biodiversity, Natural Products, and Environmental Issues

Peter Rudolf Seidl, EDITOR
Centro de Tecnologia Mineral (CETEM)

Otto Richard Gottlieb, EDITOR
Instituto Oswaldo Cruz (FIOCRUZ)

Maria Auxiliadora Coelho Kaplan, EDITOR
Universidade Federal do Rio de Janeiro

Developed from the First International Symposium on Chemistry
and the Amazon sponsored by the Associação Brasileira de Química,
American Chemical Society, Centro de Tecnologia Mineral,
and Instituto Nacional de Pesquisas da Amazonia
Manaus, Amazonas, Brazil,
November 21–25, 1993

American Chemical Society, Washington, DC 1995

Library of Congress Cataloging-in-Publication Data

Chemistry of the Amazon: biodiversity, natural products, and environmental issues / Peter Rudolf Seidl, editor, Otto Richard Gottlieb, editor, Maria Auxiliadora Coelho Kaplan, editor.

p.   cm.—(ACS symposium series; 588)

"Developed from the First International Symposium on Chemistry and the Amazon sponsored by the Associação Brasileira de Química, American Chemical Society, Centro de Tecnologia Mineral, and Instituto Nacional de Pesquisas da Amazonia, Manaus, Amazonas, Brazil, November 21–25, 1993."

Includes bibliographical references and index.

ISBN 0–8412–3159–1

1. Biogeochemistry—Amazon River Region—Congresses. 2. Natural products—Amazon River Region—Congresses. 3. Biological diversity—Amazon River Region—Congresses.   I. Seidl, Peter Rudolf, 1941–   . II. Gottlieb, Otto Richard.   III. Kaplan, Maria Auxiliadora C. IV. Associação Brasileira de Química. V. International Symposium on Chemistry and the Amazon (1st: 1993: Amazonas, Brazil) VI. Series.

QH343.7.C48   1995
574.5'0981'1—dc20   95–5473
                                                                     CIP

This book is printed on acid-free, recycled paper.

PRINTED IN THE UNITED STATES OF AMERICA

1995 Advisory Board

ACS Symposium Series

M. Joan Comstock, *Series Editor*

Foreword

THE ACS SYMPOSIUM SERIES was first published in 1974 to provide a mechanism for publishing symposia quickly in book form. The purpose of this series is to publish comprehensive books developed from symposia, which are usually "snapshots in time" of the current research being done on a topic, plus some review material on the topic. For this reason, it is necessary that the papers be published as quickly as possible.

Before a symposium-based book is put under contract, the proposed table of contents is reviewed for appropriateness to the topic and for comprehensiveness of the collection. Some papers are excluded at this point, and others are added to round out the scope of the volume. In addition, a draft of each paper is peer-reviewed prior to final acceptance or rejection. This anonymous review process is supervised by the organizer(s) of the symposium, who become the editor(s) of the book. The authors then revise their papers according to the recommendations of both the reviewers and the editors, prepare camera-ready copy, and submit the final papers to the editors, who check that all necessary revisions have been made.

As a rule, only original research papers and original review papers are included in the volumes. Verbatim reproductions of previously published papers are not accepted.

M. Joan Comstock
Series Editor

Contents

Preface

NEWS FROM THE FIRST RECORDED EXPEDITION down the Amazon reached Europe in the mid-16th century, and since then this mysterious and fascinating region has been associated with many different myths. Scientists have helped dispel or clarify many of these myths, but some large-scale projects have clearly shown the pitfalls that await those who try to explore the area without knowing enough about its many particularities and complexities.

Amazonia has been the source of much interest. Considerable research on its fauna and flora has been published since the beginning of the past century. There has recently been an important change in the focus of this research. Instead of limiting work to characterization and identification of species, considerable attention is now directed toward understanding the ways in which species interact among themselves and with their surroundings. A direct consequence of this trend is the inclusion of the human element in these studies and the realization that research on Amazonia has much to gain if the local inhabitant is made a willing partner. Consideration of measures that lead to his well-being are an essential component of any project aimed at preserving the enormous variety of species that share his habitat through judicious use of its natural resources.

Chemistry plays an important role in understanding the interaction of Amazonian wildlife and teaching humankind how to preserve fauna and flora. It provides a basis for investigating natural phenomena related to competition among species, their spatial radiation, and their response to external factors, and it generates the knowledge required for correct management of natural ecosystems. It may also reveal valuable substances and materials that, if exploited rationally or used as models for synthesis, could provide local populations with an important source of income for many generations.

Extensive research has been carried out on the geochemistry, hydrochemistry, environmental chemistry, and chemistry of natural products of the Amazon region. However, most of these studies were carried out in laboratories far from where the samples originate, and the results, although published in national and international periodicals, rarely found their way back to or aroused the interest of decision-makers responsible for the preservation and rational development of the region. When our association held a national meeting on "Chemistry of the Amazon" just after the Rio Summit in October of 1992, it became readily apparent that economic activities continue to destroy this valuable resource instead of taking advantage of its potential. This situation was taken as a clear indication of the high degree of ignorance about the region and what it has to offer.

Investigators of the geochemistry, hydrochemistry, environmental chemistry, and chemistry of natural products of the Amazon as well as specialists on chemical processes that determine biodiversity were invited to take part in the First International Symposium on Chemistry of the Amazon. Combining and comparing information from all these sources aids examination of alternatives for development of the region with the least interference possible in the natural processes that occur in the ecosystem. This book presents contributions covering prospects for treating biodiversity as a renewable resource, the bioactivity of natural products from the Amazon, their ecology and evolution, and the environmental issues involved in the region's conservation.

Acknowledgments

We are indebted to the contributing authors for their commitment to address their respective subjects within the required deadlines. Maria Nazareth S. Sucupira made an invaluable contribution to the preparation of this book.

The First International Symposium on Chemistry of the Amazon provided an opportunity to bring together the contributors to this volume. We thank its sponsors and the Ministério da Ciência e Tecnologia (Brazilian Science and Technology Ministry) for this important contribution.

PETER RUDOLF SEIDL
Centro de Tecnologia Mineral (CETEM)
Rua 4, Quadra D, Cidade Universitária
21941-590 Rio de Janeiro, RJ, Brazil

OTTO R. GOTTLIEB
Departamento de Fisiologia e Farmacodinâmica
Instituto Oswaldo Cruz (FIOCRUZ)
Avenida Brasil, 4365
21045-900 Rio de Janeiro, RJ, Brazil

MARIA AUXILIADORA COELHO KAPLAN
Núcleo de Pesquisas de Produtos Naturais (NPPN)
Centro de Ciências da Saúde, Bloco H, Cidade Universitária
Universidade Federal do Rio de Janeiro
21941-590 Rio de Janeiro, RJ, Brazil

December 5, 1994

Amazon Biodiversity: A Renewable Natural Resource?

Chapter 1

Amazon Biodiversity:
A Renewable Natural Resource?

Peter Rudolf Seidl

Centro de Tecnologia Mineral, Rua 4, Quadra D, Cidade Universitária,
Ilha do Fundão, CEP 21941–590 Rio de Janeiro, RJ, Brazil

Tropical rainforests contain a disproprotionate amount of the earth's biodiversity. The Brazilian Amazon is probably the most important biotic reserve in the world and provides an opportunity to test the prospects for making conservation of biodiversity an attractive alternative to its inexorable destruction by local communities. This is no easy task and requires new types of relationships with these communities in order to access their knowledge base on the use of the region's biodiversity and evaluate its potential economic value. Chemistry may play an important role in the identification, development and commercialization of valuable natural products (especially these that reveal some form of bioactivity) and in recovery of degraded areas with native species through specialized agroforestry systems as well as in monitoring and evaluating the impacts of any new project on its particular ecosystem.

Tropical forests are known to contain a disproportionately large share of the earth's biodiversity. Estimates of how many kinds of plants and animals inhabit the planet vary considerably, but it is generally agreed upon that, for certain groups, only a small fraction of the probable number of species have been discovered. It is anticipated that most of the species that are yet to be identified inhabit the tropics (1).

Tropical rainforests cover approximately 4% of the earth's surface and are found in more than 70 countries. They are not uniformly distributed, however, and no more than seven countries harbor more than two-thirds of these forests. Brazil alone accounts for just over 30% of this total (2) and its natural products have been investigated in laboratories all over the world (3,4).

The Brazilian Amazon thus seems to be the appropriate place to examine the propects for making the conservation of biodiversity attractive to the societies that harbor it, providing a convenient venue for applying the concepts that were negotiated at the Rio Summit (5). It covers the larger part of the Amazon basin which, with its enormous area and ecosystematic complexity, is probably the most important biotic reserve in the world. Its natural base is formed by many different ecosystems which interact among themselves, establishing complex ecological processes. in terms of dynamics as well as their interdependence (2).

Around 15 million people live in the region today yet, in view of its sheer size, it is still sparsely populated (although this situation is changing rapidly). Some 15% of its economically active population depends on what is called "extractivism", here understood as a general term which includes some environmentally sound practices and others which are extremely aggressive to local ecosystems. Among the latter are small-scale mining ("garimpo"), a direct threat to important river systems and the people who live off them, and deforestation, a process which has already claimed 8% (40 million ha) of the tropical forest. In spite of a number of environmental protection measures recently adopted, the destruction of forests and river habitats of the Amazon will inevitably continue unless the type of demand for its natural resources can be substantially modified.

The traditional economy of the Amazon Region is, to a large extent, dependent on primary activities (mining, agriculture, fishing and forestry) with a low technological content and which do not lead to socio-economical improvement for the region, but which cause considerable environmental degradation. Opportunities for changing this situation to one of sustainable development and conservation of the region are centered on the use of its renewable natural resources. The most promising alternatives are based on verticalization and diversification of the productive strucuture, specially in terms of new industries based on biotechnology and chemistry of natural resources such as biotic materials, pharmaceuticals, cosmetics, and tropical agroindustrial products with high added value. Thus a strategy for sustainable development and conservation of the Amazon requires the creation of new economic sectors, based on science and technology, that take advantage of natural resources employing environmentally friendly procedures and techniques in addition to widely publicized measures such as demarcation of indigenous areas, of parks and reserves, of extractivist reserves, of units for biodiversity conservation, as well as activities related to agroeconomic zoning of the Amazon and to monitoring anthropic effects on its ecosystems, to mention just a few of the projects included under the Pilot Program to Conserve the Brazilian Rainforest, know as the G7 Program.

Commercial exploitation of natural products from the tropical forests may, in fact, be the only way to preserve this habitat and avoid its destruction by local inhabitants in search of conditions for survival or by commercial enterprises after a short to medium-term gain. A number of projects now under way indicate that rational exploitation is possible without permanent damage to the environment and scientific management of degraded areas may accelerate their recovery.

The time is particularly favorable for the development and commercialization of products from tropical forests. The tendency to substitute products of vegetable origin by synthetic chemicals has lost much of its impetus and, in some cases, may even be experiencing a reversal. Natural products have been making a strong return to the market in the past few years and,in certain areas such as food, cosmetics and drugs, there is a marked preference for "natural" ingredients over their synthetic counterparts. An "exotic" connotation that is often associated with tropical products may even allow them to command higher prices.

Any attempt at exploiting products from the rainforest must, of course, be carefully monitored. The biogeochemical cycles that provide conditions for the generation and preservation of Amazon biodiversity have been the object of extensive research in the last two decades. Techniques for analysing anthropic effects caused by population increases in larger cities; building hidroelectric plants: prospecting, drilling and transporting oil; cattle-grazing; slash-and-burn farming and, more recently, small-scale mining ("garimpo") on the Amazon's rivers (water and sediment), air and soil should be applied to any new initiative, and a careful evaluation of its impacts must be an integral part of any new development.

Valuable   Natural   Products: Identification,   Development   and Commercialization

Interest in natural substances that have biological activity is on the rise again as approaches to "ab initio" drug design using sophisticated computational and biochemical techniques turn out to be more complex and not quite as reliable as had been originally anticipated. While major pharmaceutical companies have not abandoned this approach, they are also investing heavily in the collection, identification and preparation of extracts from plants, animals (mostly insects and marine organisms) and microorganisms for screening purposes(6). Recent advances in techniques used for screening have significantly reduced the amount of sample that is required and increased the number of tests that can be run with relatively modest investments in installations and equipment as well as costs for personnel and materials (7). As a result, vigorous programs for acquision of natural products along with novel chemical structures are integrated into drug discovery programs on the part of both industry and mission-oriented government agencies such as the National Institutes of Health (NIH) in the U.S (8).

Although medicinal agents represent the largest and most visible segment of bioactive compounds that are yet to be discovered in tropical forests, they are by no means the only one. Competition among species and development of chemical defenses against predators should provide many molecules to be tested as insecticides and herbicides. While a number of substances found in tropical forests are known to have unique flavors or fragrances and many are already commercially available, the possibility of coming across new compounds with these properties should not be overlooked. The same goes for natural dyes and other of substances that are used in formulations of cosmetics, toiletries, foods and health articles. Two of a major drug company's best-selling products are an artificial sweetener and a herbicide, for example (6).

Some form of bioactivity in an extract is the starting point, but only rarely the sole element, in the process of discovery, development and commercialization of a new compound. Depending on the final application, a promising molecule (or "hit") will go through extensive tests (and possibly chemical modification) before becoming a "lead compound" that would be subjected to clinical trials . A case in point is a promising drug against AIDS developed from a compound isolated from plant materials that was being tested as an insect regulator (the native compound was too toxic to be useful as a drug, but after a number of analogs were synthesized and tested, one of these was shown to have the necessary properties and advanced into the clinic). The plant is no longer necessary to get the substance - it is now made synthetically - but without the plant it would never have been discovered (*6*).

Another important component in the search for bioactive substances is related to genetic resources. Screens should not be limited to small molecules, samples should also be extracted for proteins in which the gene responsible for its action is identified in order be cloned. Genetic engineering has opened up new possibilities of producing variants that have desirable biochemical characteristics and these methods are equally applicable to the genetic manipulation of plants as well as lower animals and microorganisms that accumulate the bioactive compounds of interest.

Plant biotechnology offers additional possibilities for in vitro accumulation of these selected substances (*9*). An integrated strategy for conservation of genetic resources followed by screening for desirable genetic properties as well as biological activity and combined with R&D on genetic engineering/plant biotechnology of selected species should be an integral part of the effort on identification, development and commercialization of bioactive compounds. This type of approach would not only remove some of the uncertainties in prices and supplies of valuable substances but could contribute to reducing the risk of highly detrimental effects on the environment should these substances be commercialized on a large scale.

Sustainable Development

The recovery of areas that have already been degraded by activities such as mining or lumbering is an important testing-ground for agroforestry projects based on species that are native to the region. Such projects can play a significant role in reversing destruction of the rainforest and may even be necessary to complement short to medium-term initiatives that add value to products that are already exported from the region (*10*) since considerable investment and human effort will have to go into biodiversity prospecting activities before any return for local populations can be expected.

The Amazon Region should not be primarily considered a source of raw materials, however. Experience from several projects supported by government agencies as well as some that were run by powerful international concerns raises serious doubts as to the viability of large-scale agricultural projects in the region. Unlike rainforests in other parts of the world, such as Southeast Asia or Central America, soils in the Amazon are poor and largely unable to neutralize the effects of constant weathering (*11*). Domestication of tropical species is essential for the development of any adequate cultivation methods for native plants, but it requires considerable time and research (It is noteworthy that, up to now, only one specie is considered completely domesticated in the Amazon (*12*)). Chances for success in agroforestry are probably closely linked to the possibility of developing specific systems that optimize symbiotic relationships among species that thrive on and protect certain types of soils, as is the case in natural ecosystems. With very few exceptions, this implies use of small-scale, diversified projects.

Many local communities in the Amazon already have a knowledge base that suffices for sustainable management of a very small fraction of the region's biodiversity and have put its bioactive compounds to good use. Access to this knowledge requires developing new types of relationships with people who have their very own standards of living and scales of values. This is, however, no simple task. Alone the elaboration of a comprehensive strategy to face the problem is an extremely complex undertaking (*13*). It covers a gamut of socio-cultural issues that range from the interaction with societies which are quite isolated from the rest of the world (and live under rules and conditions that could be interpreted as very primitive by those who are not aware of their inherent complexities) to negotiations with large corporations whose international business strategies are based on the cutting edge of technology. As these issues span scientific disciplines, technical evaluations, commercial aplications, intellectual property rights and international relations, expertise in many different areas must be combined in order to identify all the relevant aspects of negotiations among interested parties.

Chemistry plays an important role in the analysis of propects for the sustainable use of Amazon biodiversity. Not only does it provide a link to the technical jargon of the disciplines involved; but its practitioners are quite familiar with the interplay between R&D and the marketplace. It is also required to monitor the natural cycles that are responsible for Amazon biodiversity as well as anthropogenic sources of pollution that threaten it.

Literature Cited

1.   Gentry, A. H. In *Human Medicinal Agents from Plants*; Kinghorn, A. D.; Balandrin, M. F., Eds.; ACS Symposium Series 534; American Chemical Society: Washington, DC, 1993; pp 13-24.
2.   Uribe, C. C. *Situation General de la Conservacion de laBiodiversidad en la Region Amazonica: Evaluacion de las Areas Protegidas Propostas y Estrategias*; Proyeto FAO/TCP/RLA/0160; Quito, 1993; pp 19-26.
3.   Mors, W. B.; Rizzini, C. T. *Useful Plants of Brazil*; Holden-Day, San Francisco, 1966.
4.   Ribeiro, M. N. S.; Zoghbi, M. G. B.; Silva, M. L.; Gottlieb, O. R.; Rezende, C. M. M. *Cadastro Fitoquimico Brasileiro*; INPA/FUA; Manaus, AM, 1987.
5.   Roberts, L. *Science* **1992**, *256*, 1142-1143.

6.   Brewer, S. S. J. In *The Use of Biodiversity for Sustainable Development: Investigation of Bioactive Products and their Commercial Applications*; Seidl, P. R., Ed.; Associação Brasileira de Química: Rio de Janeiro, RJ, 1994; pp 37-43.
7.   McChesney, J.D. In *Human Medicinal Agents from Plants*; Kinghorn, A. D.; Balandrin, M. F., Eds.; ACS Symposium Series 534; American Chemical Society: Washington, DC, 1993; pp 38-47.
8.   Cordell, G. A. *Chem. Ind.* **1993**, 841-844.
9.   Charlwood, B. V. In *The Use of Biodiversity for Sustainable Development: Investigation of Bioactive Products and their Commercial Applications*; Seidl, P. R., Ed.; Associação Brasileira de Química: Rio de Janeiro, RJ, 1994; pp 44-52.
10.   Gilbert, B. In *The Use of Biodiversity for Sustainable Development: Investigation of Bioactive Products and their Commercial Applications*; Seidl, P. R., Ed.; Associação Brasileira de Química: Rio de Janeiro, RJ, 1994; pp 12-22.
11.   Hoppe, A. *Natural Resources Forum* **1992**, 232-234.
12.   Miranda Santos, M. In *The Use of Biodiversity for Sustainable Development: Investigation of Bioactive Products and their Commercial Applications*; Seidl, P. R., Ed.; Associação Brasileira de Química: Rio de Janeiro, RJ, 1994; pp 7-11.
13.   Barreto de Castro, L. A. In *The Use of Biodiversity for Sustainable Development: Investigation of Bioactive Products and their Commercial Applications*; Seidl, P. R., Ed.; Associação Brasileira de Química: Rio de Janeiro, RJ, 1994; pp 104-106.

RECEIVED December 5, 1994

Chapter 2

Natural Products as Medicinal and Biological Agents

Potentiating the Resources of the Rain Forest

Geoffrey A. Cordell

Program for Collaborative Research in the Pharmaceutical Sciences, Department of Medicinal Chemistry and Pharmacognosy, College of Pharmacy, University of Illinois at Chicago, Chicago, IL 60612

As we consider the devastation that is wreaked on our planet each day, and how a burgeoning global population and rising economic wealth have literally changed the appearance of the landscape during our short lifetimes, we are constantly reminded that many of the earth's resources are unknown. Perhaps nowhere on Earth is this more true than the Brazilian rainforests. Even at a time when we are exploring the planets and beyond, in truth we know relatively little about the potential of our own, rather small, biologically active corner of the cosmos we call Earth. Whether it is the marine environment, the plant kingdom or even our own biological processes, there are fundamental questions which remain unanswered.

In terms of resources, particularly those which would be both economically useful and renewable in a finite time period, for example 10-20 years, we have barely initiated our quest. Ralph Waldo Emerson, the 19th century American philosopher wrote "What is a weed? A plant whose virtues have not yet been discovered". For "plant" one could substitute insect, marine animal, fungus, etc, the relevance of the comment remains the same. Namely, that the potential is enormous.

Yet, at no other time in the history of humankind on this planet has biodiversity been so low, the population so burgeoning and global economic development so rapid. Thus, demands for more non-renewable energy, more disposable consumer products, more drugs, more biologicals and more insecticides will inevitably increase for the foreseeable future. For mankind to survive far into and beyond the 21st century, there is a growing consensus that it will be vital to take specific actions to protect and develop our environment under sustainable, renewable conditions. Failure to do so will mean that the window of opportunity for the discovery of new medicinal and biological agents for the 21st century that will be lost forever. In this rather brief presentation we will discuss the historical use of plants and plant products as drugs, the developments that are underway for the selection and the biological evaluation of plant extracts, and the criticality of potentiating the resources of the rain forest for the drug discovery programs of the early 21st century.

Every major civilization since the earliest records of human history have employed plants as their primary source of medicament, as well as for food, shelter, furniture, writing materials, cosmetics, and numerous other purposes. Traditions of healing using indigenous plants were transferred between civilizations, primarily by the

0097–6156/95/0588–0008$12.00/0

explorers, the priests, merchants, and eventually the scientists who challenged the vast oceans, mountain ranges, jungles and deserts of Earth.

The elaboration of plants for their medicinal, nutritive (and culinary) value, through trial and error, what we now call "drug discovery", has been an avid and essential pursuit of mankind since prehistoric times. At this time in the evolution of scientific thought, it is indeed pertinent to ask why the investigation of plants for their biologically-active compounds should continue? And why these efforts have increased so dramatically in the past few years? There are probably three reasons: firstly, plants have yielded, and continue to relinquish numerous, biologically important lead structures. Typically, these compounds possess structures which have yet to be synthesized or even imagined by the traditional organic chemist or are compounds evincing new biological activities for an established natural product skeletal class or individual entity. Such compounds, therefore become prime candidates for the integrated chemical and biological development of derivatives modified to possess enhanced water solubility, resistance to gastrointestinal hydrolysis, specific biological activity and/or reduced toxicity. Secondly, it is well established that globally, and for a variety of economic reasons, the tropical rain forests of the world are shrinking at an alarming and possibly perilous rate. Thirdly, less than 15% of all terrestrial plants have been investigated for more than one biological activity, and this activity was probably a cancer-related bioassay. Important though it is as a group of debilitating and deadly diseases, for most of the world, it is not the most critical disease. Thus the potential for indigenous plants to contribute to global health, and therefore the economies of nations is indeed significant, and vastly underestimated by the nations of the North, where the expertise to integrate drug discovery into a finished pharmaceutical entity typically lies.

Farnsworth has indicated that from the small portion of the 250,000 species of flowering plants that have been investigated thus far, about 120 therapeutic agents of known structure are isolated for commercial purposes from about 90 species of plant and are utilized for the treatment of a diverse array of disease states. He estimates that seventy-four percent of these 120 plant-derived therapeutic agents were discovered based on ethnomedical records and exceptional use profiles (*1*). Some examples include the notorious mandrake, *Mandragora officinarum*, and henbane, *Hyoscyamus niger* which yielded scopolamine; Coca leaf, cocaine; opium, *Papaver somniferum*, which afforded morphine and codeine; South American curare, *Chondodendrum tomentosum*, from which tubocurarine is derived; Calabar bean, *Physostigma venenosum*, used as an ordeal poison, which gave physostigmine; digitalis, *Digitalis lanata*, yielding the digitalis glycosides; cinchona, which yielded quinine; ergot, *Claviceps purpurea*, the fungus which infests rye, which yielded the alkaloids ergotamine and ergonovine; and *Rauvolfia serpentina*, which yielded reserpine and deserpidine (*2*). In spite of this apparent importance, and the incredible untapped wealth of the world flora and fauna of the world, in the United States the percentage of drugs derived from plants declined steadily during the 1930s, 40s and 50s. Consequently, prescription product surveys during the 1970s and 1980s indicate that only approximately 25% of prescriptions filled contain a plant-derived natural product (*3*).

Elsewhere, particularly in developing countries, or those where limitations on imports due to currency restrictions and high prices inhibit the acquisition of "western-made" drugs (*4*), medicinal plants continue to represent the primary source of medication.

Indeed, the World Health Organization has estimated that this situation pertains for approximately 88% of the population in the developing countries (*1*), or approximately

3.4 billion people. In China, for example, which has about one-fifth of the worlds' population, 7,295 plant species are utilized as medicinal agents. As we consider the future development of medicinal and biological agents we need to consider whether these daily experiences merit further investigation.

Tropical Rain Forests-A Potential Source of Medicinal and Biological Agents?

It is widely considered that both philosophically and pragmatically, a nation has three forms of wealth: material, cultural and biological. The Earth began forming 4.5 billion years ago and man developed on Earth approximately 3.5 million years ago, a time which coincided with the greatest biological diversity ever. While much has been written on the impact of reduced botanical diversity and increased deforestation on the environment and the ozone layer, the presumed immense, mostly untapped, chemical and biological potential of the biome (fauna *and* flora) has been largely ignored. But what is this potential? Unfortunately, since the number of species in the biome is not known, in statistical terms this becomes an impossible question. However, there are some approximations, for example it is believed that there are about 350,000 species of vascular plant, about 200,000 species of marine invertebrates, and about 1.5 million species of terrestrial animal, insect, arthropod and other living organisms. It must be emphasized that these are estimates, and that new species of organism are constantly being discovered. For example, a new whale species was discovered in 1991, three new plant families were found in Central America in the past ten years, less than 10% of the insects in some groups have been categorized, of an estimated 1.5 million fungi, only about 70,000 have been described, and the number of deep sea organisms is completely unknown (5).

At the present the tropical rain forests are the richest concentration of biota, be they plants or arthropods. For example, a one hectare plot in Kalimantan, Indonesia yielded approx. 700 species of plants (6), and a single tree in Peru yielded 43 ant species in 26 genera (7). Yet every year 42 million acres of tropical rain forest are permanently lost, or approximately one football field every second. Every day we waste non-renewable resources for very short term gains; eliminating species at an alarming rate without having previously collected the germ plasm. This wanton destruction of the tropical rain forests has been described as "one of the great tragedies of all history", and it may well be that our children and their children will see this as an unforgivable, devastating legacy. A legacy which has resulted in a biome whose diversity is at its lowest point since the Mesozoic Period 65 million years ago. Species diversity is probably our planet's most important and irreplaceable resource. Yet it is estimated that one-fifth of all species may disappear in the next thirty years (8). Once exterminated, species regeneration, if at all possible, will take 5 to 10 million years.

In addition to the disastrous loss of the rain forests there are other losses. For the loss of, or encroachment on, habitat eliminates vertebrate and invertebrate species, fungal and bacterial flora, and in some instances, the indigenous population, which over many generations has evolved an accumulation of invaluable ethnomedical knowledge. While some may say this is patronizing, the fact remains that if these tribes have not been studied and the knowledge of the shamans catalogued, the information will be lost forever. Such losses of the biome and the knowledge of the value of the local biome are irreversible and permanent.

Following the Earth Summit in Rio de Janeiro in 1992, now, more than ever before, there is a heightened global awareness about the devastation of the tropical rain forests, in Asia, in Africa and in Central and South America. This Convention on Biological Diversity commits signatories to substantive action in many areas (*9*), including: development of national plans, strategies or programs for the conservation and sustainable use of biodiversity; inventory and monitoring of components of biodiversity and of processes adversely impacting it; development and strengthening of mechanisms for biodiversity conservation; restoration of degraded ecosystems; preservation and maintenance of indigenous systems of biological resource management, and equitable sharing of benefits with such communities; and integration of biodiversity concerns into national decision-making.

Because of the structural and biological diversity of their constituents, and their long history of indigenous medicinal use, it is widely considered that terrestrial plants offer a unique and renewable resource for the discovery of potential new drugs and biological entities. However, the vast, albeit rapidly shrinking, research base from which to draw samples for biological evaluation poses an interesting challenge. For in the recent past, numerous, increasingly sophisticated, systems for the biological detection of active compounds (*vide infra*) have appeared, therefore, it is pertinent to ask "What is to be the nature of the materials to be evaluated?" and "How do we responsibly use both the renewable (i.e. the sustainable) and the non-renewable resources available to us as effectively as possible?" and "How do we find the proverbial needle (active compound) in the haystack (plant kingdom)?" Is there the potential for an intellectual step involved in providing the samples for bioassay, or is the only approach random?

Stages in Natural Product Drug Development

The discovery of a lead candidate medicinal agent from a natural source involves several discrete and unique steps, including: i) the selection, collection and unambiguous identification of the organism, ii) the preparation and evaluation of an extract in an appropriate array of automated *in vitro* test systems, iii) prioritization of the plants to be further studied, iv) bioactivity-directed fractionation, and v) structure determination of the active isolate(s). The remaining stages of additional biological, toxicological and pharmacological testing are the same for synthetic or natural products (*10*). These drug development studies are presently estimated to cost over $230 million for a given drug to reach the stage of approval for marketing. Much has been written about what costs are included in estimating this figure, but in any event is an unacceptable number. It is inappropriate for the richest developed nation, and is an absurdity for a developing nation, whose country may well have been the original source of the lead compound which eventually became the medicinal agent for therapeutic evaluation.

Before we explore that pathway though, let us examine some of the steps involved in this preliminary drug discovery process in more detail, for they are frequently given short shrift by those who have not experienced them. There are five systematic approaches for the selection of plants that may contain new biological agents from plants: the random, the taxonomic, the chemotaxonomic, the information managed and the ethnomedical. In the random approach, all available species are collected, irrespective of prior knowledge and experience. In the taxonomic approach, plants of predetermined interesting taxa are sought from diverse locations. In the chemotaxonomic approach, a particular compound class, e.g. coumarins, may be considered as having biological interest, and plants likely to produce related compounds are collected. In the information managed approach, plants of proven biological activity which are unlikely to contain known active agents are collected in the hopes of discovering novel chemical agents. Finally, in the ethnomedical approach, credence is given to clinical, oral or written information on the medicinal use of the plant, and, based on an evaluation of this information, the plant is collected.

For any of these collection strategies, the most rational procedure is to test the material in a range of bioassays. Following prioritization, the active leads are fractionated through bioassay-directed fractionation for their active principle(s). Depending on the circumstances and even at different stages in the same discovery and development process, there are merits to each of these collection strategies. For example, it may be necessary at some point if the lead compound for development is isolated from a natural source, that a strategized search for alternative sourcing of the compound becomes important. The history of the development of taxol as a clinical entity bears witness to these changing strategies. A sixth "approach", a non-systematic one, is serendipity, where collection based on one bioactivity or ethnomedical use of the plant leads to the discovery of another bioactivity of commercial significance. The bisindole alkaloids of *Catharanthus* are an exquisite example in this sense, where the ethnomedical use of the tea of *C. roseus* as an antidiabetic agent led to the discovery of its antileukemic properties through opportunistic scientific observation and interpretation. For the purposes of the discovery of new medicinal and biological agents from plants, I believe that a combination of the ethnomedical and the information management approaches will be found to be the most productive.

Targeted application of the ethnomedical approach to a discovery program requires both the collection and prioritization of existing ethnomedical, chemical, biological and clinical data. In this way, plant collection can focus on the acquisition of plants which are most likely to yield biologically interesting new compounds for potential development. Unfortunately, the requisite information needed for such decision-making is very highly scattered. The only place in the world where this information is being collected for analysis on a global basis is the NAPRALERT database at the University of Illinois at Chicago (*11*). Through this computerized system, a list of plants can be weighted for their diverse ethnomedical, *in vitro*, *in vivo*, and clinical activities in order to identify those plants which would be a high priority for collection in the initial phases of a drug discovery program. Such a final list may be achieved for a given biological activity by gathering data sets on plants which have been studied for their active principles, ethnomedically reported plants with that or a related biological activity and plants for which a biologically activity has been experimentally established, but for which no active principle has been obtained. A comparison of these data sets affords a prioritized list of plants which have, or which are reported ethnomedically to have, a relevant biological activity, but from which no active metabolite has been obtained. A systematic collection plan for the listed plants is then developed. The evolution of any kind of directed strategy for plant collection is a relatively unappreciated, and sometimes poorly understood, aspect of natural product drug development (*12*).

Extraction and Biological Evaluation of Extracts

Two critical decisions are then necessary once the plants have been collected and identified. How are the plants to be extracted and how are the extracts to be evaluated biologically? Although these decisions are reversible, the costs involved may well be substantial, particularly in the event that substantial numbers of extracts are to be screened in a random approach. Any such decisions are likely to be a compromise which reflect a variety of local factors. Most groups opt to evaluate an organic and an aqueous extract prepared following an alcohol-based extraction and partition of a dried plant sample, although some groups prefer to extract fresh-frozen plant material. If receptor and enzyme inhibition based assays are being used, it is essential that the extract(s) be tannin-free, lest a large number of false positive leads be obtained. This usually is accomplished through either the use of a resin or a complexation/precipitation technique (*13-15*). The extraction of large numbers of random samples for biological evaluation is regarded by some companies as a rate-limiting step and indeed there is substantial room for improvement in the steps involved between a plant and the preparation of 96-well (or 384-well) plates containing the extracts. Some examples include supercritical fluid extraction and the use of various resins and size exclusion materials, which might concentrate materials in ways other than polarity and solubility.

It is critical that we use our increasingly scarce germplasm resources expeditiously for drug discovery, and answer the question of what are the major national and global diseases where drug discovery efforts are needed to yield new therapeutic agents? These major diseases include cancer, heart disease, malaria, schistosomiasis and other tropical diseases, herpes, AIDS and an array of other viral diseases, diabetes, senile dementia and a variety of other neurological diseases. Numerous factors impinge on a decision as to which therapeutic target areas are to be pursued by individual companies, but in any event, a preliminary bioassay or series of bioassays is needed, and this aspect of drug discovery has recently been totally transformed by a burgeoning technology (*15,16*).

Two facets, an improved, though substantially incomplete, understanding of the molecular basis of drug interactions, as well as substantial progress in the ethical and judicious use of animals in research, have substantially altered the basic philosophies of programs for the discovery of new medicinal agents. Formerly, the so-called Hippocratic screening method was given credence for the evaluation of a wide variety of biological activities of medicinal plants (*17*). However, the method typically permits only gross responses to be observed, requires large numbers of animals and for the purposes of biodirected-fractionation, necessitates ecologically unreasonable large samples of numerous plant extracts. It is now considered more appropriate to consider such animal experimentation to be important to look for gross responses at a time when a lead candidate structure has been identified.

The enhanced availability of receptors, enzymes and cell systems has permitted these entities to be considered as viable means for the evaluation of very small amounts of a sample on an automated basis. This, in turn, has led to another important change, namely the rate at which new biologically active entities can be discovered. Whereas it previously took weeks or months to test a few hundred samples, it now takes, for some assays, only a matter of hours. Whole cell, enzyme-based, and receptor-based assays are now quite routine in many therapeutic target areas, and the emphasis is shifting rapidly to genetically-engineered assays which evaluate the ability of a compound to interfere with a biological process in an exceptionally specific manner.

Irrespective of the *in vitro* assay, a number of criteria must be met. For example, the assay should be relatively simple and straightforward, so that it can be run without elaborate preparation or training of personnel. It should be accurate, i.e. capable of meaningful quantitation with a reasonable margin for error. It should be reproducible, both on a day-to-day basis and on a lab-to-lab basis. It should be selective, i.e. that only the individual biological event or process of interest is evaluated. It should be very sensitive, probably down to about 0.0001% of an active compound in an extract, based on the dried weight of the original organism. It should be fast and not require extensive delays for results to be obtained. Finally, it should be economical in terms of preparation, reagents, and labor costs, particularly if thousands of samples are to be evaluated. Finally, an assay possessing these qualities would be limited value unless it was both at the forefront of the biological category under study, and hopefully predictive of a therapeutic activity. Once the discovery of a biologically active chemical entity has been made, the question of protecting that invention arises.

Intellectual Property and Patent Issues

Through the international system of patent law, pharmaceutical companies, institutions and individuals have been afforded the option of being able to protect and derive exclusive reimbursement from an invention, even where it is discovered through the approach of trial and error. There is indignant outrage when these inventions are infringed upon, i.e. "stolen", by others without their knowledge and without due compensation. It is now a matter of considerable discussion that these same moral and ethical questions surrounding the application of similar standards to indigenous biodiversity *and* ethnomedical knowledge. For it has been the case historically that even when the "discovery" is based on hundreds, or perhaps thousands, of years of deliberate trial and error by an indigenous society, "property" claims often are ignored.

This indigenous knowledge of the use of plants as medicinal and biological agents is regarded now as an important tool in the discovery process, and also as evidence for the sustainable development of the tropical rain forests. However, for sustainable development and appropriate compensation to become a reality, a significant change in heart must occur within the pharmaceutical community, so that the initial knowledge which yields the lead and a portion of the profit that is acquired from the discovery of say a novel anti-inflammatory agent from a traditionally used medicinal plant is adequately compensated. Training programs for local scientists to assist them in initiating their own discovery programs are also essential. A recent series of articles (*18-22*) and books (*23-25*) has brought sharp focus to these issues.

In their global collection program, the National Cancer Institute has made specific commitments with respect to compensation in the event that a commercial discovery is made, and in addition is offering to train local scientists in its intramural laboratories (*26*), and the Merck Research Laboratories-INBio agreement in Costa Rica has led the way in stimulating discussion regarding compensation and the development of indigenous scientific infrastructure (*27*). Glaxo Research and Development has also published its policies regarding the sourcing of materials for its drug discovery programs and Gore has indicated (*28*) that for developed and developing countries "Access to native resources and protection of intellectual property are complementary concerns".

Of course the protection of indigenous rights and the avoidance of uncompensated exploitation has also been an issue for many countries and groups throughout the world. As a result a number of declarations of intent are available for consideration, for example, those of Belem (*29*), Göteberg (*30*), Chiang Mai (*31*), Kunming (*29*), Bethesda (*32*), the Hipolite Unanue Agreement (*29*), and UNESCO has published the Manila Declaration and the Bukhit Tinggi Declaration. These statements arose from conferences addressing the loss of species diversity or the ethical issues and the codes of professional practice which should be followed. One hopes that all scientists and institutions working in this area are concerned that less-developed countries are not further exploited for their local resources without adequate current and future compensation. To do otherwise would be to lose the ethical and moral basis of natural product drug discovery, and would undoubtedly lead to exclusion from development rights.

The issue of what can and cannot be patented bears directly on the commercialization of discoveries made through the ethnomedical properties of a particular plant. It has been argued that the "non-obvious" requirement for a patentable invention would preclude patenting a compound derived from a traditionally used plant. However, this is not the case, since there are patents on the use of insecticides from *Azadirachta indica*, to give just one example. The argument being that while the use may be obvious, the nature of the active ingredient, the scope of activity and the pharmacophoric unit are not available through prior art.

It is generally acknowledged that a natural product cannot *per se* be granted a patent solely as a new composition of matter. Although somewhat curiously, NIH is trying to patent parts of the human genome as they are identified, even before their biological significance has been recognized. Presumably, if these patents are awarded, every new natural product would also be eligible for protection. Emerson's comment regarding plants as weeds is also relevant at this point. However, protection for a process for the isolation of the compound, and/or a new or unanticipated biological response for a new chemical entity from a natural source <u>can</u> be afforded patent rights.

For reasons which are at a minimum arcane, a patent describing a new biological response for a known compound has substantially less esteem than one which involves a new chemical entity. Long term, this is very detrimental to the costs of drug development and thus health care. It has become apparent that although protection for an investment is certainly justified in terms of patent rights, with the rapid changes that have occurred in biology in the past ten years, this protection should in the future be based more on biology than chemistry. It can be predicted that as more highly selective agents are disclosed, numerous known compounds will be discovered which have novel biological actions of potential clinical utility. Yet without protection, these known readily available and therefore cheap compounds will remain "on the shelf" in favor of a higher priced, less available entity associated with novel chemistry. The ability to protect these discoveries of biological effectiveness should not be compromised by a lack of chemical novelty, for it has a direct bearing on the cost of bringing a clinical entity to the market place and thus health care costs overall.

The 21st Century

Since it takes 10-20 years to bring a drug discovery from the bench to the pharmacy or hospital setting, we are already engaged in drug discovery for the 21st century. Is there then a future for the discovery of medicinal and biological agents from plants? In the opinion of this author, while new, biologically important, therapeutically relevant, natural products continue to be isolated and provide lead compounds for future development, and while the search continues for natural, non-polluting insecticides and herbicides, the future will remain increasingly bright. But in practice what does this mean? What are the future challenges in developing the origins, the chemistry and the biology of natural products?

There are numerous additional facets of this broad topic which could be considered in contemplating the future. Economically and socially perhaps the most important is to evaluate the potential origins of the medicinal agents fifty years from now (*10,33*). While the cost of the chemicals involved is relatively cheap, synthetic modification to existing drug entities for the purposes of enhancing and/or reducing activity remains an attractive focus of drug development. Eventually, however, as competition between developed and developing countries for rapidly diminishing oil resources increases, and as economic standards in developing countries rise, so the pressure on the prices of fine chemicals will dramatically increase. This will have an immediate effect on the price of all synthetic and semisynthetic drugs, and may result in prohibitively expensive drugs. The ethical arguments that are just beginning to surface with respect to the cost effectiveness of continuing treatment will become even more cogent.

Back to Basics. The opportunities to develop new medicinal agents or other economically useful compounds are extremely limited; indeed they are two, synthetic or natural. A compound may be totally synthetic, partially synthetic (with a naturally-derived core) or genetically engineered. In the latter category, terrestrial and marine plants, animals and microorganisms are our only available sources. As a global society, any of our actions which jeopardize these natural resources also diminish the potential availability of new medicinal and biological agents, probably permanently. Hence the fundamental need to ensure that ethnomedical information is both appropriately catalogued and compensated before the thousands of years of passing on substantially unwritten, precious remedies are also forever lost.

There is irrefutable statistical evidence that the tropical rain forests and the indigenous cultures that have tended these environments, are a potentially immense source of biologically active compounds. On the one hand, the ethical and moral precepts briefly discusses above regarding the discovery of biologically active compounds from indigenous sources are of extreme importance for the future of the global pharmaceutical industry. But the very existence of these resources remains under dire threat (*25*). Realizing this, many pharmaceutical companies are developing their own germplasm banks and racing to collect as many samples of the worlds flora. New and important drugs and biological tools will undoubtedly be found as a result of any systematic investigation of these resources, as well as from the ethnomedical leads provided from the existing literature and the local shamans. How such work is done at a particular pharmaceutical company is probably more related to strategic choices.

The tools needed for such studies of our ethnomedical and biological inheritance are available, there is an adequate supply of expertise, and some outstanding laboratories in several countries, and there is no shortage of pharmaceutical companies to convert a drug discovery to a finished product.

However, if we are to *optimize* the development of the existing biome, a quite different approach is required, for this beautiful little corner of the cosmos has extremely limited resources. Much has been said and written about the potential of the rain forests to yield new crops and new biological and medicinal agents based on the established biological diversity of species in a given area (*23,34*). But relatively little has actually been done; thus the proof of such potential is lacking.

Globally, what is needed is a multinational, interdisciplinary program in which a multitude of plant, animal and insect species could be evaluated for their medicinal and biological potential and their germplasm stored. Coordination by a major international funding agency and sponsorship by governments, pharmaceutical and agrochemical industry, and philanthropic foundations around the world would be needed. For the long term health of mankind, I believe the question is not whether this will be done, but rather when will it start.

What are some of the challenges then for the 21st century? I believe that they include: preservation of the rain forests and the oceans; maintaining and systematizing the biodiversity; cataloging the ethnobotanical information; automating the high throughput screening; developing in-field bioassays; assuring intellectual property rights for indigenous peoples and maintaining and enhancing germplasm banks. It is apparent that we have a window of opportunity available while biodiversity is still substantial and not completely outpaced by testing capacity. It is essential that we use this period in the history of Man wisely.

As a unique community in the cosmos, we must carefully consider the environmental, medical and economic options that we have available. In this regard, the circumstance of pharmacognosy and the biome today are well described by the immortal words of Charles Dickens from *A Tale of Two Cities* come to mind: "It was the best of times, it was the worst of times, it was the age of wisdom, it was the age of foolishness, it was the epoch of belief, it was the epoch of incredulity, it was the season of light, it was the season of darkness, it was the spring of hope, it was the winter of despair."

In presenting and discussing the future of plant-derived drug discovery I have chosen to see the best, the wisdom, the belief, the light and the hope. Not to do so would truly be inhumanity to our future generations. Let me close, as I began, with a quote from Ralph Waldo Emerson, "We did not inherit the earth from our forefathers; we are borrowing it from our descendants".

Literature Cited

1. Farnsworth, N.R.; Akerele, O.; Bingel, A.S.; Soejarto, D.D.; Guo, Z. *Bull. WHO* **1985**, *63*, 965-981.

2. Tyler, V.E.; Brady, L.R.; Robbers, J.R. *Pharmacognosy*, 9th Edition; Lea & Febiger: Philadelphia, PA; **1988**, pp. 519.

3. Farnsworth, N.R.; Morris, R.W. *Am. J. Pharm.* **1976**, *148*, 46-52.

4. An interesting account of an analogous situation in the former USSR is presented in an article by F.X. Clines in *NY Times*, January 1, **1991**, p. 2.

5. Raven, P.H.; Wilson, E.O. *Science* **1992**, *258*, 1099-1100.

6. Ashton, P.; cited by Wilson, E.O. In *Biodiversity*, Wilson, E.O., Ed.; National Academy Press: Washington, DC, **1988**, pp. 3-18.

7. Wilson, E.O. *Biotropica* **1987**, *2*, 245-251.

8. Norton, B.J. *The Preservation of Species*, Princeton Univesity Press: Princeton, **1986**, pp. 305.

9. Anon (1992) *Biodiversity Conservation Strategy Update, WRI/IUCN/UNEP* **3**(2), 1-4.

10. Cordell, G.A. *Pharmacia* **1990**, *30*, 169-181.

11. Loub, W.D.; Farnsworth, N.R.; Soejarto, D.D.; Quinn, M.L. *J. Chem. Info. Comp. Sci.* **1985**, *25*, 99-103.

12.   Soejarto, D.D.; Farnsworth, N.R. *Persp. Biol. Med.* **1989**, *32*, 244-256.

13.   Wall, M.E.; Taylor, H.; Ambrosio, L.; Davis, K. *J. Pharm. Sci.* **1969**, *58*, 839-841.

14.   Loomis, W.D., Battaile, J. *Phytochemistry* **1966**, *5*, 423-438.

15.   O'Neill, M.J.; Lewis, J.A. In *Human Medicinal Agents from Plants*; Kinghorn, A.D.; Balandrin, M.B., Ed.; ACS Symposium Series No. 534, American Chemical Society: Washington, DC; **1993**, pp 48-55.

16.   *Methods in Plant Biochemistry, Vol. 6, Assays for Bioactivity*; Hostettmann, K., Ed.; Academic Press: New York; **1991**, pp. 350.

17.   Malone, M.H.; Robichaud, R.C. *Lloydia* **1962**, *25*, 320-332.

18.   Cunningham, A.B. *Cult. Surv. Quart.* **1991**, Summer, 4-8.

19.   Elisabetsky, E. *Cult. Surv. Quart.* **1991**, Summer, 9-13.

20.   Kloppenburg, J., Jr., *Cult. Surv. Quart.* **1991**, Summer, 14-18.

21.   King, S.P. *Cult. Surv. Quart.* **1991**, Summer, 19-22.

22.   Posey, D. *Cult. Surv. Quart.* **1991**, Summer, 29-35.

23.   *Biodiversity*, Wilson, E.O., Ed.; National Academy Press: Washington, DC, **1988**, pp. 521.

24.   *Biodiversity Prospecting*, Reid, W.V.; Laird, S.A.; Meyer, C.A.; Gámez, R.; Sittenfeld, A.; Janzen, D.H.; Gollin, M.A.; Juma, C.; World Resources Institute: Washington, DC, **1992**, pp. 341.

25.   *Global Biodiversity Strategy*, World Resources Institute, The World Conservation Union and The United Nations Environment Programme, Washington, DC, **1992**, pp. 244.

26.   Cragg, G.M.; Boyd, M.R.; Cardellina II, J.H.; Grever, M.R.; Schepartz, S.A.; Snader, K.M., Suffness, M. In *Human Medicinal Agents from Plants*; Kinghorn, A.D.; Balandrin, M.B., Ed.; ACS Symposium Series No. 534, American Chemical Society: Washington, DC; **1993**, pp. 80-95.

27.   Joyce, C. *New Scient.* **1991**, *132* (1791), 36-40.

28.   Gore, Jr., A. *J. NIH Res.* **1992**, *4*(10), 18-19.

29.   *International Traditional Medicine Newsletter*, (1992) Soejarto, D.D.; Gyllenhall, C., Eds.; **1992**, *4*(2), 1-3.

30.   Eisner, T.; Meinwald, J. *Chemoecology* **1990**, *1*, 38-40.

31.   *Conservation of Medicinal Plants*, Akerele, O.; Heywood, V.; Synge, H., Eds.; Cambridge University Press: Cambridge, **1991**, pp. 362.

32.   Schweitzer, J.; Handley, F.G.; Edwards, J.; Harris, W.F.; Grever, M.R.; Schepartz, S.A.; Cragg, G.M.; Snader, K.; Bhat, K. *J. Nat. Cancer Inst.* **1991**, *83*, 1294-1298.

33.   Cordell, G.A. *Amer. Druggist* **1987**, 96-98.

34.   Shultes, R.E. In *Plants in the Development of Modern Medicine*; Swain, T., Ed.; Harvard University Press: Cambridge; **1972**, p. 103-124.

RECEIVED December 16, 1994

Chapter 3

Economic Plants of the Amazon
Their Industrial Development in Defense of the Forest

Benjamin Gilbert

CODETEC Technology Development Company, Caixa Postal 6041, Campinas, 13081–970 São Paulo, SP, Brazil

In order to slow or halt rainforest destruction, high value non-timber products from native species offer perhaps the only source of income for local communities capable of competing with timber and cattle-raising. Such products include special nutrients (especially β-carotene and vitamin E), perfumes and aromatherapy materials, animal ration from wastes, toilet soap of rainforest origin, natural insecticides, exudates and balsams for mainly topical medicinal use, pharmaceutical products of natural origin.

If local populations are to benefit at an economic level which induces them to prevent further degradation of their resources then manufacture of finished products has to be conducted locally so that an appreciable proportion of the final value remains in the region.

The role of sustainable industries in the preservation of the rainforest has long been recognized but scarcely ever implemented on a scale sufficiently large to halt the ongoing destruction of these vast natural air-conditioners. The causes of forest destruction are many but the important ones in present-day Brazil can be summarized under two heads:

- slash-and-burn agriculture practiced mainly by immigrants from the south of the country;
- highly lucrative industrial operations which exploit some natural resource such as wood for timber (Note 1) or for charcoal, these often practiced by large companies whose headquarters are outside the region, or by their local suppliers. One such lucrative operation, "garimpo" gold-mining, is individual rather than company controlled but in this case too most of the financial gain leaves the region.

It is very difficult to halt either of these classes of activity because the first is a plain necessity of existence for people who have come from outside the region

0097–6156/95/0588–0019$12.00/0

Note 1

Timber, if worked rationally, could be the most important sustainable forest product. However in present practice all trees of economic trunk diameter, usually greater than 50 cm measured above the broad, often buttressed base, are felled. Some 55% of the forest, in terms of trunk sectional area, survives (Uhl, C; Vieira, I.C.G. Ciência Hoje (Rio de Janeiro), special vol. "Amazônia", Dec. 1991, 108-115). to replace the felled timber. In Juruena municipality, Mato Grosso, a rich source of high value woods, it has been estimated that commercial timbers will be exhausted in 10-20 years at the present rate of exploitation (Van Leeuwen, J.; Silva, Y.T. Pronatura-ICI, Juruena Project Report. "Agroforestry and other research proposals for sustainable land use in Juruena, MT, 1992, p. 12), whereas if hardwoods such as are being exploited grow at a rate of 0.5 cm/yr. trunk diameter then at least a century will be needed to replace those cut even if replanting is conscientiously undertaken. The present deficit in replacement of timber stocks has been described by many authors (see Jansen, M.R.A.; Alencar, J. da C. Bases Científicas para Estratégias de Preservação e Desenvolvimento da Amazônia: Fatos e Perspectivas, vol. I, Inst. Nac. de Pesquisas da Amazônia, Manaus, 1991, 187-195).

Note 2

CACEX, Brazilian Government Overseas Trade Office, published data show that copaiba oil was exported to the United States at $ 3.34/k fob (12.6 tons); to the UK at $ 4.05/k fob (7 tons); to Germany at $ 5.06/k fob (4 tons) and to France at $ 4.22/k fob (5 tons), in 1992 (Jan-Nov). The total export of 25 tons, in this period had decreased from 51 tons in 1990. Most cosmetic and over-the-counter pharmaceuticals sold in the UK in 1992 contained only very small amounts of tropically derived balsams - the total terpenoid content of a typical ointment, of which about 0.2% was collected natural exudate, costs about $ 800/k to the public. On an average there seems to be a mark-up of about 200 times between Brazilian exporter and the first world public, and there is an estimated mark-up of 10 times from collecter to exporter through three to four intermediaries. Brazil nuts were exported for 81 cents/k to $ 1.05/k in 1992, or when shelled and dried, for $1.49/k to $1.98/k, with a total export of nearly 24,000 tons ($ 32 million). The price to the collectors themselves for unshelled nuts was $ 0.07/k [Brooke, J. quoting Clay, J. (Cultural Survival Enterprises). New York Times, April 30, 1990, p. D12] and the author's estimate of the net cost to the American public of shelled nuts in candies derived from the same source is around $ 250/k. Over 99% of the added value is generated outside the country and a further 92% of the remaining 1% does not reach the collector.

Note 3

The estimate of the world cosmetic market at retail level as $ 40 billion annually is based on, for example, the Chem. Eng. News yearly Product Reports which give the US cosmetic consumer market at around $16-18.5 billion in 1990-1993 consuming over $100 million of natural raw materials (see, for example, Chem. Eng. News, April 20, 1992, 31-52 and April 26, 1993, 36,38). For the insecticide and related agrochemical market at $7 billion see, for example, von Szczepanski, Ch. v. in Crombie, L. editor Recent Advances in the Chemistry of Insect Control II, Roy. Soc. Chem., Cambridge, UK, 1990, 1-16).

and know virtually nothing of its resources, and the second is so profitable to those engaged.

Thus in suggesting sustainable industries for the Amazon region these must both satisfy the immigrants' necessities and provide a profit margin that is competitive with the extractive operations that are already going on. Low priced "commodities" such as natural rubber do not satisfy these criteria, nor do extractive activities such as copaiba oil and Brazil nuts when conducted in their traditional manner, which yields only a few cents per kilo to the collector for products that are often retailed to the first world public by "natural product" retailers at a price a thousand times higher (Note 2*).

Natural products including those just mentioned, when elaborated and packed in their final retail form do provide a sufficient margin to satisfy the criteria formulated above *so long as the operation is carried out locally* as far as this is feasible. If the world cosmetic market at retail level is $ 40 billion annually, and the insecticide and related agrochemical market $ 7 billion (Note 3), then the capture of 1% of these two markets with natural products from Amazonian producers would provide $ 470 million, sufficient to stimulate a real on-going effort for sustainable agroforestry in the region. The present article describes how this may be done.

It will be appreciated that the Amazon basin is about as big as Europe, excluding European Russia and the Ukraine, or about half as big as the United States. Just as in Europe or the USA there are many habitats in the Amazon and for each of these a distinct sustainable industrial project based on the products native to that particular habitat must be elaborated. Some broad classifications given below help one to design projects that are diversified and economically viable.

Classes of Natural Products

The classification of commercializable or potentially commercializable natural products can be made in three ways.

Classification I. Origin:
(i) Extractive products, that is, products whose exploitation requires neither clearing nor plantation.
(ii) Products of forest management, whose production requires some kind of agroforestry operation when continuing supply is envisaged.
(iii) Products derived from the plantation of devastated or marginal areas.
This classification determines the nature of the primary operation.

Classification II. Nature of the product:
(i) Vegetable oils and fats, usually obtained by expressing or extracting fruits or seeds.
(ii) Essentials oils, obtained by steam distillation, extraction or direct distillation of plant material or by a combination of such processes.

(iii) Non-glyceride, non-terpene oils or liquids, usually obtained by extraction and concentration.

(iv) Waxes, resins and latexes.

(v) Gums.

(vi) Crude extracts, obtained by the extraction with water, ethanol or other medium, of vegetable material followed by concentration to a paste or dry powder.

(vii) Pure or semi-pure products, usually solid, normally obtained from the extracts by physical procedures.

(viii) Powders obtained mechanically from the plant material or without extraction.

The subdivisions (vi), (vii) and (viii) include the majority of the colours, aromas, medicinals and pesticides which are classified below. Classification II determines the nature of the industrial installation.

Classification III. Use:

(i) Food and Drink additives
. vegetable oils and fats;
. aromas and flavours;
. colours;
. antioxidants and conservative agents;
. rheological agents (thickeners and gels);
. surfactants (emulsifiers and humectants);
. sweeteners or bitter principles;
. vitamins,provitamins,nutrients,non-nutrients;
. others.

(ii) Cosmetic and Perfume materials
. oils and fats;
. fragrances;
. pigments;
. rheological agents;
. others.

(iii) Medicinals.

(iv) Insecticides or products for agricultural use or human and veterinary disease control.

(v) Specialities for industrial use.

(vi) Starting materials for industrial chemical transformation.

(vii) Animal feed
. cellulose containing biomass;
. high nutrient wastes such as oil-cake;
. high protein native and cultivated plants;

This classification is directly associated with marketing and with recognized local needs.

Market is of course a basic factor that must be borne in mind when designing a project for a particular region. For a product elaborated and brought to retail condition in a Brazilian Amazonian locality (there is nothing unrealistic about this; one needs only to visit the sophisticated Manaus industrial estate) there are essentially three markets: the local one, the national one, that is, Brazil outside the Amazon, and the export market comprising chiefly the first world. These markets represent not only different types of demand, they have different regulatory laws and this means that a product adequate for, say, the local and national markets may not be exportable. This is particularly true of natural medicinals which are so often presented as the economic salvation of the rainforest but which, at the moment, cannot figure in any short term economically based development project because of the very high cost of taking a medicinal plant's active component through the regulatory requirements of a first world country. Natural medicinals however do figure very importantly in the solution of local health needs, and, as a first step, the identification of active components, efficacy, side-effects and practical application in public health is a research priority.

Location must also be taken into account when a project is elaborated. For example, it is quite feasible to produce a toilet soap based exclusively on Amazonian materials in Serra do Navio, Amapá and put it in a container on board ship at Santana for perhaps 15 cents a wrapped 90 g tablet, because there is a railroad from the forest location to the port. The same bar of soap produced in Juruena, Mato Grosso - a most important location since it is right on the frontier of the advancing immigrant and lumbering communities - would have to travel 910 km by road - mostly unpaved and passable only with difficulty in the rainy season - to reach Cuiabá, itself more than 1600 km from the port of Santos. For such a location a higher priced speciality is needed. A number of concrete possibilities will now be examined.

Vegetable Oils and Fats

There are a number of species whose fruits contain glyceride oils either in a pulpy mesocarp or in a, usually, nut-like kernel. In many cases these oils or fats occur in practical amounts and the plant source is sufficiently abundant naturally to make plantation unnecessary (*1-4*). There are some species which have been improved by selection, often by indigenous peoples, and have been adapted to plantation. Among the first the buriti or miriti palm (*Mauritia flexuosa*) exists in stands that may cover a 1000 km^2 at a time in the wet areas along the main river from Iquitos in the west to Marajó island in the east. *Astrocaryum* species like tucumã are also exceedingly abundant. Both these latter yield red pulp oils of the oleic type containing from 1000 to 2000 ppm of β-carotene, about double the concentration found in palm oil from *Elaeis guineensis* (*5*, Mambrim, M.C.T.; Barrera Arellano, D. *Oleagineux*, in press.). Babaçu (*Orbygnya martiana*) appears in even bigger stands in south-eastern marginal areas of Amazonia. Here the oil, of the lauric type (*6*), is in the kernel and is already an internationally sold commodity. A genetically improved palm is pupunha or peach palm (*Bactris gasipaes*) which would probably be the best native species for propagation in areas which have been cleared and sustain a rural population (*1,7*). Over the years much research has been carried out and published on the pulp and kernel oils of palms, particularly in Belém and Manaus, and their industrial and commercial potentialities can be assessed from the results of these studies (*1-13*).

Non-palm oil species include piquiá (*Caryocar villosum*) which is quite abundant in the "terra firme" primary forest and, like the related piqui (*Caryocar brasiliensis*) which grows in the "cerrado" or poor soil areas found in various parts of the Amazon, possesses a workable content of a "butter" containing, like buriti oil, a high percentage of β-carotene (*8,11*). A really hard fat, mostly trimyristin, is obtained from the seeds of the abundant *Virola surinamensis* and *V. sebifera*, the ucuúba trees (*3*) At the other end of the scale the Brazil nut tree (*Bertholletia excelsa*), also adapted to relatively dense stands, yields a highly unsaturated nut oil which, appropriately stabilized, provides a nutrient high in linoleic glycerides (*9,10*).

Some oils are not edible. An example is the bitter andiroba (*Carapa guianensis*) seed oil which contains oxygenated triterpenoids suspended in an oleic type triglyceride oil (*12,14*). The oil's analgesic or anti-inflammatory as well as insect-repellent properties well known locally have attracted the attention of both Brazilian and international cosmetic and over-the-counter pharmaceutical manufacturers.

The more abundant species which provide oils or fats adaptable to the needs of the food, special nutrient and cosmetic industries are listed in Table 1. Production of these and many other oils was a regional home industry in the past and still supplies communities that do not have access to commercial oils like soy oil from the big plantations of the south. Revival of the industry depends on the creation of specialist markets as commented above and can be made far more efficient by the introduction of solar pulp drying, mechanical oil-pressing and filtration, all of these simple technologies which widely available electrical power makes possible at the rural community level. Specialist markets include:

- β-carotene at its natural concentration or concentrated by, for example, freezing out of the more saturated glyceride components of buriti, tucumã or piquiá oils;
- base components for cosmetics containing defined concentration ranges for carotenes, tocoferols or tocatrienols and mono- and di-unsaturated fatty acids as glycerides, taking advantage of the wide melting ranges available from the hard ucuúba fat through the butter-like pequiá pulp oil to the low melting Brazil nut and *Oenocarpus* (patauá and bacaba) oils;
- analgesic or anti-inflammatory creams based on andiroba oil as OTC skin-care products;
- salad oils of the triolein type based on patauá pulp-oil.

Animal Feeds

Animal feedstuffs (Table 2) are a natural by-product of vegetable oil manufacture because this latter results in a variety of residues such as palm fruit pulp fibre and Brazil nut or other kernel oil-cake. Most cellulosic fibre wastes can be converted by steam explosion (sudden release of steam pressure to separate lignin from cellulose) to a cellulose digestible by ruminants. The oil-cakes often have high protein and carbohydrate contents and can be used as a major component of non-ruminant feeds.

Apart from vegetable-oil industry by-products high protein native plants such as caraparu or arumá-rana (*Thalia geniculata*, Maranthaceae) occur in some wet areas of Amazonia, such as the region around the mouth of the Jari river across to the islands of Gurupá and Marajó. The quantities available are extraordinarily large and would provide a virtually inexhaustible supply of raw material for animal feed.

The production of these feedstuffs must be tailored to the region. Cattle raising has been in the past one of the chief causes of destruction of virgin forest. The replacement of grasses planted often in soils unable to support cattle at an economic density by feeds produced from natural biomass from plants adapted to the habitat would seem a logical step. It must be borne in mind when planning production that the demand for cattle ration may be very high during the dry season (September to November in Marajó) and much smaller or non-existent during the rains. However in regions where no cattle raising and no meat or dairy product industry exists, ration for small animals, particularly pigs and poultry, may find a better year-round market and would obviate the necessity to clear forest for the cultivation of exotic plants to feed these animals.

Essential Oils

Although very many Amazonian plants contain essential oils that are marketed locally as perfumes, aromatherapy agents or flavours, none of these seems to be available on the scale that would be necessary to constitute an industry capable of sustaining a sizeable population. An example from the past is rosewood, *Aniba rosaeodora* (considered synonymous with *A. duckei*). Exploitation of the wood of this tree has caused almost complete eradication of the species and attempts to plant on a large scale (monoculture) have not met with success. However there is no question that plantation is the only solution for sustainable production of most of the species listed in Table 3 and the methods for achieving this successfully should be examined. Some, like *Aniba* will have to be replanted in a disperse manner in the natural habitat of the species until such time as adequate mixed plantation schedules are worked out that avoid attack and destruction by pests. Others such as *Hyptis* and *Lippia* will probably be susceptible to monoculture on degraded land judging by the frequent appearance of Lamiaceae and Verbenaceae along roadsides where the top-soil has been disturbed. Yet others, such as

TABLE I. Some Vegetable Oils of Abundant Amazonian Species that Lend themselves to Industrialization

Edible Oil Source[a]	% Oil in dry pulp[b]	% Fatty acids in pulp glycerides			% Oil in kernel[b]	% Fatty acids in kernel in glycerides		
		Oleic	Palmitic	Linoleic		Lauric	Myristic	Oleic
Babaçu					65	40-55	11-27	18-20
Bacaba	25	57	23	14				
Buriti	8-31	73-76	17-19	2-5				
Murumuru					28	43-51	26-37	0-10
Patauá	18-31	78-81	9-11	4-7				
Pupunha[c]	35	50-54	30-44	1-12	9	27-46	19-38	13-24
Tucumã	18	64-66	22-26	4	30	47-51	22-26	9-13
Ucuuba					60-68	11	61	6
		oleic	myristic	lauric				
Mucajá	26	20	22	44	41	49	16	17
						oleic	palmitic	linoleic
Brazil nut					62-67	31-56	14-18	26-46

(a) The non-edible andiroba seed (kernel) oil contains 18% myristic, 59% oleic acid glycerides; data for piquiá were not found, both pulp and kernel produce butter-like fats.
(b) Pulp or kernel oils are omitted when the source is industrially impracticable.
(c) Peach palm.

TABLE II. Principal Biomass Sources for Animal Feeds

Source	Raw Material Characteristics	Use
Palm & other fruit pulp fibre	Cellulose (after steam-explosion)	Ruminants
Babaçu & Brazil-nut oil-cake	Oil, starch & protein	All rations
Bacaba & patauá pulp[1]	Oil, carbohydrate, high protein	All rations
Palm nuts[2]	High mannose content	Not determined
Babaçu & pupunha mesocarp[3]	Starch, protein and oil	All rations
Palm pith	Starch	All rations
Non-toxic wood residues[4]	Cellulose (after steam explosion)	Ruminants
Arumás & herbaceous high protein plants, planted or wild	Cellulose, protein	Ruminants

(1) Brazil-nut oil cake could be developed for human consumption. The same applies to açaí, bacabá and patauá fruit pulp juices which are widely consumed in the region. For a typical analysis of patauá see Balick & Gershoff (13).
(2) Açaí nuts are eaten by cattle after they germinate, the same might apply to buriti kernels.
(3) For analyses of pupunha mesocarp see Clement (7) and Zapata (15).
(4) Amapá doce (*Brosimum potabile*), amapá amargo (*Paraharncornia amapa*), ucuúba (*Virola* spp), marupá (*Moronobea pulchra*) saw dusts are reported edible by cattle, even before steam explosion treatment

Conobea spp which grow in the wetlands of Marajó Island may be available naturally in sufficient quantities for direct exploitation.

The plants listed in the Table have been selected from a large number studied at the Amazonian research institutes, INPA in Manaus and Museu Goeldi in Belém. The criteria for choice were high content of essential oil based on leaf or whole plant material (not wood) and the predominance (>75%) of a single component which could be purified for sale as such. Essential oils are an attractive item for local community production requiring no highly sophisticated equipment and often commanding a good price especially if sold through distributors in the retail market. Support from a central chemical laboratory with good analytical equipment is necessary for quality control, a scheme that has worked out very successfully in Cochabamba at Bolivia with Canadian IDRC participation.

Toilet Soap

The availability of vegetable and essential oils provides the two main components necessary for the production of toilet soap. This simple industry needing a minimum of equipment is not new to the region. A family business based on the manufacture of glycerine soap using local vegetable oils and rosewood perfume was very successful in Belém for many decades. The industry has a secure local market to begin with and can gradually expand to the regional and eventually international spheres. In the case of a larger scale production the availability of both oleic and lauric oils (see Vegetable Oil section above) would be a factor in defining the locality of the operation. The sale of a soap made exclusively from Amazonian starting materials and produced in the region cannot fail to attract buyers and calculations show that competitive prices can be maintained for localities with easy access to a port (see comments on marketing above).

Insecticides and Related Pest Control Agents

Before the synthetic insecticides, herbicides and other agrochemicals began to appear in the 1940's there existed a world market for natural agrochemicals. From the Amazon and Central America came Derris, Quassia and Ryania, principally derived in Brazil from the species *Derris* (or *Lonchocarpus*) *urucu*, *Quassia amara* and *Ryania speciosa*. Derris root was often extracted with carbon tetrachloride or other suitable solvent to produce a rotenone concentrate containing not only rotenone, the main active component, but also other rotenoids which contributed to the insecticidal activity. Although these products could not compete on the multi-thousand ton scale which the synthetic insecticide industry attained, they did not altogether disappear and in many countries are available to the present day as horticultural insecticides. It is probable that this commerce could be considerably increased especially for the market garden, fruit, wine and nut industries where the modern public prefers products that have not been treated at any stage with petrochemical derived pesticides. This market could reach several

TABLE III. Some Non-Wood Sources of Essential Oils in Amazonia and their Reported Constituents

Plant Source	Part % aroma	Major Constituents (% of total aroma)			Ref.
		50-100%	23-50%	0-25%	
Cumaru - Tonka Bean *Dipteryx odorata*	seed up to 3.0	coumarin			11
Erva de Marajó *Lippia grandis*	leaves 2.2	carvacrol		p-cymene linalool	16
Laranjinha *Calyptranthes spruceana*	leaves 1.8	limonene	perilladehyde	pinenes citral	17,18
Pataqueira *Conobea scoparioides*	whole plant 0.6	thymol		p-cymene	19
Pau de erva doce *Calyptranthes sp.*	leaves 2.2	estragol			17
Pau rosa-rosewood *Aniba rosaeodora*	new leaves twigs	linalool			20
Piper aduncum (var. *cordulatum*)	leaves 1.4-3.5	dill-apiole			21
Piper hispidinervium	leaves 2.7	safrole		myrcene	22

TABLE IV. Some Plants Used Against Insect Pests

Plant Source	Main Active Compound	Use (in parentheses, potential use)	Ref.
Derris urucu root bark, timbó	rotenone and congeners	horticultural, market garden and veterinary (disease vector control)	23
Derris urucu aereal parts	DMDP or azafructo-furanoside	antifeedant still in experimental stage, nematicide	24
Quassia amara, whole plant, quassia	quassin, neoquassin	aphid and other sucking insect control, also medicinal	25
Pterodon pubescens fruits (seeds)	not identified (diterpenoid furans present)	cattle hornfly and ectoparasites	26
Carapa guianensis fruits (seeds)	not identified (oxygenated triterpenoids present)	mosquito and chigoe flea repellent, also medicinal	27,28,29

hundred million dollars and itself supply a major solution to the sustainable industrial development of the Amazon.

The plants mentioned are, of course, not the only species containing viable quantities of pest control chemicals. The existence of an abundant healthy and diverse flora in the region contrasts with the frequently observed failure of monocultures under devastating insect, fungal or other pest attack. It is quite difficult to maintain some conventional crops close to the forest because of the variety of predatory organisms that exist in that habitat. It is reasonable to suppose that pest control agents in plants play a rôle in the mutual protection of diverse species against predators. Some other possible species for pesticide production have been added in Table 4.

Considering the insecticidal plants one by one, the first, the rotenone containing leguminous plants which include *Derris urucu* have been thoroughly surveyed by EMBRAPA, the Brazilian State agricultural research institution (*23*), and three main facts emerge - firstly, *Derris urucu*, although not the highest yielding plant individually, is the best adapted to economic production of rotenone or of a crude powder containing it; secondly, plants of the genus are found all over Amazonia although *Derris urucu* is more abundant in the very humid low-lying ground along the north bank of the Eastern Amazon and similar habitats alongside other large tributaries; thirdly, some of the observed occurrences are in fact old plantations dormant 50 years but still capable of revival into industrial level production. Parallel chemical investigations have shown that the rotenone is in the root bark in *D. urucu*. Studies made in Britain (*24*) showed that practically all the *Derris* and related group of leguminous genera contain the insect anti-feedant DMDP, or aza-fructose, indicating that it might be advantageous not to extract the rotenone if by doing so one were to leave the water-soluble DMDP behind in the marc. In principle, therefore, the setting up of a Derris insecticide industry should begin with the revival of the old industry at the sites of the original pre-1950 plantations, and should concentrate on the production of a whole root bark powder, which is known to contain up to 25% of rotenone, finely ground and diluted with an inert mineral such as kaolin (abundant locally in the Eastern Amazon). One thus has a product entirely within the scope of locally installable manufacturing facilities assuring a reasonable financial return to rural populations.

Quassia amara is being developed as a crop in Costa Rica but the author has not yet located old plantations of this species although it is reasonably abundant in the *Derris* growing regions of Eastern Amazonia and could be undoubtedly co-cultivated with this crop. The internal use of this plant, both aqueous leaf and wood extracts, by local populations for protozoarian infections including malaria and even as a bitter principal, rather like quinine, in drinks, attest to the low human toxicity of the crude product, an observation that has been recently confirmed by pharmacological study. Quassin, the oxygenated modified triterpene, which is probably the main but not the only active constituent, has been reported to be allergenic when isolated (also true of rotenone), probably related to the presence of two α, β-unsaturated carbonyl groups present. Again there seems to be no reason to separate crystalline quassin or even to concentrate it but rather to aim at the crude aqueous extract as used popularly and which is remarkably effective against Aphides in the author's experience. The extract is at present undergoing clinical trial as an antimalarial agent. *Quassia amara* is not alone in its class. Many other Simaroubaceae and Meliaceae species of the Guayanas and North-eastern Brazil have been studied, particularly by the ORSTOM group (*28*), and shown to contain substances active against human and animal parasites (*25*).

Although *Ryania* was available before the Second World War as an insecticide and its active constituent, ryanodine was isolated and identified structurally, the relatively high toxicity, which has been reported to have resulted in fatal human intoxications, would justify exclusion of this product from the list of exploitable insecticides.

Clitoria racemosa (recently rebaptized *C. fauchediana*) and at least one other *Clitoria* spp have been shown to contain rotenoids and may well be viable sources of natural insecticides. *C. racemosa* is well adapted in South-Central Brazil growing well in impoverished and degraded habitats. The abundant legumes which contain the principles require evaluation as insect control agents. The first species would be an acceptable pioneer species in forest recovery due to the rapidity with which it produces shade.

Pterodon pubescens, yet another leguminous tree, produces a large yearly crop of oil-seeds. The oil has been known for some time as an inhibitor of *Schistosoma mansoni* cercarial penetration through the skin, an activity related to the presence of 15,16-epoxygeranylgeraniol (*26*). Recently the plant has been shown to confer a long-lasting protection against the cattle horn fly which could be a very useful addition to the already mentioned feedstuff production for local cattle ranchers (Santos Filho, D.; Sarti, S. J., Universidade de São Paulo, Ribeirão Preto, personal communication, 1994). *Pterodon pubescens* does not grow in the forest but rather on the cerrado a relatively open savanna which occurs in patches of sometimes several hundred or even thousand square kilometers over much of Amazonia. Many furanoid diterpenes have been isolated from the species and from others of the genus. It remains to show whether any of these or of the geranylgeraniol derivatives present in the seeds are responsible for the observed activity.

Balsams, Resins and Exudates

When the bark of an Amazonian tree is cut or bored it is common to observe an exudate which may be fluid, resinous, rubbery or gum-like. These exudates probably have the function of protecting the underlying living tissue from infection by micro-organisms, especially fungi, and invasion by larger parasites such as insects. It is not surprising therefore to find antifungal or antibacterial activity in the more fluid exudates and some of these such as copaiba oil are locally and internationally used as microbicidal agents. At present and in the past many of these products have been an item of commerce but have brought little profit to their collectors because of their very low international prices and the succession of intermediary dealers that handle them between collection and final purchase by the end user (Note 2). The collector of copaiba oil will not receive much more than the price of gasoline for his virtually finished product, although an average *Copaifera* sp tree yields from 2.5 to 10 litres per tapping (some species yield more), occurs perhaps no more than once per hectare and must be allowed to rest for several months between one tapping and the next. It is not surprising that the forest dweller prefers to sell his trees to the timber merchant who can sell the wood for perhaps $100 a cubic metre, rather than seek so poor, if continuing, a return. As with the essential oils the development of direct channels for the sale of such highly effective over-the-counter pharmaceutical aids through distributors directly to the public at the retail level is desirable. A bottle containing 20 ml of copaiba oil, even locally in Belém, costs about $2 in the pharmacy which represents about $100/litre, contrasting with the $4/litre export value fob Belém.

Examining examples (Table 5) of the various types of exudate one by one, one may first consider copaiba oil as the most important representative of the liquid terpenoid trunk exudates. The oil contains volatile neutral constituents such as β-caryophyllene whose composition varies greatly from species to species. Dissolved in these are sesquiterpene acids which appear to be responsible for part of the antibiotic activity. Copaiba oil is locally used as a treatment for skin infections and for some of these it is certainly effective. The oil also has anti-inflammatory action (30) but the affirmation that it cures skin cancer is anecdotal.

TABLE V. Examples of Medicinally Used Trunk Balsams or Resins (11,28,21)

Plant Source & Process[1]	Product and Important Constituents	Use	Ref.
Copaifera reticulata and other species of copaíba by boring trunk	oil containing sesquiterpenes and diterpene acids of eperuic and hardwickiic groups	anti-inflammatory, wide range skin antiseptic, also used internally	30
Hymenaea courbaril and other species of jutaí or jatobá, collection around base or by tapping	balsam or copal resin containing diterpene acids of eperuic group	similar to the above	31
Symphonia globulifera, anani, bark extract or exudate	yellow liposoluble resin containing prenylated and hydroxylated xanthones	skin disease treatment	32
Vismia cayennensis and other species of pau lacre, bark, leaf or fruit extract	red or orange resin containing C-methyl and C-prenyl hydroxylated anthraquinones and ring A or ring C reduced analogs	skin disease and wound treatment	33

(1) - The species cited are from two families - Caesalpinaceae and Clusiaceae. Both families feature other trees of similar use, such as, for example, *Eperua falcata* (Caesalpinaceae) whose resin is used in the preparation of a similar use, such as, for example, *Eperua falcata* (Caesalpinaceae) whose resin is used in the preparation of a fixative in perfumery. Other families with commercially important exudates are Burseraceae, especially the genus *Protium*; Sapotaceae, especially the genera, *Achras* (chiclé) and *Manilkara* (balata); Apocynaceae, especially the genus *Couma* (sorva); and Moraceae, notably the genus Brosimum (amapá doce)

A trunk resin that has similar local uses but must be applied in solution in alcohol or other solvent, is the resin from *Vismia* species. *Vismia cayennensis* resin, for example, is used in North-eastern Amazonia both against bacterial and fungal skin infections and on wounds in general as a "healing" agent. It contains, as do many other *Vismia* species, yellow to red hydroxylated and prenylated anthraquinones as well as compounds of similar structure, known as vismiones or ferruginins with one or other of the three rings aliphatic (*33*).

Conclusion

Of the groups listed under the heading Classification above, only a selection have been described in some detail. Others, which include gums, rubbers, and particularly medicines add to the list of opportunities for economic and social development of Amazônia, with simultaneous preservation or restoration of forest cover. It is to be hoped that the present fragmentary efforts can be coordinated to develop industries that benefit the region directly at a level which is competitive with present predatory activities, before it is too late.

Literature Cited

1. Clement, C.R.; Arkoll, D.B. In *Symp.: Workshop on Promising Oil-Plants (Seminário-Taller sobre Oleaginosas Promisorias)*; Forero, L.E., Ed.; PIRB, CIID, Colciências, Funbotânica, ACAC & SECAB, Bogotá, Nov. 4-6, 1985; pp 160-179.
2. Altman, R.F.A. *Publ. INPA, Ser. Quim. 4*: Instituto Nacional de Pesquisas da Amazônia, 1958; pp 1-24.
3. Serruya, H.; Bentes, M.H.S. *Proc. Intern. Mtg. Fats & Oils Tech.*; Campinas, 1991, pp 206-208.
4. Lleras, E.; Coradin, L. In *Symp.: Workshop on Promising Oil-Plants (Seminário-Taller sobre Oleaginosas Promisorias)*; Forero, L.E., Ed.; PIRB, CIID, Colciências, Funbotânica, ACAC & SECAB, Bogotá, Nov. 4-6, 1985; pp 92-143.
5. Marx, F.; Maia, J.G.S. *Acta Amazônica*, **1983**, *13*, 823-830.
6. Gunstone, F.D.; Harwood, J.L.; Padley, F. B. *The Lipid Handbook*; Chapman & Hall, London, 1986, p 99.
7. Clement, C.R. *Ciência Hoje* (Rio de Janeiro) **Dec. 1991**, *special vol. "Amazônia"*, 66-73.
8. Leal, K.Z.; Costa, V.E.U.; Seidl, P.R.; Campos, M.P.A.; Colnago, L.A. *Cienc. Cult.* **1981**, *33*, 75-84.
9. Pereira, P.L. *Dissertation;* Campinas State University, Campinas, SP, 1976.
10. Pinto, G.R. *Bol. Tec. Inst. Agron. Norte* (*Min. Ag., Belém, Brasil*) **1953**, *31*, 209-273.
11. Mors, W.B.; Rizzini, C.T. *Useful Plants of Brazil*; Holden-Day, S. Francisco, 1966; pp 16-36.
12. Pinto, G.R. *Bol. Tec. Inst. Agron. Norte* (*Min. Ag., Belém, Brasil*) **1953**, *31*, 195-206.

13. Balick, M.J.; Gershoff, S. N. *Econ. Bot.* **1981**, *35*, 261-271.
14. Ollis, W.D.; Ward, A.D.; de Oliveira, M.M.; Zelnik, R. *Tetrahedron* **1970**, *26*, 1639-1645.
15. Zapata, A. *Econ. Bot.* **1972**, *26*, 156-159.
16. Leão da Silva, M.; Maia, J.G.S.; Mourão, J.C.; Pedreira, G.; Marx, M.C.; Gottlieb, O.R.; Magalhães, M.T. *Acta Amazônica*, **1973**, *3*, 41-42.
17. Correa, R.G.C.; Leão da Silva, M.; Maia, J. G. S.; Gottlieb, O.R., Mourão, J.C.; Marx, M.C.; Alpande de Moraes, A.; Koketsu, M.; Moura, L.L.; Magalhães, M.T. *Acta Amazônica*, **1972**, *2*, 53-54.
18. Leão da Silva, M.; Luz, A.I.; Zoghbi, M.G.B.; Ramos, L.S.; Maia, J.G.S. *Phytochemistry*, **1994**, *23*, 2515-2516.
19. Araujo, V.C. de; Correa, G.C., Gottlieb, O.R.; Leão da Silva, M.; Marx, M.C.; Maia, J.G.S.; Magalhães, M.T. *Anais. Acad. Bras. Ciênc.* **1972**, *44*, suppl., 317-319.
20. Alpande de Moraes, A.; Mourão, J.C.; Gottlieb, O.R.; Leão da Silva, M.; Marx, M.C.; Maia, J.G.S.; Magalhães, M.T. *Acta Amazônica* **1972**, *2*, 45-46.
21. Gottlieb, O.R.; Koketsu, M.; Magalhães, M.T.; Maia, J.G.S.; Mendes, P.H.; Rocha, A.I. de; Leão da Silva, M.; Wilberg, V.C. *Acta Amazônica* **1981**, *11*, 143-148.
22. Maia, J.G.S.; Leão da Silva, M.; Luz, A.I.R.; Zoghbi, M. das G.B.; Ramos, L.S. *Química Nova* 1987, *10*, 200-204.
23. Lima, R.R., *EMBRAPA-CPATU, Doc. 42, Informações sobre Duas Espécies de Timbó, Derris urucu (Killip et Smith) Macbr. Derris nicou (Killip et Smith) Macbr. como Plantas Inseticidas*; Belém, 1987.
24. Evans, S.V.; Fellows, L.E.; Shing, T.K.M.; Fleet, G. W. J. *Phytochemistry* **1985**, *24*, 1953-1955.
25. Robins, R.J.; Morgan, M.R.A.; Rhodes, M.J.C.; Furze, J. M. *Phytochemistry* **1984**, *23*, 1119-1123.
26. Mors, W.B.; Santos Filho, M.F.; Monteiro, H.J.; Gilbert, B.; Pellegrino, J. *Science* **1967**, *157*, 950-951.
27. Le Cointe, P. *Amazônia Brasileira III, Árvores e Plantas Úteis*; Livraria Clássica: Belém, 1934; p. 23.
28. Grenand, P.; Moretti, C.; Jacquemin, H. *Pharmacopées Traditionelles de Guayane.*;ORSTOM: Paris, 1987; pp 289, 290, 396-405.
29. Pio-Corrêa, M. *Dicionário das Plantas Úteis do Brasil*; Min. Agric., I.B.D.F., Rio de Janeiro, 1984; vol. 2, pp 370-375.
30. Basile, A.C.; Sertié, J.A.A.; Freitas, P.C.D.; Zanini, A.C. *J. Ethnopharmacol.* **1988**, *22*, 101-109.
31. Nakano, T.; Djerassi, C. *J. Org. Chem.* **1961**, *26*, 167-173.
32. Locksley, M.D.; Moore, I.; Scheinmann. *J. Chem. Soc. Org. (C)* **1966**, 2265-2269.
33. Delle Monache, F.; Torres, F.F.; Marini-Bettolo, G.B.; de Lima, R.A. *J. Nat. Prod.* **1980**, *43*, 487-497.

RECEIVED December 10, 1994

Chapter 4

Agroforestry Strategies for Alleviating Soil Chemical Constraints to Food and Fiber Production in the Brazilian Amazon

Erick C. M. Fernandes[1,3] and João Carlos de Souza Matos[2]

[1]Department of Soil Science, North Carolina State University and Empresa Brasileira da Pesquisa Agropecuaria, Caixa Postal 319, CEP 69047−660, Manaus, AM, Brazil
[2]Centro de Pesquisa Agroflorestal da Amazônia Ocidental, Caixa Postal 319, CEP 69047−660, Manaus, AM, Brazil

The main soil chemical constraints on agricultural production systems in the Amazon include soil acidity, phosphorus (P) deficiency, and low effective cation exchange capacity (ECEC). About 80% of the Amazon has acid soils with pH values of less than 5.3 in the topsoil. Associated with soil acidity, is the problem of aluminum toxicity and phosphorus deficiency. The low soil ECEC of most Amazonian soils is also a major soil chemical constraint to plant productivity since the leaching of mobile nutrients (such as potassium) increases as ECEC decreases. Agroforestry systems have the potential to control erosion, maintain soil organic matter and soil physical properties, augment nitrogen fixation, and promote efficient cycling of the scarce nutrients in Amazonian soils. The rationale for supposing that agroforestry systems can alleviate the soil chemical constraints to food and fibre production is that the nutrient exports via the harvest of crop and animal products, erosion, leaching, volatilization, and the deterioration in soil physical properties due to cropping or grazing can be counteracted by the tree component via: 1) nutrient uptake by deep rooted trees allowing for capture and surface deposition via tree litter, of nutrients beyond the reach of crop roots (more efficient nutrient cycling), 2) increased amounts of organic (shoot and root) inputs to the soil help to maintain soil organic matter and thus improve soil structure and nutrient status, and 3) increased nutrient additions to the soil via nitrogen fixation.

[3]Current address: Department of Soil, Crop, and Atmospheric Sciences, Cornell University, Bradfield Hall, Ithaca, NY 14853

Agroforestry refers to a range of land use systems in which trees are managed together with crops and/or pastures. The trees may be present on a given land unit at the same time as crops or pastures (zonal agroforestry) or may be rotated with the herbaceous species (rotational agroforestry). Traditional agroforestry systems have played a significant role in sustaining populations in a variety of tropical soil and climate conditions (*1*). Apropriate agroforestry systems have the potential not only to maintain the productivity of currently cropped lands, but also to rehabilitate abandoned crop and pasture lands. As many forest products (fuel wood, fruits, nuts, medicinal plants) commonly gathered by rural populations are now increasingly in short supply due to deforestation, agroforestry systems could substitute for these natural forest production systems. The objective is to establish agroforestry systems on already deforested lands for the production of food, fruits, and other forest products currently being extracted in an uncontrolled manner from primary forests.

The chemical and physical properties of soils determine the species composition, above and below ground biomass production capacity, and hence the nutrient cycling and soil conservation potential of the vegetation. The focus of this paper is to identify ways in which the inherent biological and structural diversity of agroforestry systems can be harnessed for the sustained production of food and fibre on already deforested land without the necessity to clear additional forest. As the bulk of the deforested land in the Brazilian Amazon is currently poor quality or abandoned pasture (*2*), we also envisage a major role for agroforestry systems in the rehabilitation and conversion of abandoned pasture land to diversified tree-crop-livestock (agrosilvopastoral) systems. The land use policy changes that will be required for the socially just distribution of appropriate incentives and subsidies to promote the conversion of abandoned pastures to productive agroforestry are beyond the scope of this paper.

Soil Chemical Factors Constraining Amazonian Food and Fibre Production Systems

Nearly 75 percent of the Amazon basin contains acid, infertile soils classed as oxisols and ultisols. Upland ("terra firme") soils such as Oxisols, Ultisols, and some Entisols and Inceptisols, are characterized by low nutrient reserves, low effective cation exchange capacity (ECEC), high aluminum toxicity, and low phosphorus availability (*3*). The "varzea" or river bank soils of the white water rivers are generally more fertile due to the replenishment of nutrients via sediments deposited by flooding. Dark water varzeas are generally infertile sands and not normally used for agriculture. While Oxisols are likely to have very low levels of potassium, calcium and magnesium, Ultisols may present greater problems of Al toxicity because of higher exchangeable levels of Al. The high levels of exchangeable aluminum in Ultisols, some Inceptisols, Oxisols, and Spodosols can severely restrict root growth, nutrient uptake, and hence nutrient cycling (*4*). Phosphorus fixation is usually high and hence P availability is low on Oxisols. Sandy soils are especially low in nitrogen, although phosphorus, calcium and magnesium may also be low.

Of the 482 million hectares in the Amazon Basin, 81% of the area had native pH values in the topsoil less that 5.3 and 82% had native pH values less than

5.3 in the subsoil. Associated with these low pH values is aluminum toxicity. Cochrane and Sanchez (3) reported that 73% of the soils in the Amazonian basin have an aluminum saturation of 60% or more in the top 50 cm. Ninety percent of the soils in the Amazon have topsoil P levels less than 7 mg/kg (3). Assuming a critical P level of 10 mg/kg, these soils will not support crops without additions of P. Fortunately, only 16% of the soils are estimated to be strong P fixers, that is, they have over 35% clay and a high percentage of iron oxides. The remaining soils can be managed by prescribing small P applications on a crop by crop basis.

The low ECEC is considered to be a soil constraint (5). The susceptibility of leaching of mobile nutrients increases as ECEC decreases. This is of major importance in an environment where rainfall exceeds potential evapotranspiration most of the year and where nutrients are in short supply to begin with. It is critical that mobile nutrients added to the soil remain in the soil as long as possible giving the plant adequate opportunity to utilize them. For example, potassium is considered to be a constraint on 56% of the land area (3). It is important to note that these critical values are for traditionally grown crops and some pasture legumes. Almost nothing is known about the critical soil nutrient contents for the highly adapted tree and shrub species that are components of agroforestry systems.

Slash and Burn Agriculture and Agroforestry in the Amazon

Traditionally, farmers in the tropics have temporarily overcome soil acidity and low soil nutrient contents by slashing and burning the forest vegetation. Farmers typically clear about a hectare of primary or secondary forest, burn it and then plant crops for one or more years taking advantage of the nutrients released in the ash. The quantities of nutrients accumulated in the forest biomass are typically in the range 100-600 kg/N/ha, 10-40 kg/P/ha, 200-400 kg/K/ha, 150-1125 kg/Ca/ha, and 30-170 kg/Mg/ha (6,7). Between 20% to 40% of the biomass burns to ash. Data from measurements of the nutrient contents of ash at various sites in the Amazon are presented in Table I. Large losses of nutrients occur during the burn probably via the physical removal of ash via air currents generated during the burn. Approximately 88-95% of N, 42-51% of P, 30-44% of K, 33-52% of Ca, and 31-40% of Mg contained in the above-ground biomass was reported lost during burning (8,9). After burning, further loss of nutrients from the site continues via rainfall run-off and soil erosion, leaching, volatilization, and crop harvests. Once yields decline as a result of decreasing soil fertility and/or increasing weed pressure, the site is abandoned. The accumulation of nutrients and shading out of weeds by the regenerating forest vegetation, and the action of roots and associated microorganisms and fauna are the processes by which the potential soil productivity of the abandoned site is gradually improved to a state approaching that of a primary rainforest.

Due to increasing population pressure and new land use policies that prohibit deforestation and burning of primary forests, regenerating forest fallows are under increased pressure from farmers. The fallows are being slashed and burned well before soil productivity has been recovered which results in rapid and

Table I. Nutrient Contribution of Ash upon Burning of Primary Forests and Fallow Regrowth at Different Sites in the Amazon

Location and Soil	Vegetation	Ash Dry weight	Nutrient Additions								
		t ha^{-1}	N	Ca	Mg	K	P	Zn	Cu	Fe	Mn
							kg ha^{-1}				
Manaus, Brazil Xanthic Hapludox	Primary forest	9.2	80	82	22	19	6	0.2	0.2	58	2.3
	Secondary forest (12 years)	4.8	41	76	26	83	8	0.3	0.1	22	1.3
	Abandoned pasture (5 years regrowth)	2.2	18	58	14	40	3	--	--	--	--
Yurimaguas, Peru Typic Paleudult	Secondary forest (25 years)	12.1	127	174	42	131	17	0.5	0.2	4	11.1
	Secondary forest (17 years)	4.0	67	75	16	38	6	0.5	0.3	8	7.3
	Secondary forest (11 years)	1.1	10	217	51	81	8	0.7	0.1	2.7	3.4

Source: Ref. 23-26.

sometimes permanent soil degradation. Farmers urgently require land use alternatives that fit diverse socioeconomic and ecological conditions and sustainably produce enough food and wood without degrading the soil resources required for future production. We suggest that such potentially sustainable systems must have as many of the following characteristics as possible:

- Biological and structural diversity to minimize biophysical (pests, drought) and economic (volatile markets) risk and provide resilience so that both the farmer and the system survive the occasional years of severe drought or pest attack.
- A high degree of soil cover via plant canopies or plant residues left on the soil surface.
- High value but low biomass of products to minimize nutrient exports.
- The return of crop residues (or the animal manure) to the cropped land to maximize nutrient recycling.
- The application of adequate levels of organic and/or inorganic fertilizers to balance nutrients removed in harvests.
- The use and optimum management of improved crop and animal species or varieties.

Many of the above characteristics are commonly found in existing agroforestry practices in the tropics (10). These systems generally have low to medium capital requirements and produce a range of food, wood and other economically useful goods. In addition, agroforestry systems are characterized by a variety of potential service roles such as soil conservation and the maintenance of soil fertility (11-13). The key difference between agroforestry and sole cropping systems lies in the potential for farmers to incorporate tree species that are considerably more tolerant to acid soil conditions than common annual crops. The main functions of these adapted tree species is to protect the soil, conserve existing soil nutrients, enhance the uptake and the recycling of added organic or inorganic fertilizer nutrients. A discussion on the various tree-soil improvement hypotheses using data from the tropics was undertaken by Fernandes et al., (2). A key issue that requires further research is the search for appropriate tree species, their improvement, and the development of suitable management strategies to optimize the role of trees as soil improvers.

Tree Characteristics for Maintaining Soil Productivity Under Agroforestry

Fast-growing, leguminous tree species are major components of agroforestry technologies (14). Adaptability to soil chemical and physical conditions is an obvious but vital requirement for any tree species to contribute to soil improvement. Based on studies from the Amazon and other regions in the humid tropics (2,4,15), we suggest the following criteria for judging the suitability for soil improvement of a tree species or provenance:

- Shoot biomass production: 8 to 10 Mg/ha/yr from one to four prunings a year (Table II).

Table II. Biomass Production by Tree Species in Alley Cropping on Fertile and Infertile Soils in the Humid Tropics (Adapted from Ref. 16)

Species	Trees/ha	Tree Age (mo)	Prunings (No/yr)	Dry Matter (Mg/ha/yr)
Yurimaguas, Peru; rainfall 2200 mm/yr; Ultisol, pH 4.2-4.6, P = 8 ppm (Olsen)				
Inga edulis	8888	11	3	09.6 l+w
Gliricidia sepium 14/84	5000	11	3	•08.1 l+w
Gliricidia sepium 34/85	5000	11	3	01.8 l+w
Onne, S.E. Nigeria; rainfall 2400 mm/yr; Ultisol, pH 4.0, P = 50 ppm (Bray-1)				
Acioa barteri	2500	48	n.d.	13.8 l+w
Alchornea cordifolia	2500	48	n.d.	14.9 l+w
Cassia siamea	2500	48	n.d.	12.2 l+w
Gmelina arborea	2500	48	n.d.	12.3 l+w
Sumatra, rainfall 2575 mm/yr; Oxisol, pH 4.1, P = 4.8-6.8 mg/kg (Melich I)				
Paraserianthes falcataria	19900	09	4	04.9 l+w
		21	4	09.7 l+w
Calliandra calothyrsus	19900	09	4	06.8 l+w
		21	4	10.7 l+w
Gliricidia sepium	10000	09	4	00.6 l+w
		21	4	01.4 l+w
Costa Rica, rainfall 2640 mm/yr; Inceptisol, pH 4.3-4.8, P = 8-15 ppm (Olsen)				
Gliricidia sepium	6666	24	2	09.6 l+w
		60	2	15.2 l+w
Erythrina poeppigiana	555	24	2	07.4 l+w
		60	2	11.1 l+w
Western Samoa, rainfall 3000 mm/yr; mod. fertile Inceptisol, no soil data				
Calliandra calothyrsus	5000	48	3	12.1 l+w
	3333	48	3	07.6 l+w
Gliricidia sepium	5000	48	3	10.7 l+w
	3333	48	3	06.5 l+w

l = leaves and green shoots, w = woody material.

- Nitrogen fixation potential of 10 to 50 kg/N/ha/yr.
- Most of the tree fine roots (<2mm diameter) concentrated below the depth (15-20 cm) at which the bulk of annual crop roots are found.
- Capacity to form effective mycorrhizal associations with native populations of Vesicular-Arbuscular (VA) mycorrhizal fungi or ectomycorrhizal fungi in order to enhance the utilization efficiency of low levels of native soil P and small amounts of P added via fertilizers.
- Moderate to high nutrient concentrations in leafy biomass (e.g. 2.0-3.5% N, 0.2-0.3% P, 1-3% K and 0.5-1.5% Ca). Interpreting data for micronutrient concentrations in tree and shrub biomass is still difficult. Leafy biomass derived from vigorously growing trees of *Inga edulis* grown on an Ultisol were as follows: Mn 112, Cu 13, Zn 35, and Fe 95 mg/kg (*17*).
- Rapid litter decay (1 to 3 weeks) where tree biomass is used to provide nutrients to associated crops (e.g. *Leucaena leucocephala, Sesbania sesban, Gliricidia sepium*) or slow decay (2 to 6 months) when tree biomass is used as mulch for weed suppression and soil protection (e.g. *Inga edulis* and *Flemingia macrophylla*).
- Absence of toxic substances in the foliage or root exudates. For example, in the acid savannas (*cerrado*) of Brazil some species have been shown to accumulate from 4,000 to 14,000 mg/kg of Al in the foliage (*18*). Non-accumulator species have Al concentrations < 200 mg/kg.

Several promising tree and shrub species have the potential to improve soils. Most commonly preferred tree species tend to be nitrogen-fixing. For acid soils in the Amazon, the following tree or shrub species appear to have good potential: *I. edulis, Calliandra calothyrsus, Flemingia macrophylla, Gliricidia sepium, Paraserianthes falcataria,* and *Senna reticulata* (*2*).

Indigenous and Introduced Agroforestry Systems in the Amazon. Amerindian peoples of the Amazon have long planted and managed trees for a variety of products and services in close association with annual and perennial food crops (*19,20*). The Kayapó create "resource islands" of trees, shrubs, herbs, and root crops at the forest margin and also in open grasslands. These species are generally collected as seedlings in the forest and transplanted to clearings and campsites. Over a hundred species have been encountered in these "agroforestry" islands (*21,22*). A mosaic of different aged, multistoried plots established and managed by the Arara Indians in the vicinity of the Transamazon Highway contained at least 19 varieties of plants belonging to 13 species. The crop and tree species included squash (*Cucurbita* spp.), sweet potato (*Ipomoea batatas*), banana (*Musa* spp.), ginger (*Renealmia occidentalis*), pineapple (*Ananas comosus*), cotton (*Gossypium hirsutum*), annato (*Bixa orellana*), and araticum (*Anona nitida*).

The great majority of these traditional systems involve managing the fallow vegetation (for a variety of products and services) following cropping and subsequent abandonment to permit regeneration of forest species. Most of the Amazonian tree crops in use today were probably domesticated via these traditional systems. In addition to a high species diversity, indigenous agroforestry systems are characterized by a variety of species associations of different age classes spread over several sites. These system characteristics maximize labor efficiency per unit area of land, minimize tree and crop failure due to drought or severe pest attack, and guarantee the availability of food at relatively modest levels of species productivity. Traditional agroforestry systems are generally for subsistence and support low population densities.

Migrants to the Amazon and their descendants have introduced and developed a great variety of agroforestry systems that attempt to combine the domesticated species and structural diversity of Amerindian systems with new species and rotations to meet the household food and cash needs. Most such implanted systems also start off with annual and semi-perennial crops (rice, cassava, banana) following slashing and burning of the forest vegetation. The next phase generally involves the establishment of a variety of tree crops and incorporation of livestock. By themselves, these tree crops do not constitute agroforestry. The planting of tree crops, however, may follow annual crops (rotational agroforestry) or the tree crops may be present at the same time as annual crops and or livestock (mixed or zoned agroforestry). Given the ecological and economic interactions between the first phase annual crops and the second phase tree crops and/or livestock, however, most of these systems can be classified as agrisilviculture, silvopasture or agrisilvopasture.

Tree-based homegardens are typical agrosilvopastoral practices involving a variety of both native and exotic species for fruit, timber, shade, medicines, spices, and forage (Table III). As many as 190 plant species at various stages of domestication have been recorded in these agrosilvopastoral systems in the tropics (*1*). The high species diversity and sustainability of homegardens make them ideal for use in buffer zones around extractive reserves and protected forests and improves the chances for gene flow from wild to semi-domesticated populations of selected food and fruit species. In the Amazon, systems involving around 30 perennial and annual plant species has been reported from Para, Brasil (*27*) and over 70 species from Peru (*28*). Multi species, tree-based homegardens have a high degree of ecological and biological sustainability coupled with good social acceptability. The factors that promote sustainability include diversified production, reduced risk of crop failure, enhanced labor efficiency, continuous production thereby minimizing post harvest losses, good nutrient cycling and reduced erosion because of good ground cover. The species diversity is greatest close to the house and declines as one moves further away. In most cases, homegardens contain food producing species and other high value species where the farmer can protect them more easily. In Africa, where deforestation is resulting in significant loss of biodiversity, homegardens have been identified as important *in situ* germplasm banks of semi-domesticated species (*29*). Amazonian homegardens could also play a vital role in helping to preserve native, semi-domesticated species in areas where the pressure of deforestation is high.

Table III. Tree and Crop Species Encountered in Surveys of Agroforestry Homegardens in the States of Acre, Amazonas, Para, Rondonia and Roraima (Fernandes, E. C. M., Unpublished Data)

Common Name	Scientific Name	Uses
Abacate	*Persea americana*	fruit
Coco	*Cocos nucifera*	food, oil, cash crop
Guaraná	*Paulinia cupana*	drink, cash crop
Tucumã	*Astrocaryum aculeatum*	fruit, fibre
Fruta pão	*Artocarpus altilis*	seeds
Jaca	*Artocarpus heterophyllus*	fruit, seeds
Goiaba	*Psidium guajava*	fruit
Limão	*Citrus aurantifolia*	fruit, cash crop
Manga	*Mangifera indica*	fruit
Pupunha	*Bactris gassipaes*	fruit, palm heart
Caju	*Anacardium occidentale*	fruit, nut cash crop
Abacaxi	*Ananas comosus*	fruit
Cupuaçu	*Theobroma grandiflorum*	fruit, cash crop
Annatto	*Bixa orellana*	seeds for dye
Acerola	*Malpigia glabra*	fruit
Pimenta do reino	*Piper nigrum*	spice, cash crop
Cacao	*Theobroma cacao*	Seeds, cash crop
Banana	*Musa* spp.	fruit, cash crop
Café	*Coffea canephora*	drink, cash crop
Tapereba	*Spondias mombin*	fruit
Ingá	*Inga edulis*	fruit, fuel wood
Biriba	*Rollinia mucosa*	fruit, cash crop
Graviola	*Anona muricata*	juice, ice cream
Açai	*Euterpe oleracea*	fruit, palmito
Araça boi	*Eugenia stipitata*	juice
Jambu	*Eugenia jambos*	fruit
Pitanga	*Eugenia uniflora*	fruit, juice
Mamão	*Carica papaya*	fruit
Caimito	*Pouteria caimito*	fruit
Sapotilla	*Manilkara zapota*	fruit, chewing gum
Bacuri	*Platonia insignis*	fruit
Genipapo	*Genipa americana*	fruit, wood
Araticum	*Anona montana*	fruit
Bacaba	*Oenocarpus bacaba*	wine, wood, leaf baskets
Mandioca	*Manihot esculenta*	tubers for starch
Maracujá do mato	*Passiflora nitida*	fruit
Maracujá caiano	*Passiflora macrocarpa*	fruit
Umari	*Poraqueiba sericea*	fruit
Seringa	*Hevea brasiliensis*	latex
Mapati	*Pourouma cecropiaefolia*	fruit
Cubiu	*Solanum sessiliflorum*	fruit
Pitomba	*Talisia esculenta*	fruit
Carambola	*Averrhoa carambola*	fruit
Buriti	*Mauritia flexuosa*	fruit

Agroforestry Practices with Potential for the Amazon

We envisage that agroforestry systems are more appropriate than monocropping systems for already deforested land and not as alternatives to primary forest. Several agroforestry practices, that are not common in the Amazon, have the potential to significantly reduce deforestation while permitting food and wood production without soil degradation. On already degraded sites and abandoned pastures, these same systems can play a major role in the rehabilitation of site productivity. The key factor for abandoned pastures will be providing a modest input of chemical nutrients to prime the nutrient cycling pump of agroforestry systems.

Managed Tree Fallows. Traditionally, after cropping a patch of cleared land for two to four years, farmers abandon it to a natural forest fallow. We argue, however, that promoting the establishment and managing fast-growing leguminous and other tree species in fallows may significantly reduce the time taken for soil productivity to recover to original forest levels (*30*). This is due to the rapid growth rates and specialized nutrient accumulation strategies that enable these species to take up and concentrate nutrients in biomass even at very low soil concentration levels. Prinz (*31*) identified the following factors for successful management of tree fallows: a soil that has not been completely depleted, good regeneration of natural forest species and/or natural forest adjacent to the fallow area, and a sufficiently long fallow period to permit complete regeneration of the soil. Ideally, tree species for fallows should be capable of fast growth, high biomass accumulation, possess extensive root systems, and high nutrient uptake capacity. In addition, the rapid establishment of tree species in fallows could significantly improve not only the seed dispersal of secondary fallow species due to increased habitats for perching and foraging by birds and bats, but also create favorable microhabitats for the germination of other fallow species. To make the concept of tree fallows attractive to farmers, it is important that at least some of the fallow species yield economically useful products.

Fallows with High Nutrient Accumulating Species. In a study in the Peruvian Amazon, Szott (*30*), found that eight months after establishment, managed fallows with leguminous species had higher nutrient stocks than the natural forest fallow control. By 29 months, treatments with tree species had significantly higher nutrient stocks than treatments with natural forest species or no tree species. In another study on degraded pastures in the Brazilian Amazon, the concentrations of N, P, and K in leaves of *Laetia procera* were double those found in leaves of various species from primary forest vegetation on acid infertile soils (*9*). For example, *Cecropia* spp. have been reported to accumulate Ca and P on acid soils (*32*). Other tree species reported to colonize degraded and abandoned pastures include *Vismea* spp., *Goupia glabra, Bellucia grosulariodes, Dipteryx odorata,* and *Zanthoxylum procerum* (*33*). Based on these observations, it is possible that by enriching natural forest fallows with these high nutrient accumulating species, the time to recover soil productivity could be significantly reduced.

Fallows with Economically Valuable Species. The inclusion of species with economically valuable products (flowers, fruits, homeopathic medicines, essences, resins) provides an economic return to tide farmers over until the fallows biological, ecological, and site rehabilitation potential has been realized. For example, fruit species (*Anona muricata, Bactris gassipaes, Eugenia stipitata, Rollinia mucosa, Theobroma grandiflorum*) and a variety of plant species with medicinal compounds are likely to provide earlier and more substantial monetary returns (relative to natural fallows). It is likely that several potentially valuable species await discovery in the Amazon. The strategy with economically enriched fallows is to use species that produce a low volume but high value product so as to avoid excessive export of nutrients from the site. Mixing species that produce an economically valuable product with nutrient accumulating species gives the farmers a good incentive to maintain a forest fallow for the minimum time required to recover soil productivity.

Alley Cropping. This system has potential to sustain crop productivity via improved soil protection, nutrient cycling, and reduced weed pressure (*34*). The system involves the growing of annual food crops in the alleys formed by hedgerows of fast-growing, nitrogen-fixing trees. The hedgerows are pruned periodically to provide green manure or mulch for the crops in the alleys and to minimize shading and root competition by the hedgerows. While experimental results from several studies on fertile soils (mainly Alfisols and Entisols) show that alley cropping can sustain crop yields, maintain soil nutrient status and prevent soil organic matter decline (*34*), unsatisfactory short to medium term crop yields have been reported from alley cropping trials on acid infertile soils of the Amazon (*35*). Low nutrient contents of Oxisols and Ultisols (the dominant soils in the Amazon) result in high competition between hedgerows and annual crops. Nutrient cycling by the hedgerows is often at the expense of the crop in the alley. The labor required for the frequent pruning of hedgerows in order to reduce tree-annual crop competition is prohibitive.

We suggest that alley cropping on acid soils in the Amazon should be used for the production of perennial rather than annual crops. Perennial crop species that could grown in such a system include *Bactris gassipaes, Theobroma grandiflorum, Bertholettia excelsa, and Eugenia stipitata.* The chief function of the hedgerows would be to minimize runoff and erosion and control weeds via the provision of mulch. Suitable hedgerow species for acid soils include *Inga edulis, Gliricidia sepium, Cassia reticulata, Flemingia congesta, Calliandra calothyrsus,* and *Paraserianthes falcataria* (*16*). Since weed control is often a major factor reducing successful establishment and yields of many perennial crops, selection of commonly known species such as *Inga edulis*, a species with slowly decomposing litter, could significantly reduce weed pressures (*15*) and increase chances of the system being adopted by farmers. The traditional model of alley cropping for annual crop production may only be justified on sloping land where the danger of soil erosion and rainfall runoff is high.

Alley cropping could be used for the establishment of high value forest species as well. Hedgerows could be established with annual crops in the alley and after a year or so, forestry species planted in the alleys with the hedgerows serving as shade ("nurse") trees for the next 5 to 10 years. Thereafter, periodic lopping of the hedgerows would provide mulch to aid in weed control and recycle nutrients in hedgerow biomass.

Tree-Pasture (Silvopastoral) Systems. Despite the beneficial effects of perennial crop plantations on soil conservation and nutrient cycling (*36*), sustainable production of tree crop plantations (mainly rubber and oil palm) in the Amazon has rarely been possible due to a combination of pest problems and poor economic returns. The addition of a cover crop and an animal component provides additional flexibility with respect to markets, economic returns and the purchase of required inputs. In addition to soil protection, leguminous cover crops (such as *Centrosema macrocarpum, Desmodium ovalifolium, Pueraria phaseoloides*) can contribute to improve tree root growth and soil N via nitrogen fixation (*37*). Even in the absence of leguminous cover crops, the integration of sheep, poultry and bees in tree crop plantations can be not only socially and economically attractive, but also result in a significant reduction in the cost of weed control in the plantations.

Given the vast national and international market potential for timber, the establishment of timber species in pastures would considerably improve the long-term economic returns and justify incentives and subsidies in the short term to help establish improved (grass-legume) pastures. An example of such a system in Paragominas, involving *Schizolobium amazonicum, Bagassa guianensis, Eucalyptus tereticornis* and various pasture grasses was reported by Veiga et al., (*38*). Other potentially useful tree species for pastures include *Carapa guianensis, Cedrelinga catenaeformis, Cordia goeldiana, Swietenia macrophylla* and other economically important tree species. Fast-growing, "nurse species" (e.g. *Schizolobium amazonicum, Sclerolobium paniculatum, Inga edulis*) can be used to yield economic returns in the short term while protecting the more valuable timber species from wind and pest damage in the first six to ten years.

Live Fence Posts and Hedges. In all systems involving livestock, the use of fencing for animal control requires a large number of fence posts. The continuous removal of young trees from primary and secondary forests for establishing and maintaining fences is a serious and largely unnoticed form of deforestation. An agroforestry alternative involves the planting of large (1.5 - 2 m) woody cuttings that have the capacity to take root and continue growing, thereby producing live fence posts. Planting of such cuttings more densely can result in hedges. Species that are commonly used for live fence posts and hedges include *Gliricidia sepium, Erythrina* spp., *Spondias* spp., *Pithecellobium dulce* (*39*). Live fence posts and hedges could have a significant impact against deforestation, help reduce soil erosion from cropped fields, and help contain nutrient loss in rainfall runoff.

Modified "Taungya". The system was developed as an agrisilvicultural (involving crops, trees and livestock) practice for the recovery of deforested and degraded land via plantings of timber trees and food crops (*40*). Shifting cultivators are attracted to establish forest plantations by various incentives such as housing, clinics, schools, and tenure over permanent agricultural plots. The farmers are required to help establish and maintain forest plantations, in which they are permitted to raise agricultural crops during the first three to four years of establishment. In the Amazon, the modified "Taungya" scheme could be used to reforest the large areas of degraded uplands while providing employment and settlement for landless migrants. Brienza and Yared (*41*) reported satisfactory establishment at reduced costs of *Cordia goeldiana, Jacaranda copaia,* and *Bagassa guianensis* via intercropping the trees with cow pea (*Vigna unguiculata*). Incentives and credit facilities to permit intensification of land use on the farmer's permanent field plots would help towards self sufficiency in food production.

Fiscal incentives and capital costs for the modified "taungya" approach can be justified on the grounds of social equity, the fact that shifting cultivators would become agents of reforestation, the reduced need for timber harvests from primary rain forests if forest plantations are successfully established. These forest plantations would also play an active role in reducing atmospheric levels of the greenhouse gas CO_2 since the carbon accumulation by growing forest plantations is 20 to 100 times greater than that of degraded pastures (*42*). Plantations of timber species have the potential for fixing atmospheric carbon over a significantly longer period than plantations for fuel wood, biomass or pulp. The carbon stored in soils is nearly three times that in above-ground biomass and approximately double that in the atmosphere (*43*). Numerous studies have shown that forest plantations on degraded soils can increase soil organic matter contents via enhanced production of above-ground litter (*44*) and high turnover of fine roots (*45*).

Management Strategies for Sustaining Soil Productivity Via Agroforestry Systems

On acid, infertile soils (Oxisols, Ultisols, Dystropepts, Psamments, and Spodosols) the potential of trees to sustain soil productivity lies in their role for maximizing soil cover (via canopies and mulch) and minimizing the loss of scarce nutrients via rainfall run off, soil erosion, and leaching (*4*). Due to reduced nutrient losses in systems with trees, the nutrient use efficiency of relatively small amounts of chemical fertilizers can be significantly improved.

We suggest the following strategies to enhance the potential of agroforestry systems to sustain soil productivity on recently deforested acid, infertile soils in the Amazon:

- Selection and use of acid-soil tolerant tree (wood, forage, medicines, resins, dyes), food and cover crop germplasm and compatible rhizobia and vesicular arbuscular (VA) mycorrhizal fungi.

- .Use of a variety of leguminous crop and tree species to harness the potential of biological nitrogen fixation. Given the low levels of available P in most acid soils, enhancement of available soil P status and the uptake of P by the plant will be important since both nodulation and N_2 fixation require phosphorus. Improving P availability and uptake can be achieved by ensuring effective mycorrhizal associations and the addition of 20 to 50 kg of P ha^{-1} since mycorrhizal fungi do not manufacture P but only help in P uptake by the plant.
- If possible, establish rapid soil cover via a fast-growing, herbaceous leguminous cover (e.g. *Mucuna pruriens, Canavalia brasiliensis*) crop immediately after slashing and burning of the original vegetation. The function of the cover crop is to protect the soil surface, take up the nutrients released in the ash, and control weed growth. At the appropriate time during the planting season, the herbaceous cover crop can be slashed and the agroforestry system established.
- Nutrients removed in grain, leaf, and wood biomass should be replaced via organic and inorganic fertilizers otherwise the system ceases to be sustainable. On recently deforested sites, additional nutrients may not be required for several years before productivity declines set in. Abandoned pastures, however, show a combination of soil chemical and physical constraints to crop production and will require inputs to alleviate soil compaction and nutrient deficiencies (principally phosphorus and calcium).
- On degraded and abandoned pastures, we hypothesize that a modest input of chemical fertilizers (25 kg N, 50 kg P, 50 kg K, 25 kg Ca, and 25 kg of Mg per hectare) will be essential to prime the nutrient cycling pump of agroforestry systems. Ideally, these nutrients should be added to the first crop established.

Conclusions

The major advantage of agroforestry systems on acid, infertile soils of the Amazon lies in the superior soil conservation characteristics relative to traditional slash and burn or introduced sole cropping systems. The potential to use highly acid-soil tolerant tree species for biomass and nutrient accumulation, nitrogen fixation, to form physical barriers against soil erosion, and to produce a diverse range of economically valuable products makes these systems naturally attractive to farmers. Since the bulk of new land clearings for agriculture are projected for the Amazon region with acid, infertile soils, work is urgently needed on optimizing the sustainability and performance of agroforestry systems on these marginal soils.

Key research issues that need systematic and long term attention include: 1) identifying potential social and economic barriers to the adoption by farmers of improved and potentially sustainable agroforestry systems, 2) the selection and improvement of tree, crop, and pasture germplasm both for tolerance to acid soils and for growth in mixed species associations, 3) the simultaneous selection of effective symbiotic partners (rhizobia and/or mycorrhizal fungi) for these plant species, 4) characterizing the role of management (spacing, rotation, pruning, mulching, and minimum chemical fertilizer applications) on the productivity, nutrient cycling potential and sustainability of agroforestry systems.

Agroforestry systems that involve combinations of annual crops with a variety of tree species and in some cases livestock are being evaluated in an on-going project, for their potential to rehabilitate abandoned pastures. The joint project between the Department of Soil Science and the Brazilian Enterprise for Agricultural Research (EMBRAPA) is being funded by the Rockefeller Foundation. Several agroforestry systems involving annual crops, perennial crops, leguminous, herbaceous and tree species, and valuable timber trees are being studied for their potential to rehabilitate soil productivity of abandoned pastures in the western Amazon. Besides the biological productivity of these systems, the economics of establishing and maintaining these systems and their impact on greenhouse gas emissions are also being studied. Local farmers are also contributing to the evaluation of similar system prototypes via on-farm trials. Data from the first two years suggest that rotations and/or combinations of acid-soil-tolerant crop, tree, and pasture germplasm is an important strategy to establish a rapid and productive vegetative cover for the soil. Nitrogen-fixing species together with modest P applications can significantly improve plant biomass production, and nutrient recycling.

Large numbers of publications debating deforestation rates, greenhouse gas emissions and highlighting the speculative nature of the process of deforestation have done little to provide concrete solutions to countering the effects of deforestation or more importantly improve the livelihood of the people currently slashing and burning forests for their daily subsistence. It is worth noting that the bulk of the deforestation for subsistence agriculture by small farmers has been brought about by such "sophisticated" technological interventions as the machete, the axe, and the matchbox! Pending major land policy reform, the deforestation undertaken by small holder farmers in the Amazon will continue to increase for the foreseeable future. Agroforestry systems are not the miraculous solution that will halt all future deforestation by subsistence farmers. Rather, such systems offer the very real possibility that farmers practicing them may need to use "the machete and the matchbox" much less frequently because of better soil conservation, higher nutrient recycling, diversified products and hence minimized risk of food and/or cash crop failure. The cumulative effect on reducing deforestation and improving the standard of living of the "farmers in the forests" will be significant pending the results of endless ecological and political debates.

Literature Cited

1.   Fernandes, E. C. M.; Nair, P. K. R. *Agric. Syst.* **1986**, *21*, 279-   310.
2.   Fernandes, E. C. M.; Neves, E.; Mattos, J. C. In *Forestry for Development: Policy, Environment, Technology and Markets. Proceedings of the 1st Panamerican Forestry Congress.* Brazilian Society of Silviculture and Brazilian Society of Foresters, São Paulo, 1994; pp 96-101.

3. Cochrane, T. T.; Sanchez, P. A. In *Amazonia: Agriculture and Land Use Research*;. Hecht, S. B., Ed.; CIAT, Cali, Colombia, 1982; pp 137-209

4. Szott, L. T.; Fernandes, E. C. M.; Sanchez, P. A. *Forest Ecology and Management* **1991**, *45*, 127-152.

5. Cassel, D.K.; Lal, R. *Soil Sci. Soc. Am. Special Publication* **1992**, *29*, 61-89.

6. Sanchez, P. A.. *Properties and Management of Soils in the Tropics*; John Wiley; New York. 1976; 618 p.

7. Andriesse, J. P. *Monitoring Project of Nutrient Cycling in Soils Used for Shifting Cultivation under Various Climatic Conditions in Asia.* Final Report: Part I; Royal Tropical Institute: Amsterdam, The Netherlands, 1987; 141 p.

8. Coutinho, L. M. *Ciência Hoje* **1990**, *12*, 23-30

9. McKerrow, A. J. *M.Sc. Thesis*, North Carolina State University, Raleigh, NC. 1992.

10. Nair, P. K. R.. *An Introduction to Agroforestry*; Kluwer, Dordrecht, Netherlands, 1993.

11. Lundgren, B. L.; Raintree, J. B. In *Agricultural Research for Development: Potentials and Challenges in Asia*; Nestel, B., Ed.; ISNAR, The Hague, 1982; pp 37-49

12. Nair, P. K. R. *Agrofor. Syst.* **1984**, *3*, 97-128.

13. Young, A. *Agroforestry for Soil Conservation. Science and Practice of Agroforestry*; 1989, No. 4, C.A.B. International/ICRAF. Wallingford, UK.

14. Nair, P. K. R.; Fernandes, E. C. M.; Wambugu, P. A. *Agrofor. Syst.* **1984**, *2*, 145-163.

15. Szott, L. T.; Palm, C. A.; Sanchez, P. A. *Adv. Agron.* **1991**, *45*, 275-301.

16. Fernandes, E. C. M.; Garrity, D. P; Szott, L. A.; Palm., C. A. In *Tropical Trees: The Potential for Domestication*; Leakey, R. R. B.; Newton, A. C., Eds.; Proc. IUFRO Centennial Year (1892-1992) Conf. Edinburgh,. HMSO: London, 1993; pp. 218-230.

17. Fernandes, E. C. M.. *Ph.D. Dissertation*, North Carolina State University, Raleigh, USA. 1990

18. Haridasan, M. *Plant and Soil* **1982**, *65*, 265-273.

19. Denevan, W,M. and Padoch, C. In *Swidden-Fallow Agroforestry in the Peruvian Amazon. Advances in Economic Botany*; Denevan, W. M.; Padoch, C., Eds; 1987, Vol. 5; pp 1-7.

20. Posey, D. A. In *Change in the Amazon Basin*; Hemming, J., Ed.; Manchester University: Manchester, Ukm 1985, Vl. 1; pp 156-181.

21. Kerr, W. E.; Posey, D. A.. *Interciência* **1984**, *9(6)*, 392-400.

22. Posey, D. A.. *Advances in Economic Botany* **1984**, *1*, 112-126.

23. Seubert, C. E.; Sanchez, P. A.; Valverde, C. *Trop. Agric.* **1977**, *54*, 307-321.

24. Smyth, T. J.; Bastos, J. B. *Rev. Bras. Sci. Solo* **1984**, *8*, 127-132.

25.   Sanchez, P. A. In *Management of Acid Tropical Soils for Sustainable Agriculture;* Sanchez,. P. A.; Stoner, E. R.; Pushparajah, E., Eds.; International Board for Soil Research and Management: Bangkok, Thailand, 1987; pp 63-107.

26.   Smyth, T. J.; Alegre, J. C.; Palm, C. A.. In *Manejo de Suelos Tropicales en Latinoamerica*; Smyth, T. J; Raun, W. R.; Bertsch, F., Eds.; North Carolina State University, Raleigh, NC, 1991; pp 39-47.

27.   Subler, S.; Uhl, C. In: *Alternatives to Deforestation. Steps toward Sustainable Use of the Amazon Rainforest*; Anderson, A. B., Ed.; Columbia University: New York, 1990; pp 152-166.

28.   Paddoch, C.; Jong, W. *Econ. Bot.* **1991**, *45(2)*, 166-175.

29.   Okafor, J. C.; Fernandes, E. C. M.. *Agrofor. Syst.* **1987**, *5*, 153-   168.

30.   Szott, L. *Ph.D. Dissertation*; North Carolina State University, Raleigh, USA, 1987

31.   Prinz, J. M. *Plant Res. and Dev.* **1986**, *24*, 31-56.

32.   *A tropical Rainforest*; Odum, H. T.; Pigeon, R. F., Eds.; Office of Information Services, U.S. Atomic Energy Commission, Washington, DC, 1970; Vol. III.

33.   Uhl, C., Buschbacher, R.;. Serrão, E. A. S. *J. Ecology* **1988**, *76*, 663-681.

34.   Kang, B. T.; Reynolds, L.; Atta-Krah, A. N. *Adv. Agron.* **1990**, *43*, 315-359.

35.   Fernandes, E. C. M.; Davey, C. B.; Nelson, L. In *Technologies for Sustainable Agriculture in the Tropics*; Ragland, J.;. Lal, R, Eds.; ASA Special Publication 56, ASA, Madison, WI., 1993; pp 77-96.

36.   Alvim, P. T. *Agrotrópica (Centro de Pesquisa do Cacau,* Ilheus) **1989**, *1(1)*, 5-26.

37.   Broughton, W. J. *Agro-Ecosystems* **1977**, *3*, 147-170.

38.   Veiga, J. B.; Brienza, S. Jr.; Marques, C. L. T.; Serrão, E. A. S.; Yared, J. A. G., Bastos, J. B.; Costa, M. P.; Kitamura, P. C. In *Relatório Técnico Anual do CPTU*; EMBRAPA-CPTU: Belém, AM, 1988; pp 61-62.

39.   Budowiski, G. *Agroforestry: A Decade of Development*; Stepper, H. A.; Nair, P. K. R., Eds.; ICRAF: Nairobi, Kenia, 1987; pp 69-88.

40.   Boonkird, S. A.; Fernandes, E. C. M.; Nair, P. K. R.. *Agrofor. Syst.* **1984**, *2*, 87-102.

41.   Brienza, S. Jr.; Yared, J. A. G. *For. Ecol. and Managemente* **1991**, *45*, 319-323.

42.   Houghton, R. A. *Ambio* **1990**, *19(4)*, 204-209.

43.   Eswaran, H.; Van den Bergh, E.; Reich, P. *Soil Soc. of Am. J.* **1993**, *57*, 192-194.

44.   Lowry, J. B.; Lowry, J. B. C.; Jones, R. *Nitrogen Fixing Tree Research Reports* **1988**, *6*, 45-46.

45.   Montagnini, F.; Sancho, F. *Ambio* **1990**, *19*, 386-390.

RECEIVED January 10, 1995

Chapter 5

Secondary Compound Accumulation in Plants—The Application of Plant Biotechnology to Plant Improvement

A Proposed Strategy for Natural-Product Research in Brazil

Marcia Pletsch, Antônio Euzébio G. Sant'Ana,
and Barry Victor Charlwood

Departamento de Quimica—Centro de Ciências Exatas e Naturais,
Universidade Federal de Alagoas, Campus Universitário,
57072–970 Maceció, AL, Brazil

The vast flora of Brazil is not only an invaluable source of supply of secondary compounds but also provides an almost limitless bank of genetic information concerning the formation, transport and storage of these metabolites within the plant. The recent expansion in demand for natural products will certainly continue in the foreseeable future, but commercialisation of these compounds from natural sources can easily lead to detrimental effects on the environment. Whilst production methods involving chemical synthesis are available for a number of secondary compounds, many molecules are too complex to be commercialised through this route. Plant biotechnology offers an alternative strategy through *in vitro* accumulation of selected products, although this may not be a widely applicable approach nor, arguably, will it be one that is necessarily in the best commercial and/or political interests of Brazil. Plant genetic engineering has opened up wide possibilities to produce crop plant variants that have desirable biochemical characteristics, and these methods are equally applicable to the genetic manipulation of plants that accumulate secondary compounds. Examples of the application of plant biotechnology to both the *in vitro* and the *in planta* approach to secondary compound production are considered in this paper.

The extensive and diverse flora of Brazil is a natural asset of immense potential value for the provision of secondary compounds. Such materials already receive wide commercial application in the manufacture of, for example, pharmaceuticals, food additives, cosmetics and agrochemicals. The world-wide

0097–6156/95/0588–0051$12.00/0
© 1995 American Chemical Society

production of secondary compounds will be an area of significantly increasing industrial activity in the next decade as the demand for an even greater range of "natural" products continues.

The possible application of the plant resource within Brazil may not have been fully recognised by the competent authorities thus far. The exploitation of secondary compounds has often been left to entrepreneurs from USA, Europe and Japan with the profits accruing to these countries accordingly. With the advent of the modern techniques of plant biotechnology there is a window of opportunity for Brazil to take a major role in the development and application of these new methodologies for the commercialisation of natural products in Brazil.

Resource Factors

Extensive Flora.   It is reported that at least 16% of the estimated 500,000 species of plants that exist in the world are to be found in the Amazonian basin of Brazil alone. Further, less than 10% of all known plants have been studied chemically and far fewer have been studied with respect to their biological activities. And yet at least 3000 well characterised, biologically active natural products are already known (1). These factors taken together mean that a vast number of highly active compounds are waiting to be discovered and the majority of them are to be found in the plants presently growing wild in Brazil in areas such as the Cerrados, the Pantanal, Caatinga, the Atlantic Forest and, of course, the Amazon itself. Unfortunately, however, this magnificent resource is rapidly diminishing, not only because of land clearance and agricultural practice, but also because of the interest of (commercial) research groups outside of Brazil. Any strategy for plant biotechnology research must take into account both the aspects of learning about the valuable plant materials that are available to us and also of conserving those species which, whilst presently not directly exploitable, may very well be of value in the years to come.

Agricultural Base.   Brazil has a large, fertile land mass much of which is still available for agricultural exploitation. Naturally, the majority of Brazilian agriculture is aimed at primary food-crop plants, but there are currently a number of good examples of cash-crop farming for natural products (such as stevioside production by Ingá Stevia Industrial S.A., Maringá, Paraná). As the export markets for many primary crops (for example, sugar-cane) are becoming more difficult, it will be increasingly important for Brazil to bring on stream a more mixed farming with greater reliance on novel crop plants. It can be reasonably argued that plants yielding high value natural products would be excellent alternative crops in years to come but the preparation for this should commence immediately.

Labour Force and Capital Financing. Brazil has a large rural labour force skilled at working on the land. It is axiomatic that any application of plant biotechnology should enable the employment of an increased work force which can be adequately remunerated and hence enhance support for the social fabric of the rural areas. It is further recognised that many facets of fermentation biotechnology are extremely expensive in terms of plant and equipment. Any application of plant biotechnology should be able to be brought on stream with a minimum of capital expenditure which must be well within the means of State or Federal resources.

Taken together these two considerations mitigate against a strategy for plant biotechnology research involving the ultimate aim of production of secondary compounds in large scale bioreactors, particularly where a novel crop plant would be a direct alternative, or where advanced chemical engineering problems (such as the provision of specially designed bioreactors) would be necessitated. There are, however, a number of cheap "alternative technologies" which may very well enable *in vitro* technology to be operated on a commercial scale in rural areas in Brazil and applications in this area will be considered below.

Strategies for Commercialisation of Secondary Compounds

Determination of Plants with Appropriate Biological Activities. The essential pre-requisite to any form of commercialisation, namely plant screening, is dealt with in detail elsewhere in the present volume but, if Brazil is to retain some degree of control over the use and ultimate exploitation of its floral resource, it is important that this screening should be carried out as far as possible within Brazilian Institutes.

Conservation of the Gene Source and Build-up of Biomass for Further Exploration. Following the determination of a potentially valuable plant, it will be necessary to obtain considerable biomass in order to carry out activity-focused isolation and structure elucidation of the active components. The traditional method would be to harvest large quantities of the plant material, both fresh tissue and, preferably also, seed stock. It may be noted that collectors typically aim to harvest in the region of 4 kg of seed material in order to maintain a seed bank to conserve a specie. The effect of such collection could easily result in irreparable damage to the survival of the specie and also to the ecosystem which supported it. Plant biotechnology may readily assist in the conservation and production of large amounts of plant material even for the most sensitive of species through cryopreservation and micropropagation techniques.

Until the present time, the germplasm banks that maintain tissue culture collections do so in order to provide breeders and scientists with genetic material of agricultural and horticultural plants and their wild relatives. These technologies may be applied also to plants which produce commercially valuable secondary compounds in order: (i) to conserve and protect rare or inaccessible plants which accumulate secondary compounds of known biological activities; (ii) to retain genomic material of endangered plants whose biological properties are presently either unknown or not exploited; and (iii) to store transgenic plants, or their progenitor forms, that may be impaired with respect to their ability to reproduce by the normal means. The relatively high expense involved in this form of germplasm storage implies that its application must initially be limited to the storage of material for which there is a defined rather than a speculative use. The most obvious application is in the universally difficult area of obtaining fresh material of an unusual plant for biochemical and biomedical studies, and of retaining the original line for future experiments. In this case, *in vitro* storage of the parent plant, coupled with *in vitro* clonal propagation, could provide the continuity of material required. Interest in the micropropagation of medicinal plants has recently emerged with studies concerning the *in vitro* multiplication of, for example, *Pilocarpus microphyllus*, *Maytenus ilicifolia*, *Baccharis trimera* and *Artemisia annua* currently underway in a number of laboratories in Brazil.

Chemical Elucidation of Biologically Active Principles. Brazil has a world-recognised history of excellence in classical phytochemistry and many Institutes in Brazil are already adequately equipped and employ highly qualified personnel to carry out basic structural elucidation. It is now of paramount importance that research funding be available specifically to those groups wishing to undertake phytochemical studies where the separation of compounds is monitored by the increase in a target biological activity and not merely by chemical class. Clearly such a radical alteration in primary objective of phytochemical research requires that the mechanisms for obtaining a wide range of biological tests should be set up immediately.

Technologies for the Large Scale Production of Natural Products. The appropriate technology for the commercial scale production of a natural product depends on the characteristics of the target compound and could involve full chemical synthesis, biotechnological processes or a combination of both. Ideally, a potential target should be a single molecule or a small group of closely related molecules. For any other situation, chemical synthesis would be inappropriate and plant biotechnology is not sufficiently advanced to enable the manipulation of multi-pathways. At present it is not possible to enhance productivity of complex flavour, aroma and cosmetic compounds by biotechnological methods, and traditional cropping techniques would remain the most suitable form of exploitation. However, the yield of the producing plant could still be significantly improved by protecting the plant from, for example, microbial attack through genetic engineering as outlined below.

When an identified target molecule has a relatively simple chemical structure, it may be possible to carry out a total synthesis and purification of the compound on an economical basis. This is not usually the case, however: out of 50 purified products from plants, all of which can be chemically synthesised, only 9 may be manufactured economically through total chemical synthesis (2). A specific example of this is to be seen in the case of the anti-malarial sesquiterpenoid artemisinin found in *Artemisia annua* which is now being considered as a potential crop plant in Western Europe, India, USA and Brazil. The active principle is effective against chloroquinine-resistant, chloroquinine-sensitive and cerebral malarias (3); indeed, the ß-ethyl ether of artemisinin has been selected by the World Health Organisation for the treatment of severe and complicated forms of *Plasmodium falciparum* malaria (4). *A. annua* also produces several sesquiterpenoid endoperoxides which have potential as natural herbicides (5), and a highly aromatic essential oil, the major component of which is artemisia ketone (6), which may possess insect repellent and anti-microbial activities. The various synthetic routes to artemisinin that have been described (7 and papers quoted therein) are multi-step, complex and not commercially plausible.

Where total chemical synthesis is not viable then the active principle must be obtained either completely or partially from a natural source. Often, however, active principles are present only in minute amounts in material from the source plant. The classical example of this situation refers to the anti-leukaemic dimeric alkaloids vincristine and vinblastine from *Catharanthus roseus*. These compounds may be obtained from the leaves of periwinkle plants grown in the field, but the yield of active principle is less than 0.001% (w/w) making these amongst the most expensive of drugs. Similarly, the content of artemisinin in *A. annua* has been shown to be rather low and variable within the range 0.003 - 0.21% w/w (8).

Even when an active compound is present in extractable quantities in the plant, the natural source itself may still not be appropriate for large scale production. Thus, whilst the anti-cancer diterpenoid taxol may be prepared from the bark of the yew tree (*Taxus brevifolia*), harvesting source material destroys the plant (12 trees are required per patient) and seedlings take 50 - 60 years to reach maturity. Where the plant source is rare, endangered or essential for the preservation of the natural ecosystem, then use of the natural source as such is clearly inappropriate.

Application of Plant Biotechnology to Secondary Compound Production

Plant biotechnology offers two distinct strategies for the production of secondary compounds on a commercial basis. When a product is required in a pure form, in small quantities (less than 5000 kg per year) and at a high price (minimum $3000 per kg) then the possibility to use fermentation technology becomes available. This may be through *de novo* synthesis in a culture system, or by a two-step process involving both chemical synthesis and a biotrans-formation stage. Alternatively, the yield of the compound(s) of interest may be increased *in planta* by the manipulation of the intrinsic biosynthetic pathways through genetic engineering. The latter technology offers many exciting possibilities and would seem to be the most appropriate in the context of resources available in Brazil. Indeed, this strategy is the only available route when a product is required to act within the plant itself (ie transferring a specific antimicrobial activity to a crop plant).

Fermentation Processes.   Following two decades of intensive research, there are currently only a handful of potentially, commercially viable processes for the production of secondary products - including the well known examples of shikonin and berberine (*9, 10*) - through *de novo* fermentation of unorganised plant cells. Since the fermentation of cell cultures has produced little success in the past, and has a relatively poor prognosis for the future, research effort in this area would seem to be better directed towards the use of fully differentiated organ cultures, the biosynthetic capacities of which are preserved *in vitro*. In this context, transformed root organ cultures which can be grown in the dark, have low aeration and nutrient requirements and show fast growth rates, present very good possibilities for the development of a commercial process.

Transformed root organ cultures may often be formed readily by infection of a sterile plant, explant or callus with the soil-borne bacterium *Agrobacterium rhizogenes*, or by direct insertion, using a biolistic technique, of the *rol* genes (which are responsible for rhizogenesis) isolated therefrom (*11*). The neoplastic outgrowths which thus form typically within 2-3 weeks, can be removed and cultured in either solid or liquid medium as axenic root organs. These transformed roots (often referred to as "hairy-roots") commonly exhibit very high growth rates (sometimes 10 times higher than those of their non-transformed counterparts) when cultured in very low strength medium and in the absence of growth regulators, the presence of which may be detrimental to secondary compound accumulation *in vitro*. The cultures are genetically and biochemically very stable and usually possess the same biosynthetic capacities as the roots of the intact plant. Table I provides examples of the accumulative capacities of some transformed root cultures. In view of their rapid growth rate, their minimal bioreactor requirements, and the relative ease with which products may be isolated from these cultures, it is anticipated that transformed root cultures will be amenable to commercial exploitation for the production of high value fine chemicals. Low technology, low cost bioreactors of the roller-bed type are already available for the production of high quantities of biomass

Table I Examples of Production of Secondary Compounds in *Agrobacterium rhizogenes*-transformed Root Cultures

Plant Species	Compounds Accumulated	Productivity cf. Normal Root[*]
Atropa belladonna	scopolamine, hyoscyamine	6.6 fold increase
Calystegia sepium	scopolamine, hyoscyamine	26 fold increase
Centranthus spp.		
Fedia spp.	valepotriates	up to 25 fold increase
Valerianella spp.		
Hyoscyamus spp.	scopolamine, hyoscyamine	8.2 fold increase
Lippia dulcis	hernandulcin, monoterpenes	no hernandulcin produced by non-transformed roots
Nicotiana tabacum	nicotine	10 fold increase
Panax ginseng	ginsenosides	13 fold increase

[*] The amount of secondary compound produced (per litre of culture per day) by the transformed culture compared with its non-transformed counterpart.

from transformed roots, and such systems could be attractive for natural compound production in Brazil.

Two-step Synthesis. Whilst unorganised cultures may not accumulate secondary compounds in significant amounts, they typically maintain the capacity to catalyse individual steps in the biosynthetic pathways elaborated by the parent plant. Cell-free systems or partially purified enzyme preparations derived from tissue cultured cells are often able to transform both natural precursors and foreign substrates, the latter yielding novel natural products which should be suitable for pharmaceutical screening (*12*). It is anticipated that a mixture of technologies (ie chemical synthesis coupled with *in vitro* techniques) might be utilised either by the initial, facile (ie cheap) chemical synthesis of an intermediate followed by biotransformation to the end product, or by the formation of a complex compound through *in vitro* culture which can be extracted and submitted to minimal chemical transformation to yield a biologically active derivative.

The work that perhaps demonstrates most significantly the potential application of this area of plant biotechnology for the synthesis of medicinal compounds is that of Kutney and his co-workers in Vancouver, Canada (*13*) concerning the synthesis of etoposide. This lignan, originally isolated from

Podophyllum peltatum, has been shown to be efficacious in the treatment of myelocytic leukaemia, testicular, bladder and small-cell lung cancers. As is the case for many complex natural products, total synthesis of the active principle is not commercially feasible: the most attractive process developed so far involves the relatively facile modification of the pre-formed skeleton of podophyllotoxin which must also be obtained from *P. peltatum*. However, dependence on a plant source for a major feed-stock may be disadvantageous in terms of quality, quantity and continuity of supply. A synthesis of etoposide has been developed (*13*) that uses both traditional chemical reactions involving readily available starting materials, together with further elaboration of the complex structure using an enzyme system derived from plant cell cultures. Starting with a readily available aldehyde, a precursor of podophyllotoxin could be obtained by a short reaction sequence which afforded very high overall yields of product. Peroxidases present in cell free extracts from cultures of *Catharanthus roseus* are, apparently, not only capable of bringing about C-C bond linkages in indole alkaloids normally present in the parent plant, but can also accept foreign substrates. Such a cell free system was able to cyclize the podophyllotoxin precursor in excellent yields and the product could be readily converted into podophyllotoxin for further elaboration to etoposide.

Similar studies are underway in a number of laboratories and it is to be expected that many more "hybrid" synthetic routes will be developed as the potential of *in vitro* systems becomes more apparent.

Genetic Engineering. A key and exciting avenue of plant biotechnology which is currently receiving extensive research interest is the improvement of higher plants through gene transfer. With such technology it is possible:
(i) to increase the protein content (and hence activity) of a regulatory enzyme in a pathway already extant in a plant and thus to up-regulate the accumulation of compound(s) on the target pathway - clearly application of this procedure should result in plants with increased yields of biologically active principles;
(ii) to introduce a novel biosynthetic pathway into a plant - an example application might be to transfer chemical defence mechanisms between plants or even between micro-organisms and plants;
(iii) to suppress partially or completely the production of unwanted compounds by inhibiting the formation of the enzyme(s) responsible for that portion of the pathway - an alternative application of this strategy could be to block a particular branch of a pathway hence diverting carbon flux (precursors) along desirable branches of the pathway.

The introduction and stable incorporation of a foreign DNA sequence into the chromosome of a commercially important plant has, until recently, been commonly brought about using *Agrobacterium*-based vector systems. However, such systems have one major disadvantage namely the limited host-range of infection. For example, in the case of *Glycine max* (soybean) infection with *Agrobacterium tumefaciens* is restricted to one variety only and this, unfortunately, has no commercial value. The latest technology for the delivery of foreign genes into plant material employs direct gene transfer using biolistic devices. This method involves bombardment of cells or tissues with high velocity metal particles coated with DNA, and it has been used in numerous laboratories for genetic transformation of diverse plant species (*14*). The advantages of the biolistic system are that it permits transformations independent of genotype and allows a large number of transformants to be recovered in order to assess the levels of gene expression. Whether foreign DNA is introduced into unorganised cells by a biolistic technique or by using disarmed *Agrobacterium* vectors, an efficient regeneration protocol must be established in order to obtain an intact, transgenic plant.

Details of examples of transfer of specific genes encoding for key enzymes in biosynthetic pathways to secondary compounds are now becoming available (Table II). Hamill and co-workers (*15*) used an *A. rhizogenes* vector system to introduce copies (3 to 7 inserts per genome) of a yeast cDNA sequence coding for ornithine decarboxylase into roots of *Nicotiana rustica*. The transformed roots showed a 3-fold increase in activity of this enzyme compared to control cultures and were also enhanced in their capacity to accumulate putrescine (the product of ornithine decarboxylation) and nicotine (which derives from putrescine), the latter being increased by 2-fold. The authors point out that, although the increase in enzymatic activity in transgenic roots was lower than had been anticipated, it was "demonstrated that flux through a pathway to a plant secondary product can be elevated by means of genetic manipulation". On the other hand, Berlin and co-workers (*16*) obtained a 10-fold increase in tryptophan decarboxylase (TDC) in roots of *Peganum harmala* which had been transformed with *A. rhizogenes* containing a DNA sequence from *C. roseus* coding for the decarboxylase. In this case, although non-transformed roots showed undetectable levels of seratonin (a product of tryptophan decarboxylation), the transgenic roots accumulated 15 - 20 mg per gram dry weight of this proto-alkaloid. A similar large increase in TDC (more than 40-fold) was obtained following insertion of a cDNA encoding for this enzyme into regenerated plants of *N. tabacum* (*17*), and in this case tryptamine levels were stimulated by up to 260-fold compared with control plants.

A further successful report of the up-regulation of an existing pathway concerns the enhanced formation of the anticholinergic drug scopolamine from the less active hyoscyamine in *Atropa belladonna* (*18*). A cDNA sequence from *Hyoscyamus niger* coding for hyoscyamine 6-ß-hydroxylase (an enzyme which catalyses both hydroxylation and epoxide formation in hyoscyamine), placed under the control of the CaMV 35S promoter, was transferred using a disarmed *Agrobacterium tumefaciens* vector to yield transgenic plants of *Atropa belladonna*. Selfed progenies from one transgenic line yielded up to 1.25% scopolamine (on a dry weight basis) in leaf tissue compared with the

Table II Regulation of Secondary Products by Genetic Manipulation

Enzyme Coding Sequence	Source of DNA	Plant Transformed	Secondary Product	Reference
ornithine decarboxylase	yeast	*Nicotiana rustica*	putrescine, nicotine	15
tryptophan decarboxylase	*Catharanthus roseus*	*Peganum harmala*	seratonin	16
tryptophan decarboxylase	*Catharanthus roseus*	*Nicotiana tabacum*	tryptamine	17
hyoscyamine 6-β-hydroxylase	*Hyoscyamus niger*	*Atropa belladonna*	scopolamine	18
strictosidine synthase	*Catharanthus roseus*	*Nicotiana tabacum*	enzyme activity enhanced	19
lysine decarboxylase	*Hafnia alvei*	*Nicotiana tabacum*	cadaverine	20
stilbene synthase	*Arachis hypogaea, Vitis vinifera*	*Nicotiana tabacum*	resveratrol	21, 22
trichodiene synthase	*Fusarium sporotrichioides*	*Nicotiana tabacum*	trichodiene	23
isochorismate hydroxymutase	*Escherichia coli*	*Rubia peregrina*	alizarin	29
3-hydroxymethylglutaryl CoA reductase	hamster	*Nicotiana tabacum*	sitosterol, stigmasterol, campesterol, cycloartenol	30
chalcone synthase anti-sense	*Petunia hybrida*	*Petunia hybrida*	anthocyanins	24
polygalacturonase anti-sense	*Lycopersicon esculentum*	*Lycopersicon esculentum*	enzyme activity reduced	25
3-hydroxymethylglutaryl CoA reductase anti-sense	*Arabidopsis thaliana*	*Fedia cornucopiae*	enzyme activity reduced	26
caffeic acid 3-O-methyltransferase anti-sense	*Medicago sativa*	*Nicotiana tabacum*	lignin	27

wild-type plant which accumulated only 0.29% alkaloid, and of this approximately 96% was hyoscyamine.

Assuming that the correct substrate is available for a pathway not presently operating within a specific plant, it should be possible to introduce a novel gene the expression of which would lead to a novel product. A modified cDNA sequence coding for strictosidine synthase, a key enzyme in the pathway to vinblastine and vincristine in *C. roseus*, has been transferred to *N. tabacum* (*19*). The construct was efficiently expressed in the transgenic plants with the enzyme being correctly targeted to the vacuole: interestingly, tobacco produced two separate isoenzymic forms of the synthase whereas *C. roseus* possesses just one form. No strictosidine was produced in the transgenic tissue since tobacco lacks the appropriate substrate. Herminghaus and co-workers (*20*) transformed *N. tabacum* using a disarmed binary vector harbouring a plasmid containing the lysine decarboxylase coding sequence from the bacterium *Hafnia alvei* fused to DNA encoding a chloroplast-targeting peptide. Transformants showed higher chloroplast decarboxylase activity and also accumulated cadaverine at 0.3 - 1.0% dry weight in leaf tissue: non-transformed plants and those transformed with a non-targeted sequence contained barely detectable levels of cadaverine.

Some workers have succeeded in transferring genes associated with the formation of secondary compounds which may have important roles in plant-microorganism or plant-insect defence systems. In one case, a genomic sequence from *Arachis hypogaea* coding for stilbene synthase has been introduced into tobacco protoplasts (*21*). This synthase is responsible for the formation of the phytoalexin resveratrol (3,4',5-trihydroxystilbene) which is induced following attack by, for example, *Botrytis cinerea* on *Vitis vinifera*, *Picea sitchensis* or *A. hypogaea*. Stilbene synthase activity could be induced, by UV light or fungal elicitor, in all parts of the transgenic plants regenerated from protoplasts, and up to 50 ng of resveratrol per gram fresh tissue was accumulated within 24 hours of induction. In an extension to these studies (*22*), two full length genes from *V. vinifera* for stilbene synthase were introduced into tobacco through direct protoplast uptake, and the transgenic plants regenerated from them were able to produce up to 400 μg resveratrol per gram fresh weight following infection with *B. cinerea*. These plants showed a significantly higher resistance to the fungal infection than did non-transgenic controls. This work is of considerable importance since it is the first to show that the introduction of a foreign phytoalexin gene under the control of its own inducible promoter can confer a degree of disease resistance in the recipient plant.

In another example, a gene, isolated from the fungus *Fusarium sporotrichioides*, which encodes for trichodiene synthase (the enzyme responsible for the production of the toxic sesquiterpenoid trichodiene) has been introduced into tobacco (*23*) and trichodiene was accumulated at low concentrations (5 to 10 ng per gram fresh weight of tissue) in the transgenic plants. The constitutive synthesis of this fungal toxin may very well bring about an alteration in insect feeding habit.

Introduction of a specific cDNA sequence in an inverse orientation ("anti-sense") can often lead to inhibition of expression of the endogenous gene. For example, a chalcone synthase gene was introduced into *Petunia hybrida* in an anti-sense orientation causing a reduction in chalcone synthase mRNA and resulting in plants with a reduced flower pigmentation (*24*). Transgenic tomatoes that stably expressed RNA anti-sense for polygalacturonase showed a significant reduction in the activity of this enzyme and, consequently, a remarkable delay in the onset of ripening of the tomato fruit (*25*). Similarly, when a cDNA sequence coding for 3-hydroxymethylglutaryl coenzyme A reductase (HMGR: a key enzyme in the regulation of the early isoprenoid pathway) was introduced in the anti-sense mode into roots of *Fedia cornucopiae*, HMGR activity was reduced by 11-fold (*26*).

A further example of the use of anti-sense technology is the constitutive expression in the anti-sense mode of a caffeic acid 3-*O*-methyltransferase gene from *Medicago sativa* in transgenic tobacco leading to a reduction (of approximately 50%) in lignin content (*27*). The down-regulation of lignin production would be highly advantageous to the paper pulp industry and could also be employed to improve the digestibility of forage crops.

Genetic manipulations involving both the up-regulation of key enzymatic activities through gene-dosage and the inhibition of competing or unwanted pathways by anti-sense RNA should provide a successful strategy by which secondary compound yield in biologically active plants might be improved. An example of a strategy involving both sense and anti-sense technologies in order to regulate pathway genes is provided in our current studies involving the enhancement of accumulation of artemisinin in *Artemisia annua* in which we are attempting to up-regulate HMGR, farnesylpyrophosphate synthase and germacranadiene cyclase in order to enhance carbon flux towards sesquiterpenoid synthesis, and partially to down-regulate squalene synthase in order to reduce synthesis of steroids. It is of interest to note that this is the very strategy employed by tobacco cells when the synthesis of 5-*epi*-aristolochene (a sesquiterpenoid phytoalexin) is induced following treatment with cellulase: the mechanism of induction involves an increase in sesquiterpenoid cyclase activity (correlated with the synthesis of mRNA coding for this enzyme) and a decrease in squalene synthase activity (*28*).

Conclusion

As can be seen from the examples provided above, Brazil could readily employ these new methodologies in plant biotechnology for the exploitation of bioactive compounds to be found in the diverse floral resource of the Country.

Clearly the technology is now in place through which both primary and secondary metabolism may be perturbed by manipulation of the plant genome. Whilst our knowledge of the natural control of secondary metabolism is still insufficient to enable us fully to predict the magnitude of the effect of foreign gene insertion, it has been shown already that it is possible to regulate segments of pathways. The control of entire metabolic pathways may be available to us in the future, possibly through the manipulation of genes which control developmental processes. Even with our present state of knowledge, however, the prospects for designing transgenic root organs and regenerated plants with higher yields of specific secondary compounds are extremely promising. On the other hand, our ability to produce plants in which unwanted side-products are eliminated is virtually definite.

Acknowledgements

The authors are very grateful to the Conselho Nacional de Desenvolvimento e Pesquisa (CNPq/RHAE) and the Fundação de Amparo à Pesquisa do Estado de Alagoas (FAPEAL) for financial support of the Plant Biotechnology programme at the Universidade Federal de Alagoas. One of us (BVC) wishes to thank the Associação Brasileira de Química (ABQ) for financial support in attending the 1st International Symposium on Chemistry and the Amazon.

Literature Cited

1. Harborne, J. B.; Baxter, H. *Phytochemical Dictionary*; Taylor and Francis: London, **1993**.
2. Phillipson, J. D. In *Secondary Products from Plant Tissue Culture;* Charlwood, B. V.; Rhodes, M. J. C., Eds.; Clarendon Press: Oxford, **1990**; pp 1-21.
3. Klayman, D. L.; Lin, A. J.; Acton, N.; Scovill, J. P.; Hoch, J. M.; Milhous, W. K.; Theoharides, A. D.; Dobek, A. S. *J. Nat. Prod.* **1984**, *47*, 715-717.
4. ElSohly, H. N.; Croom, E. M.; El-Feraly, F. S.; El-Sherei, M. M. *J. Nat. Prod.* **1990**, *53*, 1560-1564.
5. Chen, P. K.; Leather, G. R.; Klayman, D. L. *Plant Physiol.*, **1987**, *83S*, Abstr. 406.
6. Banthorpe, D. V.; Charlwood, B. V. *Nature (New Biol.)* **1971**, *231*, 285-286.
7. Ravindranathan, T.; Kumar, M. A.; Menon, R. B.; Hiremath, S. V. *Tet. Lett.* **1990**, *31*, 755-758.
8. Charles, D. J.; Simon, J. E.; Wood, K. V.; Heinstein, P. *J. Nat. Prod.* **1990**, *53*, 157-160.

9.   Fujita, Y.; Tabata, M. In *Plant Tissue and Cell Culture*; Green, C. E.;
      Somers, D. A.; Hackett, W. P.; Biesboer, D. D., Eds.; Alan R. Liss: New
      York, NY, **1987**; pp 169-186.
10.   Fujita, Y. In *Plant Tissue Culture: Applications and Limitations;*
      Bhojwani, S. S., Ed.; Elsevier: Amsterdam, **1990**; pp 359-275.
11.   Kodama, H.; Irifune, K.; Kamada, H.; Morikawa, H. *Transgenic Res.*
      1993, *2*, 147-152.
12.   Suga, T.; Hirata, T. *Phytochemistry* **1990**, *28*, 2393-2406.
13.   Kutney, J. P. *Synlett* **1991**, *(1)*, 11-19.
14.   Sanford, J. C. *Physiol. Plant.* **1990**, *79*, 206-209.
15.   Hamill, J. D.; Robins, R. J.; Parr, A. J.; Evans, D. M.; Furze, J. M.;
      Rhodes, M. J. C. *Plant Mol. Biol.* **1990**, *15*, 27-38.
16.   Berlin, J.; Rügenhagen, C.; Dietze, P.; Fecker, L. F.; Goddijn, O. J. M.;
      Hoje, J. H. C. *Transgenic Res.* **1993**, *2*, 336-344.
17.   Songstad, D. D.; De Luca, V.; Brisson, N.; Kurz, W. G. W.; Nessler,
      C. L. *Plant Physiol.* **1990**, *94*, 1410-1413.
18.   Yun, D-J.; Hashimoto, T.; Yamada, Y. *Proc. Natl. Acad. Sci. USA.* **1992**,
      89, 11799-11803.
19.   McKnight, T. D.; Bergey, D. R.; Burnett, R. J.; Nessler, C. L. *Planta*
      1991, *185*, 148-152.
20.   Herminghaus, S.; Schreier, P. H.; McCarthy, J. E. G.; Landsmann, J.;
      Bottermann, J.; Berlin, J. *Plant Mol. Biol.* **1991**, *17*, 475-486.
21.   Hain , R.; Bieseler, B.; Kindl, H.; Schröder, G.; Stöcker, R. *Plant Mol.
      Biol.* **1990**, *15*, 325-335.
22.   Hain, R.; Reif, H-J.; Krause, E.; Langebartels, R.; Kindl, H.; Vornam,
      B.; Wiese, W.; Schmelzer, E.; Schreier, P. H.; Stöcker, R. H.; Stenzel, K.
      Nature **1993**, *361*, 153-156.
23.   Hohn, T. M.; Ohlrogge, J. B. *Plant Physiol.* **1991**, *97*, 460-462.
24.   Van der Krol, A. R.; Lenting, P. E.; Veerstra, J.; Meer, I. M.; Van der
      Koes, R. E.; Gerats, A. G. M.; Mol, J. N. M.; Stuitje, A. R. *Nature* **1988**,
      333, 866-869.
25.   Smith, C. J. S.; Watson, C. F.; Ray, J.; Bird, C. R.; Morris, P. C.;
      Schuch, W.; Grierson, D. *Nature* **1988**, *334*, 724-726.
26.   Munasinghe, K. *PhD Dissertation*; University of London, **1992**.
27.   Ni, W.; Paiva, N. L.; Dixon, R. A. *Transgenic Res*. **1994**, *3*, 120-126.
28.   Vögeli, U.; Chappell, J. *Plant Physiol.* **1990**, *94*, 1860-1866.
29.   Lodhi, A. *PhD Dissertation;* University of London, **1994.**
30.   Chappell, J.; Proulx, J.; Wolf, F.; Cuellar, R. E.; Saunders, C. *Plant
      Physiol.* **1991**, *96*, 127.

RECEIVED November 2, 1994

NATURAL PRODUCTS: BIOACTIVITY AND CHEMISTRY

Chapter 6

The Promise of Plant-Derived Natural Products for the Development of New Pharmaceuticals and Agrochemicals

James D. McChesney

Research Institute of Pharmaceutical Sciences, University of Mississippi, University, MS 38677

Plant-derived natural products hold great promise for discovery and development of new pharmaceuticals and agrochemicals. Careful consideration of the process of discovery and development - a "systems" approach - will be required to bring this great potential to realization. A "systems" analysis is outlined and ilustrated in this paper.

This article covers the approach at the University of Mississippi, Research Institute of Pharmaceutical Sciences on the process of discovery and development of new pharmaceuticals and agrochemicals starting with natural product preparations. Pharmaceuticals will be emphasized since those are often more easily understood in a general context. It must be pointed out, however, that all the things that are said about pharmaceuticals similarly apply to agrochemicals. The key issue is the particular and specific biological question that is being addressed: Does the substance being investigated have utility to treat disease in the human thus making the agent a pharmaceutical or, alternatively, does the substance have biological activity to control an agricultural parameter such as inhibit the growth of weeds, control insect damage or provide medical treatment for domestic animals. Those would all represent agrochemical applications. To summarize, pharmaceutical development will be utilized as the example here but it should be emphasized that development of pharmaceuticals and agrochemicals utilize very similar processes.

Drug Development

Initially, the pipeline concept of drug development should be discussed. This is shown in Figure 1. There are very distinct periods during the time of development of new pharmaceuticals; a period of discovery and once certain criteria are met, those discoveries become development candidates and in turn, once safety and efficacy have been demonstrated, approval to market the substance as a pharmaceutical can be gained from various regulatory agencies (1). The usual periods that are associated with the development of new pharmaceuticals are shown in Figure 1. There is a period of conceptualization where it is decided that new lead compounds with specific desirable biological activity will be sought; a decision is made about how to seek those compounds, whether from synthetic, organic

0097–6156/95/0588–0066$12.00/0

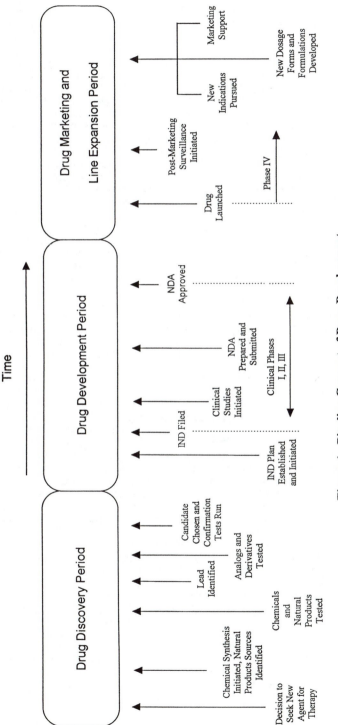

Figure 1. Pipeline Concept of Drug Development.

chemistry libraries or from natural product sources. The next and most important step is the development of a specific and appropriate biological assay procedure and then with that biological assay, various chemical substances are evaluated and ultimately chemical leads are discovered. Those leads represent specific compounds which have promise to bring about the particular biological response that is being sought. As lead compounds are identified, then effort is made to expand upon those through standard medicinal chemistry approaches including molecular modeling. From medicinal chemistry effort will flow a suite of compounds, so-called chemical cousins, and the evaluation of their biological activity gives an idea about structure activity relationships to elicit this biological activity. Ultimately from that family of compounds, certain specific "candidates" are identified based upon their potency and other desirable properties. Of particular importance at this stage, is initial evaluation of the breadth of pharmacological activity, i.e. how specific is the substance in eliciting the particular biological response and what are the parameters that might be associated with toxicity of the compound. Once candidates have been chosen and various additional tests evaluated, then a program of development is defined. An investigational new drug development plan is formulated and the substance evaluated such that it may ultimately enter human clinical evaluation or trial. When sufficient information is in hand, an investigational new drug (IND) application can be filed with the regulatory agency. That investigational new drug application, when approved, provides the authorization to enter the substances under study into human clinical evaluation. Those evaluations in the clinic are divided ordinarily into three or more phases. Initially evaluation of safety is made; how toxic or how well tolerated is the agent in healthy volunteers? (Phase I). For compounds which pass this early phase, then they are evaluated in patients wherein efficacy to treat the particular disease is initially evaluated. (Phase II). Finally with sufficient safety and efficacy demonstrated in phases I and II, then a broad Phase III clinical evaluation involving large numbers of patients can be initiated. As clinical trials progress, information is being developed in support of a new drug application (NDA) which will be submitted to the regulatory agency seeking approval to market the substance for treatment of a particular disease or disorder. That approval process, when finished, which allows the pharmaceutical company to enter a marketing phase, also requires that the company monitor for any unexpected toxicities which might show up once the agent is being used broadly in a diverse population of patients.

How long does all this take? The usual range of time required to accomplish the development processt ranges from 4 years at the very shortest to as much as nearly 20 years. On average more than one decade is required. To validate the numbers in this statement, it should be pointed out that in 1992 the Food and Drug Administration of the United States approved between 25 and 30 new chemical entities as pharmaceuticals. The average time in development for those new drugs was 12.6 years. The costs associated with the various stages of this process total between to 30-35 million dollars. This amount seems to be a very low compared to what one sees quoted in newspapers and hears reported in television news shows dealing with drug development. Those quoted numbers of 300 million dollars seem to be quite large compared to what is reported here (2). However, I would emphasize that the lower numbers relate just to the specifics of development of a single successful agent. They do not account for all of the failures that occur during the process nor they account in any way for the "lost opportunity" that one would have simply by taking money and investing it in government bonds, for example, rather than risking the money by investing it in research and development for which there is no certainty of outcome. So the 30-35 million dollars is an estimate of the costs associated with the successful agent. It does not in any way account for the costs associated with all of the trial and error leading to the discovery of that first successful candidate and its subsequent development and the many cases where agents have failed because of toxicity or other unexpected problems.

Discovery/Development of New Pharmaceuticals

If it takes so long and costs so much, how is it that new pharmaceuticals are found at all? Well, there are various ways that new pharmaceuticals are discovered and developed. One of the most direct is the observation that drugs that have been in utilization for a period of time show up in the clinic as having potential for other uses. A particular example is the agent levamasol which has been utilized for a very long time as a veterinary product and more recently has been shown to possess selective and specific immune stimulatory activity and so is now incorporated into cancer chemotherapy regimens to enhance the body's own resistance to cancer and thereby enhance chances of remission following surgery or radiation therapy. So old drugs are observed in clinical use to have potential new uses. A second way in which substances are shown to have drug utility is through the evaluation of normal or usual physiological substances. In fact, this approach is the basis upon which much of the biotechnology era of discovery and development of pharmaceuticals is based. Hormones or regulatory agents, enzymes, etc. have been identified, their role in the physiology of man defined, and as a consequence, control of their concentration and thereby their physiological effects provides pharmaceutical benefit. This is the basis for the development of many biotechnology companies. A third approach that might be utilized is the synthesis of new chemical substances. One may look at lead compounds or compounds known to have some desirable biological activity and use those as models or templates - the so-called "designed" chemical synthesis approach. As was mentioned, for example, in the initial discussion of the discovery and development of pharmaceuticals, once a lead substance has been identified, then chemical cousins are prepared and evaluated. That's a type of design chemical synthesis that is utilized extensively for the discovery and development of pharmaceuticals. An alternative is to go into the laboratory as an organic chemist and have fun just making new compounds - compounds that have never been made before. It's a little bit like the mountain climber who climbs the mountain "just because it's there" and no one else has already climbed it. Since a chemist visualizes this particular molecule, it's not one which has been made and so the challenge is to make it and then once it has been made it, to evaluate it for potential utility. Numerous examples might be quoted in this area; perhaps the most significant being the discovery of the benzodiazepene class of minor tranquilizers. That discovery arose from just such an approach; preparation of an interesting chemical structure in the laboratory and the subsequent submission of that chemical substance for pharmacological evaluation. Finally, new pharmaceuticals can be discovered and developed through evaluation of natural product preparations. Those natural product preparations may derive from animals, for example, marine animals; they may derive from micro-organisms, and here antibiotics can be used as an example of a particularly significant group of pharmaceuticals that have been discovered and developed from micro-organism fermentation broths. And finally, plants might serve as a source for new substances that can be evaluated for pharmaceutical application. And, in fact, historically, plants have been a very significant source of new pharmaceuticals, perhaps the most important source.

Role of Natural Products in Discovery of New Pharmaceuticals

So natural products have a role in discovery. This role derives from the observation that all organisms interact among themselves and with their environment by chemical means. Organisms protect themselves, communicate with others of their own kind, etc. by means of chemical substances (3). Those interactions of organisms are very specific and effective. One may wonder how it is that certain insects are ever able to identify their opposite sex and thereby propagate their kind. That communication takes place largely by chemical means through various specific substances which attract either the male or female to the opposite gender. Many organisms protect themselves by secreting chemical defense substances (3). The function of these chemicals is to enhance the competitiveness of the organisms in their environment, making them more successful, in turn creating the opportunity for them to propagate offspring. Because there are literally millions of different organisms, all communicating among and between themselves with chemical means, that means that there are literally millions of chemical structure types to select from for evaluation for pharmaceutical or agrochemical application.

This has not gone unnoted historically since natural product preparations have been the major source of pharmaceutical agents. Indeed, more than 90% of current therapeutic classes derive from a natural product prototype and interestingly, even today, roughly two-thirds to three-quarters of the world's population rely upon medicinal plants for their primary pharmaceutical care. Those medicinal plants represent natural product substances or preparations which have utility as pharmaceutical agents.

There have been some perceptions which have limited interest within the pharmaceutical industry for plant derived natural products. Some years ago there were significant problems sourcing authenticated plant materials. It was easy to collect plants and demonstrate that they had interesting biological potential but then when researchers went back to recollect and confirm those potentials and ultimately to carry out the development and commercialization process; then failure often resulted because their original plant collections had not been adequately identified and were consequently lost. Another area that has created problems has been the ability to measure desirable biological activity. Natural product preparations ordinarily are very complex mixtures of materials and to measure a desirable biological activity utilizing a mixture of materials creates some serious problems. Interactions of the components of the mixture, either antagonism by one material of another's activity or addition or even synergy of activities, can give very misleading results. So in many cases this has created significant problems. A third perception has been that one could not quickly and cost effectively purify and identify active principals from complex natural product mixtures. Those mixtures containing dozens to hundreds of different chemical substances, often of quite similar chemical and physical properties, historically have created great challenges for the separation and purification of the active principles. The time required, as well as the effort involved, has been quite significant and that has contributed to the perception that this is a serious limitation, particularly with regard to plant derived natural products.

Another important consideration is that natural products often are poor pharmaceuticals: their chemical stability may be marginal; they may have poor solubility characteristics or poor bioavailability characteristics; they may not formulate well, etc. All of these pose serious limitations but as will be shown, they can be overcome. A very important consideration, which in my judgment has perhaps limited interest most in plant derived natural products for pharmaceutical discovery and development, has been concern over the availability of quantities of pure plant derived chemical substances. These quantities are required for development, i.e., the generation of the various kinds of information needed to understand and assess the real potential of the substance for pharmaceutical application. Ultimately, perhaps the most limiting is how to deal with the quantities required to meet market demand should a pharmaceutical become a successful drug in the market place. That market demand can reach the hundreds to thousands of kilograms per annum scale.

A number of changes in capability are fostering a renaissance in natural products research. Foremost among these are the advances which have been made in bioassay technology over the last several years. We now have highly automated, very specific and selective bioassays in which complex mixtures of materials can be rapidly and cost effectively evaluated. Indeed, advances in bioassay technology have been so great that the availability of materials to evaluate has become more limiting than the ability to carry out those evaluations. Once biological activity has been demonstrated in an appropriate bioassay or primary screen we now have available, based upon advances in separations and structure elucidation technology, the capability to isolate, purify and determine the chemical structure of the active principle in a few weeks, or, at most, a few months. Those separation advances are particularly associated with high performance chromatography methodologies including high performance liquid chromatography, high performance supercritical fluid chromatography and capillary electrophoresis. Most recently, improved methodologies in counter current partition chromatography have further expanded the capabilities for separations. Structure elucidation technology has evolved particularly with the development of high field NMR spectrometry as well as high resolution technologies in mass spectrometry. Most important are the two dimensional NMR techniques that have been developed which allow very rapid and straightforward assignment of structure to complex natural products. Additionally, the technologies of coupled liquid chromatography-mass spectrometry and similar techniques provide very potent or powerful methodologies for separation and structure elucidation. Further as an increased understanding of biological and physiological pathways in all organisms is reached, much more specific and selective questions with regard to potential drug application can be formulated, that is, for example, the investigation of substances which interact only with a very specific receptor rather than with a family of receptors and with the advances that have been made in biotechnology those receptors can be cloned and "constructs" prepared in which cloned receptors become a component of a created cell line which then ultimately forms the basis of a high throughput very selective and specific bioassay. In this way, the advances in several areas can be put together to focus upon the discovery of new substances as lead compounds for pharmaceutical development. Natural products represent the most important source of unique chemical substances for evaluation with these new assaying strategies for potential pharmaceutical utility.

Another issue that is fostering a renewance of interest in natural products research is the recognition that the biological diversity of the earth is diminishing rapidly. Indeed, one cannot pick up a newspaper or news journal without encountering some article dealing with the rate, consequences, cause, etc. of loss of biological diversity. However, it must be emphasized that it is the loss of chemical diversity represented by those organisms that represents the true loss, the possible utilization of those chemicals for the benefit of humankind will be lost. Even the foodstuffs, the building materials, the fibers that are utilized to make clothing, etc. are chemicals derived from nature and it is, indeed, the complex and unique natural product chemicals in the various kinds of organisms that create the chemical diversity that are to be evaluated for pharmaceutical or agrochemical discovery and development. The loss of those organisms and, in turn, the loss of the chemical diversity represented by those organisms, is a very important issue stimulating natural products research.

Another important consideration is that an era of worldwide economic competition is beginning. Indeed, in the U.S. at the moment the pharmaceutical industry still represents one of the important areas of industrial leadership and the recognition that the discovery and development of new pharmaceuticals and agrochemicals maintains the competitive position of that industry leads to interest in natural products research to increase the efficiency of discovery and development. So worldwide competition does motivate a certain portion of the new interest in natural products research that can be observed. And finally, the historical success of the approach of evaluating natural product preparations for pharmaceutical and agrochemical discovery and development must not be overlooked. Indeed, one can demonstrate that nearly all pharmaceutical classes were derived from natural product prototypes. Relatively few exceptions exist to that general statement. So the historical success of evaluating natural products for pharmaceutical discovery and development and increased capabilities that have been reviewed, contribute greatly to a renaissance in this area for new discovery and development activities.

Now let me turn for a moment to our general utilization of plants and then more specifically to the utilization of plants for the discovery of pharmaceuticals. It is estimated that approximately 300,000 species of higher plants exist in nature. In some cases this number reported to be 250,000. In other cases it may be as high as 500,000. The differences in those numbers reflect partly a difference in philosophy among systematic botanists, but also I would emphasize that as we begin to explore more aggressively unusual environments, or particularly diverse environments, such as the tropical rainforests new species of higher plants are in fact being encountered continually. Of the 300,000 or so species of higher plants, about one percent, or roughly 3,000, have been utilized for food and of those 3,000 or so about 150 had been commercially cultivated. In today's marketplaces throughout the world unusual fruits and vegetables are beginning to appear because there is an increasing desire on the part of the world's populations for more "exotic" foodstuffs. However, the vast majority of caloric intake derives from about 20 species of plants. These 20 represent the basis upon which the world's population is fed. This represents a very narrow foundation upon which to support the human populations of the world.

Turning to plants as sources of medicine, it is verified that approximately 10,000 of them have a documented and recorded utilization for medicine - considerably more than the 3,000 or so that have been utilized for food materials. Looking specifically at the utilization of plant derived materials in western medicine, the U.S., Western Europe, etc., it is found that roughly 150 to 200 of such agents are incorporated. This is still a very small percentage of all higher plants and thus there are many more important discoveries in the plant kingdom which may be exploited for pharmaceutical or agrochemical application.

How does the process of discovering a natural product with potential for pharmaceutical/agrochemical application and carrying that natural product through the process of assessment and development so that it ultimately becomes a successful pharmaceutical or agrochemical occur?. At the University of Mississippi Research Institute of Pharmaceutical Sciences, we have a general approach outlined in Figure 2. This general approach specifies the various steps or stages of the discovery and development process. The specific steps of that process are outlined

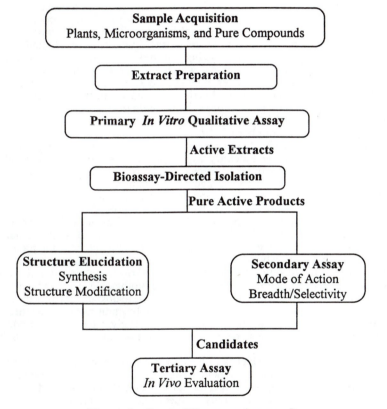

Figure 2. General Program Approach.

in the following paragraphs. Samples are acquired for evaluation in the discovery program. Those samples are acquired from collections of higher plants, from micro-organism fermentation broths, and more recently in the Mississippi program, from marine organism collections. When the opportunity is presented, synthetic substances are evaluated for potential utility. All of those sources represent samples to be evaluated.

In the case of natural product materials, a suitable extract must be prepared. Ordinarily that means to grind or homogenize the biomass and then to carry out a series of selective solvent extractions to remove from the biomass the small molecule natural products which will be subsequently evaluated for their potential application as pharmaceuticals or agrochemicals.

Evaluation takes place initially in a primary *in vitro* qualitative assay or bioassay. Two different approaches are utilized in the Mississippi program. One is based upon an *in vitro* cell or whole organism assay. This is easily described in the context of anti-infective agents where extracts would be evaluated for their antimicrobial activity utilizing growing cultures of select micro-organisms as indicator or assay organisms against which specific or selective antimicrobial activity is being sought. A number of cell based assays are utilized in the primary *in vitro* qualitative assay strategy. In addition mechanism based assays are being utilized; looking at receptor binding, enzyme inhibition, and other mechanism of action based assays. In the latter case, plant preparations, and plant extracts particularly, contain a number of non-specific inhibitors or toxins, that cause a general inhibition of protein based assays. This creates a real challenge to assay or bioassay these materials and identify within the extracts substances with selective and specific biological activity in mechanism of action based assays against the background of these non-specific agents; the tannins, the phenols, etc. Some strategies to overcome this limitation for the evaluation of plant extracts in mechanism of action based assays at Ole Miss, have evolved.

Once the primary *in vitro* qualitative assays have been carried out and active extracts have been identified then a strategy of bioassay-directed isolation is undertaken which leads to the isolation of active constituents from the preparation. This strategy is vigorously adhered to since it is not cost effective to fractionate extracts and isolate inactive substances no matter how scientifically interesting or challenging that process may be. Rather it is the biological activity which justifies the investment of time and effort to identify novel substances from these plant extracts. Methodology utilized draws extensively upon modern techniques of chromatography and solvent partitioning, etc. Ultimately, a pure active product will be isolated and its homogeneity demonstrated by various qualitative and quantitative assay technologies.

Once the active product is in hand, a series of secondary assays are then carried out to quantitate the biological activity of the substance; i.e. determine its potency, begin developing some specific information about mode and mechanism of action, evaluate breadth of activity or specificity or selectivity of activity, and importantly, if these agents are to be carried into further development, a view of their pharmacological effects is desirable. Thus some preliminary evaluations are made at the whole organism level, particularly for toxicity and general pharmacological evaluation. This allows one ultimately to identify substances which show selective enough activity to become candidates for *in vivo* evaluation.

While the secondary assays are being carried out, the pure active product is also undergoing structure elucidation. The application of high field NMR technology is particularly important. As the structure elucidation progresses and one is able to identify the specific chemical structure, then some preliminary structure modification work will be initiated to gain some information about the importance of various functional groups and their spacial relationships for biological activity. Ultimately from this, then there will evolve a small list of candidate molecules which will be evaluated in a tertiary *in vivo* assay, one which begins to measure specifically the efficacy and the toxicity of the compound. This tertiary assay is ordinarily a murine model of disseminated disease. In the case of anti-effectives, it is easy to understand that a mouse model of the infectious condition would be available and infected mice would be treated with candidate drugs and either the suppression of the infection, or alternatively, the cure of the infection would be observed.

In the case of plant derived natural products, one of the perceptions that was mentioned earlier which has limited interest has been the availability of the quantity of chemical required to carry out all of these studies. To identify the biological activity and to isolate the active product and determine its chemical structure requires only about a total of 50 mg of the chemical substance. In order to obtain sufficient chemical substance and to carry out this entire process, an understanding of the quantity of plant material required for the preparation of the necessary quantities of chemical substance is needed. To isolate and carry out full chemical characterization requires approximately 50 mg and therefore about 5 kg of dry plant material would be required if it is assumed that the active product were present at the concentration of only 0.001% of the dry weight of the biomass. This may be considered a "worst case" scenario. It is clear that the technologies are capable of identifying potential utility in the bioassay and isolating and characterizing the natural product which occurs in plant biomass at this low concentration. Once the chemical substance is isolated and characterized and its biological properties determined, a decision point is reached: Is the chemical structure novel? Does this substance represent a potential new prototype? The circumstance where the answer to these questions is "yes"; i.e. we have a substance, a natural product, that we wish to carry forward into development is first evaluated. What is the quantity of material required? The next step is to assess the real potential of the substance. Some confirmatory bioassays to make sure that the suspected biological activity is actually present must be carried out. Some secondary biological assays are undertaken to gain a full understanding of the breadth and selectivity of the biological activity. Some preliminary toxicology of the substance be determined, i.e. if it cures a particular infection but kills the host then it is not really likely to be a drug substance. Once all of that information is available, some initial *in vivo* evaluation must be carried out to determine that the agent has real promise both in terms of its efficacy and toxicity in a "real world" situation. To carry out those assessments in the program at the Research Institute, about a half gram, 400-500 mg, of pure active product is needed. That represents a ten-fold increase in the plant material required, so as much as 100 kilograms of dry weight of biomass will need to be processed to gain this additional information. This is not particularly daunting and it is very probable that one would carry forward to this stage. Success at this stage then would suggest that one would wish to go into preclinical evaluation of the agent. The quantity of pure chemical substance required for full preclinical development and a subsequent clinical trial is roughly two kilograms of pure active product. In a "worst case" scenario of active product concentration of only 0.001%, 200,000 kilograms of dry plant biomass would be required to produce the approximately two kilograms of pure active product. This aready begins to look like a very daunting quantity of dry biomass to process. What is the quantity of material that would be required to meet market need should one have carried through the development and shown that the agent was truly of value as a pharmaceutical? Assuming that the agent would be utilized to treat an acute condition and that a relatively small patient population of only about 10,000 patients per year existed and that approximately two grams of the agent were required for a course of therapy then 20 kilograms per year of bulk active drug would be required to meet that market need. At a "worst case" scenario of 0.001% of active product present in the biomass this would then lead to a requirement of two million kilograms of dry biomass per year. That seems like a truly daunting or impossible quantity of material to collect and process. However, if we put it in the context of a plant based commodity; something that is more easily understood, this represents roughly 2200 tons of biomass which is the equivalent of about 75,000 bushels of wheat or corn or soybeans or any other commodity. Indeed an average American farmer produces roughly this 75,000 bushel quantity each year. Consequently, in that context one is not talking about the necessity to chop down and process entire tropical rainforests to obtain the two million kilograms per year of dry plant biomass. Now let us turn to an alternative scenario where the agent would be used to treat a chronic condition and assuming the patient population is considerably larger - 100,000 patients per year, the agent would have reasonable potency so that only 50 mg or so per patient per day would be required to treat the condition. Under those conditions, 2,000 kilograms of bulk active drug would be required to meet the market need. For our worst case scenario of 0.001% concentration of active product in the biomass, 200,000,000 kilograms of dry weight of biomass would be required for processing to produce the 2,000 kilograms per year of bulk active substance. This number of 200,000,000 kilograms appears to be very large and it is but when placed in the context of a plant-based commodity such as wheat or corn or soybeans, it is again obvious that this represents a modest production level. Indeed many agricultural counties of the United States produce as much or more than the 7,500,000 bushels of a commodity the 200,000,000 kilograms represent. In this context it is clear this quantity of biomass is readily obtainable. Two examples that were picked deliberately to emphasize this point may now be analized. In 1990, information on the worldwide production of marijuana and cocaine led to the following observations: at least 30,000 tons of marijuana were produced worldwide with 5,000 tons of that being US production (4). At a price per less than $2,000 per pound and considering all of the criminal penalties that would be paid if one were convicted of producing marijuana, it is clear that if there is market for a plant biomass there will be an entrepreneurial effort to meet that market. Indeed, considering the billions of dollars spent each year to suppress drug plant production then it is easily appreciated that giving producers an alternative, profitable and ethical crop to produce will lead readily to the production of the necessary quantities of biomass for drug production to meet pharmaceutical application. Cocaine is utilized as a nearly pure chemical entity and indeed a thousand tons, roughly one million kilograms, were produced worldwide in 1990, greatly in excess of the examples required to meet an ethical pharmaceutical market. Clearly, capability is there if there is a stable and bona fide market for the plant-derived chemical substance.

If plant derived natural products are to be produced for utilization in the pharmaceutical or agrochemical industry, then there will be certain criteria that a system of production must meet. It clearly must be economic. After all, if the drug costs hundreds or thousands of dollars per dose, then there is no viable product. It must be sustainable and reliable. Patients will need the drug this year, next year and perhaps a decade from now and a source of that agent must be available to meet those medical needs. Clearly in today's society, production of plant derived natural products must be environmentally safe, non-environmentally impacting. One cannot propose to cut down the rainforests or denude the earth of a particular species for the production of a plant derived natural product.

In order to meet these criteria and establish a viable production system, one must systematically evaluate all the steps of production of a plant derived natural product. The first order of business is to identify a superior source of that substance. A strain or variety of the species must be discovered which has a high and consistent concentration of the natural product or a precursor of the natural product that can be converted economically to the final bulk active product.

Once that superior source has been identified then one must secure an uninterruptible and stable supply of that material. That ordinarily means that one must develop an agronomic system for biomass production, i.e. one must bring the source into cultivation. This allows then full expression of the genetic capability of the cultivar. One endeavors to match climate and soil types to the requirements of the plant, understand the impact of fertilization, irrigation and the like on the production of the biomass and its chemical constituents and learn generally how to grow and cultivate the material economically. In the future one may well be able to develop control over production of secondary substances and plants through the use of growth regulators of one sort or another.

After one have learned to produce the biomass, one must learn to harvest that biomass appropriately. When during its growing season does the drug concentration reach its highest concentration? How does one handle the freshly harvested biomass to retain drug content? And finally, in order to maintain an economic system of production, one must consider mechanization of the harvest process.

Once the biomass is harvested in order to build an economic processing facility, one wishes to level processing of the biomass for isolation of the active product over an entire calendar year. One does not wish to build or be required to build a processing facility which would process all of the biomass immediately after its harvesting. So technology must be developed to stabilize the biomass so that it retains drug content during storage prior to its ultimate processing. This usually involves developing an appropriate drying process.

Once processing of the biomass is initiated that extraction/purification system must be economic; it must be efficient in its recovery of the natural product from the biomass. It must be safe in its operation and the generation of waste products must be minimized so that there is no deleterious environmental impact from the processing of the biomass material. If one carries out a systematic evaluation of a production strategy for plant derived natural products, there is evidence that the quantity of material should not become a limitation either in development or ultimate commercialization of pharmaceuticals or agrochemicals derived from plant derived natural products.

This point may be illustrated with the example of taxol - a recently introduced anticancer agent of plant origin. Taxol was initially discovered through the National Cancer Institute program for evaluation of plant preparations for anticancer activity (5). In 1962, USDA botanist Arthur Barkley collected *Taxus brevifolia* and submitted that biomass to the cancer institute anticancer evaluation effort. In 1964 an extract of the bark was shown to be highly cytotoxic *in vitro* to cancer cells. This material then was recollected and the biological activity was confirmed in certain animal models of cancer. By 1967 3,000 pounds of bark were collected and processed, leading in 1971 to the structure elucidation of taxol by Wall and coworkers. Efforts by Susan Horowitz in the early 1970s showed that taxol had a unique mechanism of action in its suppression of the growth of cancer cells. This led in 1977 to its evaluation in animal models of cancer where it showed high activity which led to its designation for development. In 1983 taxol entered human clinical trials. By 1988 initial clinical results in ovarian cancer were very encouraging and a major effort was initiated. The development of a system for the economic production of taxol and its final development and approval for utilization in the treatment of cancer followed. In 1992 taxol was approved for the treatment of refractory ovarian cancer. In 1993 it is estimated that Bristol-Myers Squibb Company will have sales or did have sales of more than 150 million dollars of taxol. Thus one can see that this very complex natural product of plant origin has great utility in the treatment of human cancer and because of its complex chemical structure, it will not likely be economically prepared by synthesis and so one must rely on isolation of the agent from a natural source. And indeed Bristol-Myers Squibb has evolved a system of production based upon isolation of a precursor of taxol from the leaves or needles of *Taxus baccata* or *Taxus wallichiana* and the conversion of that precursor by chemical synthesis into taxol. In this way, the hundreds of kilograms of taxol required per year for the treatment of cancer patients will be made available in the future.

In summary, plant derived natural products hold great promise for discovery and development of new pharmaceuticals and agrochemicals. Careful consideration of the entire process of discovery and development - a "systems" approach - will be required to realize this great promise effectively. Such an analysis was outlined and illustrated in this paper

Literature Cited.

1.   Basara, L. R.; Montagne, M. *The Food and Drug Administration and Drug Approval*; Chapter 4, Pharmaceutical Products Press: New York, 1994; pp 77-108.
2.   *Drug Prices and Profits, 1993*; Pharmaceutical Manufacturers Association, Washington, DC, 1994.
3.   Waterman, P. G. *Secondary Metabolites: Their Function and Evolution*; In Ciba Foundation Symposium 171, Wiley, Chichester England, 1992; pp 255-275.
4.   *The Supply of Illicit Drugs to the United States*; The National Narcotics Intelligence Consumers Committee Report, Drug Enforcement Administration, Office if Intelligence, Washington, DC, 1990.
5.   Kingston, D. G. I.; Molinero, A. A.; Rimoldi, J. M. *Progress in the Chemistry of Organic Natural Products*; vol. 61, New York, 1993.

RECEIVED December 15, 1994

Chapter 7

Poisons and Anti-poisons from the Amazon Forest

Walter B. Mors

Núcleo de Pesquisas de Produtos Naturais, Universidade Federal do Rio de Janeiro, 21941−590 Rio de Janeiro, RJ, Brazil

Having been used as insecticides in the past, the rotenoids seem destined for a comeback. Easy degradability and consequent lack of persistence on the treated objects are playing in favor of these materials due to our current environmental concerns. Thus, new sources of rotenoids are being investigated. It has also been discovered that the rotenoids in *Derris* species are accompanied by pyrrolidine derivatives which are active compounds in their own right. On of them, 2,5-dihydroxymethyl-3,4-dihydroxypyrrolidine (DMDP), a powerful glucosidase inhibitor, has now been isolated from the roots of *Derris urucu*. As for anti-poisons: A number of plants active against snake bite occur in the Amazon region. Several active compounds have already been isolated. Assays in laboratory animals showed antimyotoxic, anti-hemorrhagic and anti-protéolytic activity. Both extracts and pure compounds were able to protect mice from the effects of snake venom when administrated by mouth. Development of a product for practical use by humans is in sight.

It is with pleasure that I accept the invitation of the President of the Associação Brasileira de Química and the Coordinator of the Scientific Committee on Natural Products, to speak about my experience in the Amazon region during my professional life. It is exactly fifty years now that I, then recently graduated, started to work at the Instituto Agronômico do Norte, in Belém, state of Pará. As a first task I was put in charge of the chemical laboratory where routine analyses of timbó roots were performed. Curiously, the subject "fish poisons" has accompanied me ever since.

Analyses of samples collected in different places and grown at the I.A.N. were performed on roots from 232 clones of *Derris (Lonchocarpus) urucu*

0097−6156/95/0588−0079$12.00/0

Saponins ...

...which kill

Serjanoside B

from

Serjania caracasana

(Sapindaceae)

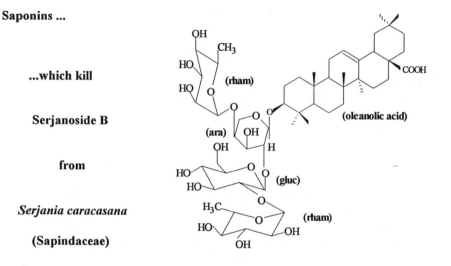

and

...which disperse

Derrissaponin

from

Derris urucu

(Fabaceae)

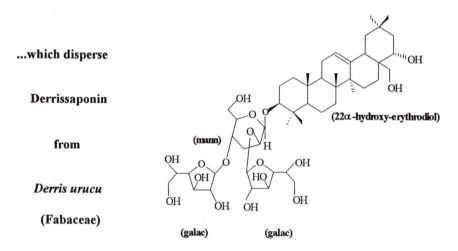

Figure 1. Fish Poisons Containing Saponins.

(Fabaceae) and 148 clones of *D. utilis*. The results were published by the U.S. Department of Agriculture in 1948 (*1*). Three years ago a similar commitment led to the evaluation of another 69 clones of fish poison plants, mainly *Derris* species from the Amazon, now at our research center (NPPN) at the Federal University of Rio de Janeiro (Lima, R. R.; Chaves da Costa, J. P.; Mathias, L.; Parente, J. P. ; Mors, W. B., unpublished data.). The difference in methodology is striking. While in the early 40's the procedure was still gravimetric, by weighing the rotenone crystallized from the root extracts, HPLC is today the method of choice - faster and, of course, more accurate.

In the years in between I had the opportunity - with the collaboration of many colleagues and students - to study fish poisons derived from other plant families: A polyacetylene, ichthyothereol, was isolated from *Ichthyothere terminalis* (Asteraceae) (*2*) and so were the active saponins from *Serjania caracasana* (Sapindaceae) (*3*).

Saponins are also present in *Derris* roots. But, contrary to those of Sapindaceae, they are devoid of toxicity. They do contribute, as dispersants, to suspend the rotenoids, which by themselves are insoluble in water (*4*). Figure 1 shows examples of the two types.

Having been used as insecticides in the past, the rotenoids seem destined for a comeback. Factors which were formerly held agains these substance, such as their degradability and consequent lack of persistence on the treated objects (plants and animals), are now playing in favor of these materials due to our current environmental concerns. Thus, new sources of rotenoids are being investigated by us; and just now, *Derris* extracts had, for the first time, their activity evaluated against larvae of *Triatoma infestans* one of the vectors of Chagas' diesease, with highly promising results (*5*).

It has been shown in the 30's that *Derris* (*Lonchocarpus*) extracts artificially deprived of rotenone still show considerable insecticidal activity (*6-8*). Our own experiments extended these findings to *Derris* species which are naturally devoid of rotenoids but nevertheless show toxicity to fish of the same order as rotenone (*9*). Chalcones present in these were inactive. Several other species which do not contain rotenoids are reported to serve as fish posions. What are, then, the active principles in these suspensions or extracts?

Nitrogen-containing sugar analogs have recently been described from several Fabaceae species. One of them, 2-5-dihydroxymethyl-3,4-hydroxypyrrolidine (DMDP), was first isolated form leaves and roots of *Derris elliptica (10)*. These sugar mimics were shown to be powerful glucosidadse inhibitors *(11)*. DMDP acts as an antifeedant for locusts and is toxic for certain caterpillars and beetles. It was isolated in our laboratory from the roots of *Derris urucu* (Mathias, L.; Parente, J. P., unpublished data.)

It is possibly responsible, at least in part, for the toxicity of *Derris* extracts, even if they do not contain rotenoids. The matter is being investigated by us at present. Figure 2 shows some of the mentioned molecules.

So much for the poisons. Now to the anti-poisons.

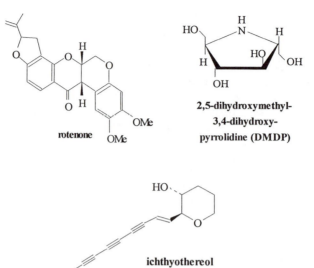

2,5-dihydroxymethyl-
3,4-dihydroxy-
pyrrolidine (DMDP)

rotenone

ichthyothereol

Figure 2. Some Molecules Discussed in the Lecture.

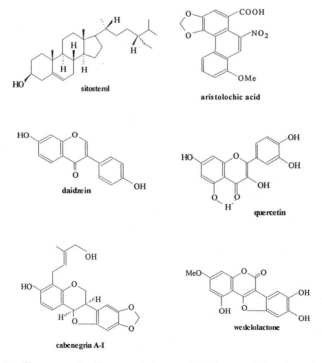

sitosterol

aristolochic acid

daidzein

quercetin

cabenegrin A-I

wedelolactone

Figure 3. Compounds Responsible or Co-Responsible for Anti-Snake Venom Activity in Plants.

Several years ago, Brazilian chemists and pharmacologists entered a completely new field of bioactive plants: those reputed by people as antidotes against snake bite. Incredible as it may look on first sight, these medicines actually do work *(12)*. Many of them have been studied recently, with the most surpirising results. Some of the most famous are listed in Table I.

Table I. Amazonian Plants Used as Snake Venom Antidotes

Family	Species	Common Names
Cyperaceae	*Cyperus corymbosus* Rottböll	priprioca
Araceae	*Dracontium asperum* K. Koch	batata-de-cobra, jararaca, milho-de-cobra, tajá-de-cobra
Moraceae	*Dorstenia asaroides* Gard.	caapiá, carapiá
Aristolochiaceae	*Aristolochia trilobata* L.	jarrinha, papo-de-peru, urubucaá
Capparaceae	*Crataeva benthami* Eichl.	catauari, catoré
Icacinaceae	*Humirianthera duckei* Hub.	mairá, surucucuina
Loganiaceae	*Potalia amara* Aubl.	erva-de-cobra, pau-de-cobra
Verbenaceae	*Aegiphila salutaris* H.B.K.	contra-cobra
Scrophulariaceae	*Stemodia viscosa* Roxb.	boia-caá, paracari
Asteraceae	*Eupatorium triplinerve* Vahl.	erva-de-cobra, japana

Figure 3 shows examples of molecules which, in our own research as well as in the international literature, were found to be responsable for anti-snake venom activity (*12*). What do these compounds have in common? Chemically, very little. But they are all low molecular weight compounds with some strong biodynamic properties detected in the past. Their targets in living organisms are polypeptides and proteins - the kinds of structures present in toxins and enzymes which make up snake venoms. *In vitro* and *in vivo* assays in laboratory animals showed antimyotoxic, anti-herrorrhagic and antiproteolytic activity. Both extracts and pure compounds were able to protect mice from the effects of snake venom when administrated by mouth. Even if the mechanism of action has not yet been ascertained, the development of a product for practical use by humans is in sight.

Having formerly yielded rubber, resins, rosewood oil, quassia, cacao, Brazil nuts and tubocurarine, the Amazon forest may still be able to offer many as yet unexploited products for the benefit of mankind.

Literature cited

1.   Higbee, E. C. *Lonchocarpus, Derris* and *Pyrethrum* Cultivation and Sources of Supply. *U.S.Dept. Agriculture Miscellaneous Publication no. 650,* **1948**,   Washington, D.C.

2.   Cascon, S. C.; Mors, W. B.; Tursch, B. M.; Aplin, R. T.; Durham, L. J. *J. Am. Chem. Soc.* **1965**, *87*, 5237.

3.   Xavier, H. S.; Mors, W. B. *Ciência e Cultura* (São Paulo) **1975**, *27 (7,supl.)*, 179.

4.   Parente, J. P.; Mors, W. B. *Anais Acad. Brasil. Ci.* **1980**, *52*, 503.

5.   Bronfen, E.; Oliveira, C. L.; Campos, A. E. S.; Mathias, L.; Mors, W. B.; Parente, J. P. *Summary of Papers Presented at Meeting on Basic Research on Chagas' Desease*, Belo Horizonte, November **1993**.

6.   Campbell, F. L.; Sullivan, W. N.; Jones, H. A. *Soap* **1934**, *10*, 81.

7.   Ginsburg, E. *J. Econ. Entomol.* **1934**, *27*, 393.

8.   Jones, H. A.; Gersdorff, W. A.; Gooden, E. L.; Campbell, F. L.; Sullivan, W. N. *J. Econ. Entomol.* **1933**, *26*, 451.

9.   Mors, W. B.; Nascimento, M. C.; Ribeiro do Valle, J.; Aragão, J. A. *Ciência e Cultura* (São Paulo) **1973**, *25*, 647.

10.   Welter, A.; Jadot, J.; Dardenne, G.; Marlier, M.; Casimir, J. *Phytochemistry* **1976**, *15*, 747.

11.   Evans, S. V.; Fellows, L. E.; Shing, T. K. M.; Fleet, G. W. J. *Phytochemistry* **1985**, *24*, 1953.

12.   Mors, W. B. In *Economic and Medicinal Plant Research*; Wagner, H.; Farnsworth, N. R., Eds.; Academic Press, **1991**, Vol 5; 353-373.

RECEIVED October 13, 1994

Chapter 8

Biologically Active Neolignans from Amazonian Trees

Massayoshi Yoshida

Instituto de Química, Universidade de São Paulo, Caixa Postal 20.780, CEP 01498–970 São Paulo, SP, Brazil

Typical magnolialean species of the Amazonian forest contain an especially characteristic class of natural products, the neolignans. Many of their isolated representatives are distinguished by important biological properties. The expectations for further successful pharmacological investigations are justified by several striking features of neolignans. With respect to their biosynthesis, one marvels that not less than 40 structural types and close to 700 derivatives can originate by oxidative dimerization of allylphenols and propenylphenols, from only two simple precursors. The second fact concerns a series of rearrangements which, possibly involving cyclic reaction paths, occur spontaneously under natural conditions.

Neolignans can be defined as oxidative dimers of propenylphenol/propenylphenol as shown in Figure 1, propenylphenol/allylphenol as presented in Figure 2 and allylphenol/allylphenol. Different coupling modes of these units and further modifying steps increase the number of neolignans (1). Based on more than 400 neolignans isolated from Amazonian species of Myristicaceae and Lauraceae, Gottlieb and Yoshida (2) proposed a model of evolution in a micromolecular system. In this scheme the Cope, retro-Claisen, Claisen rearrangements constitute an example of microscopic reversibility as pictured in Figure 3.

Neolignans from Myristicaceae

Ethnopharmacological interest in Myristicaceae species first evidenced by Schultes in 1954 (3), inspired Gottlieb and co-workers to develop phytochemical investigations on these arboreal plants dispersed over Amazonia. The occurrence of γ-lactones, flavonoids and neolignans was observed in trunk wood and a

0097–6156/95/0588–0085$12.00/0

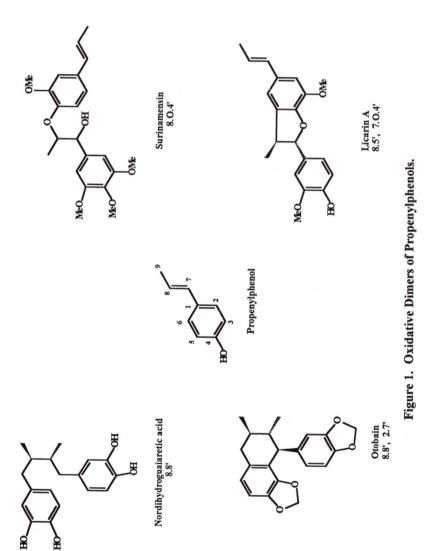

Figure 1. Oxidative Dimers of Propenylphenols.

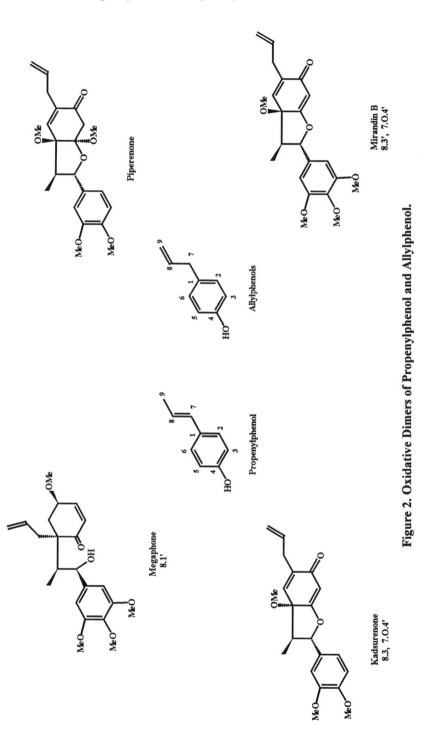

Figure 2. Oxidative Dimers of Propenylphenol and Allylphenol.

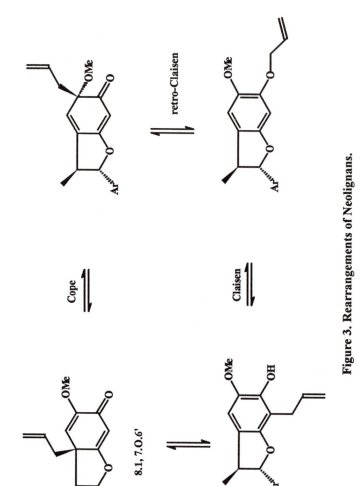

Figure 3. Rearrangements of Neolignans.

communication about isolated compounds and folk medicine of myristicaceous species was published by Gottlieb in 1979 *(4)*.

Otoba butter, sometimes called American nutmeg butter, American mace butter or otoba wax, is a popular name for the fat expressed from the seeds of *Myristica otoba*. This fat has long been used by Colombian people as a medicine for skin diseases of domestic animals *(5)*. Otobain was isolated from otoba fat after hydrolysis with KOH and chromatography on alumina of neutral water insoluble fraction *(6)*. It is still not clear if the activity of otoba fat is due only to otobain.

The fruits of *Virola sebifera* have been found to contain oxootobain *(7)*. Unusual neolignans were also found in fruits of this species and these new indanone neolignans were proposed to be biosynthesized by a pinacol-pinacolone type rearrangement of trihydroxyaryltetralones *(8)* as depicted in Figure 4.

Nutmeg and mace, respectively seed and aril of the fruits of *Myristica fragrans,* have been used in folk medicine as aphrodisiacs or in the treatment of digestive disorders, rheumatism and cholera. *Myristica fragrans* comes originally from Moluccas and is now cultivated for commercial purposes in Malaysia. Nutmeg is used for the preparation of nutmeg butter by expression and essential oil by steam distillation. Nutmeg butter contains of trimyristin (70-85%) and essential oil contains monoterpenes (85%) and aromatic ethers (8%). Essential oil obtained from powdered mace also contains monoterpenes (93%) and aromatic ethers (6.5%). The narcotic effect of nutmeg is explained by presence of myristicin, an aromatic ether *(9)*. Shulgin speculated that myristicin could be the narcotic principle of snuffs called yakee, paricá or epena, prepared from the bark resin of a *Virola* tree *(10)*. However, analysis of epena showed it to contain several hallucinogenic amines, such as derivatives of N,N-dimethyltryptamines or of β-carbolines *(11)*.

In a search for the psychoactive components present in nutmeg and mace, representatives of two different structural types of neolignans, the 8.O.4'-type and the 8.5', 7.O.4'-type, were isolated *(12)*. Surinamensin and virolin, neolignans of the 8.O.4'-type, occur in the leaves of *Virola surinamensis*, besides elemicin and the 8.8', 7.O.7'-neolignans galbacin and veraguensin *(13)*. A hexane extract of the leaves showed activity against the penetration of cercaria of *Schistosoma mansoni*. It was shown that this activity is due only to surinamensin and virolin.

Smart et al. reported the regression of malignant melanoma in a patient following continued medication with an aqueous extract of the cresote bush *(14)*. Cresote is a common name of *Larrea divaricata* (Zygophyllaceae) and the major chemical constituent of leaves and stems is nordihydroguaiaretic acid (NDGA). This compound is a potent antioxidant which has been used to prevent rancidity of vegetable and animal fats and the oxidation of vitamins A and E. Many experiments have been described about the antioxidant properties of NDGA in biological systems, such as the inhibition of eletron and energy transfer in rat liver mitochondria, reported by Dakshinamurti *(15)*. Six neolignans with the skeleton of NDGA were isolated from the chloroform extract of arils and kernels of *Virola calophylla*: four are 1,4-diaryl 2,3-dimethylbutanes, one is an 1-aryl-2,3-dimethyltetralin and the last is a 2,5-diaryl-3,4-dimethyltetrahydrofuran. Larvicide

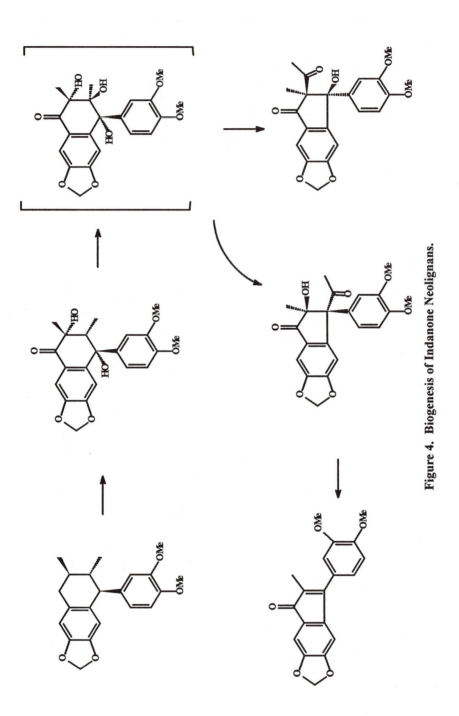

Figure 4. Biogenesis of Indanone Neolignans.

activity of neolignans was tested on *Schistosoma mansoni* and molluscicide activity on *Biomphalaria glabrata*. The most active compounds in these biological tests are the neolignans which possess phenolic hydroxyls (Brocksom, U., Universidade Federal de São Carlos, unpublished data).

Neolignans from Lauraceae

The neolignans of the 8.5', 7.O.4'-type were isolated from trunk wood of *Licaria aritu* which occurs along the Manaus Itacoatiara road. Licarin-A is identical with dehydrodiisoeugenol and its absolute configuration was deduced by comparison of ORD curves *(16)*. Licarin-A was also isolated by LeQuesne et al. *(17)* from leaves and stems of *Nectandra rigida*. The interest of investigation of the chemical constituents of *Nectandra rigida* was due the cytotoxicity of its extracts in the KB assay. it was observed that the major cytotoxic activity of the plant appears to be due to the presence of licarin-A, which displayed an ED_{50} of 7.0 mg/ml against KB cells, in comparison with a active fraction of extract with ED_{50} of 18-31 mg/ml.

The alcoholic extract prepared from dried ground roots of *Aniba megaphylla* demonstrated inhibitory activity *in vitro* against KB. Partition of an extract guided by KB activity led to isolation of three neolignans: megaphone, megaphone acetate and megaphyllone acetate, that showed cytotoxicity against KB cell culture at 1.70, 1.75 and 2.55 mg/ml, respectively *(18)*.

The trunk wood of *Nectandra miranda* contains mirandin A and B besides five benzofuran neolignans *(19)*. The structures of mirandin A and mirandin B are closely related to kadsurenone (see Neolignans from Piperaceae) and denudatin B (see Neolignans from Magnoliaceae), respectively.

Neolignans from Magnoliaceae

Magnolia fargesii, a herb, is popularly known as "hsin-i" in China, and used for the treatment of nasal congestion with headache, sinusitis and allergic rhinitis. It was found that a chloroform extract of flower buds of *Magnolia fargesii* exhibited a Ca^{++}-antagonistic activity. The bioassay-guided isolation of the active principles, by using chromatographic techniques, afforded several neolignans including denudatin B, which possessed most potent activity *(20)*.

Neolignans from Piperaceae

A benzene extract of leaves of *Piper futokadzura* showed an antifeedant activity against larvae of *Spodoptera litura*. The extract yielded the neolignan piperenone. A herbal preparation of the stem of *Piper futokadsura*, a medicinal plant which grows in the southeast of China and Taiwan, is popularly known as haifenteng. Haifenteng is used for the general relief of bronchoasthma, inflammation and rheumatic pain. A methylene chloride extract of haifenteng, after chromatographic fractionation, afforded kadsurenone, kadsurin A, kadsurin B and piperenone. The active principle of this traditional medicine, identified as receptor antagonist for

PAF-acether, is the neolignan kadsurenone. The compound shows a K_i of 5,8 x 10^{-8} M vs. a K_i of 6.3 x 10^{-9} M for PAF itself *(21)*.

It is hopped that this impressive diversity of biological activities will enhance interest in the several hundred additional bicyclooctane and benzofuran neolignans of yet unreported bioactivity, isolated and described with respect to structure by our laboratories in the past decade.

Literature Cited

1. Gottlieb, O.R.; Yoshida, M. In *Natural Products of Woody Plants*; Rowe, J.W., Ed.; Springer-Verlag: Berlin, 1989; pp. 439-511.
2. Gottlieb, O.R.; Yoshida, M. *Nat. Prod. Lett.* **1993**, *1*, 281.
3. Schultes, R.E. *Bot. Mus. Leaf. (Harvard University)* **1954**, *16*, 241.
4. Gottleib, O.R. *J. Ethnopharm.* **1979**, *1*, 309.
5. Baughman, W.F.; Jamieson, G.S.; Brauns, D.H. *J. Amer. Chem. Soc.* **1921**, *43*, 199.
6. Gilchrist, T.; Hodges, R.; Porte, A.L. *J. Chem. Soc.* **1962**, 1780.
7. Lopes, L.M.X.; Yoshida, M.; Gottlieb, O.R. *Phytochemistry* **1982**, *21*, 751.
8. Lopes, L.M.X.; Yoshida, M.; Gottlieb, O.R. *Phytochemistry* **1984**, *23*, 2021.
9. Forrest, J.E.; Heacock, R.A. *Lloydia* **1972**, *35*, 440.
10. Shulgin, A.T. *Nature* **1966**, 210, 380.
11. Agurell, S.; Holmstedt, B.; Lindgren, J.E.; Schultes, R.E. *Acta Chem. Scand.* **1969**, *23*, 903.
12. Forrest, J.E.; Heacock, R.A.; Forrest, T.P. *J. Chem. Soc. Perkin I* **1974**, 205.
13. Barata, L.E.S.; Baker, P.M.; Gottlieb, O.R.: Ruveda, E.A. *Phytochemistry* **1978**, *17*, 783.
14. Smart, C.R.; Hogle, H.H.; Robbins, R.K.; Broom, A.D.; Bartholomew, D. *Cancer Chemother. Rep.* **1969**, *53* (Part I), 147.
15. Bhuvaneswaran, C.; Dakshinamurti, K. *Biochemistry* **1972**, 11, 85.
16. Aiba, C.J.; Correa, R.G.C.; Gottlieb, O.R. *Phytochemistry* **1973**, *12*, 1163.
17. Le Quesne, P.W.; Larrahondo, J.E.; Raffauf, R.F. *J. Nat. Prod.* **1980**, *43*, 353.
18. Kupchan, S.M.; Stevens, K.L.; Rohlfing, E.A.; Sickles, B.R.; Sneden, A.T.; Miller, R.W.; Bryan, R.F. *J. Org. Chem.* **1978**, *43*, 586.
19. Aiba, C.J.; Gottlieb, O.R.; Pagliosa, F.M.; Yoshida, M.; Magalhães, M.T. *Phytochemistry* **1977**, *16*, 1547.
20. Chen, C.C.; Huang, Y.L.; Chen, H.T.; Chen, Y.P.; Hsu, H.Y. *Planta Med.* **1988**, 438.
21. Chang, M.N.; Han, G.R.; Arison, B.H.; Springer, J.P.; Hwang, S.B.; Shen, T.Y. *Phytochemistry* **1985**, *24*, 2079.

RECEIVED October 15, 1994

Chapter 9

Proteins from Amazonia

Studies and Perspectives for Their Research

Lauro Morhy

Laboratório de Bioquimica e Quimica de Proteinas, Departamento de Biologia Celular, Universidade de Brasilia, 70910–900 Brasilia, DF, Brazil

Very little is known about the proteins of the great variety of species of plants and animals present in Amazonia. Perspectives for their research and some results of studies on enterolobin, a cytotoxic protein from *Enterolobium contortisiliquum* seed, and methionine rich proteins from *Bertholetia excelsa* and *Lecythis usitata* seeds are presented.

Amazonia is the last and the largest natural reserve of biodiversity of the world today. Determination of the number of plants and animals existing in the region is a task that still must be completed, to say nothing about research at a molecular level. Phytochemists started work on small molecules some time ago, but only a few studies on macromolecules have been made so far. Ultimately, biodiversity could be described as diversity of informational molecules present in the living organisms; therefore, proteins attract special interest of scientists, and represent a wide range of scientific research and biotechnological applications.

Protein Diversity and Potentiality. Proteins are synthesized according to genetic instructions. Modern evolutionary theory assumes that genes have been modified in the course of time and that natural selection has led to evolutionary lines and to different species. If we consider that proteins are encoded by genes, according to a genetic code, we can expect that the resulting amino acid sequences reflect the evolution and the biodiversity.

Proteins or genes having a significant number of similarities are said to be homologous. They could descend from a common ancestor. Comparison between homologous proteins of a single family shows that certain amino acid positions are conserved, while others present variations. These variations should not affect the general conformation of the molecule, which determines its function. It is interesting to observe the functional invariability of hydrophobic residues in homologous proteins, as in the classical case of Cys14 and Cys17 in cytochrome c, to which the heme is attached, and in the region 70-80, which is folded in such way as to form the apolar heme pocket. Crystallography and comparative protein chemistry show that the interior of the protein globule is more resistant to evolutionary changes than surface parts. Proteins having almost identical biological activities and some amino acid sequence differences have been named isoproteins and have been assumed to result from duplication of a structural gene.

Proteins are certainly the most abundant biomacromolecules. Extremely versatile in their function, they are present in an immense range of biological systems. The biochemical catalysts (enzymes) are proteins. The main biochemical reactions that occur in living cells are catalyzed by enzymes. This also includes DNA replication, protein biosynthesis and photosynthetic reactions. Even the light produced by fireflies results from a reaction involving the enzyme luciferase. Proteins are utilized as nutrient; in transport (hemoglobin and membrane transporters); in defense against bacterial and viral infection (immunoglobulins); in blood coagulation (fibrinogen and thrombin); in defense against microorganisms (bacterial toxins), animals (snake, scorpion and other venoms) and plants (ricin, enterolobin, protein inhibitors, etc.); in contraction and motility (actin and myosin, in muscle; tubulin and dynein in flagela and cilia); to support biological structures (collagen, in tendons and cartilage; elastin, in ligaments) and in coats (keratin in hair, fingernails, and feathers; fibroin, in silk fibers and spider webs); in regulation of cellular and physiological activity (hormones). Many other proteins with unusual or exotic properties (intensely sweet, antifreezing, etc.) have been found. If we consider the number of possible combinations of the 20 amino acids out of which proteins are made, we can expect many other unknown varieties of proteins to also exist.

Proteins have been exploited for various purposes by man. However, food and medical use are certainly the most important ones. In the first case, proteins constitute our main amino acid source and are used as ingredients to prepare better dishes. Food proteins must be palatable, digestible, non-toxic, and economically available. It is important that they fulfill nutritional needs of essential amino acids. Modern technological utilizations of proteins also explore chemical modifications and physical properties (as viscosity, surface tension and solubility) to get desirable characteristics of foods as beverages, soup, sauces, bread, cakes, ice creams, desserts, egg substitutes, sausage, texturized vegetable proteins, food coatings and others products. Enzymes are used in sugar refining, oligo and polysaccharide (starch, cellulose, etc.) processing; ethanol fermentation; beer brewing; baking (growth, ripening and storage); dairy industry; amino acid production (for food supplements, medicinal agents, etc.); as antioxidant or for removing oxygen and reactive oxidants (glucose oxidase, superoxide dismutase, catalase, etc.); in protein processing (gellatine, peptones, collagen, soy and whey proteins, wheat gluten hydrolisis, yeast extracts industry, clinical analysis, conversion of porcine insulin in human insulin, tenderising meat, aspartame (a very sweet synthetic dipeptide ester); in fruit processing; as cleansing and detoxifying agents (*1*).

Enzymes, proteins, antibodies, biologically active peptides and other protein derivatives have long been used in medicinal agents. Modern biotechnology opened horizons to create new processes and products. For example insulin, interferon and other medical proteins are now being produced commercially by genetic engineering.

Perspectives for Research. Research on biodiversity is normally based on morphological aspects. But, while classification of biological species is relatively easy by descriptive visual methods, the same cannot be said in the case of molecular species. In this case, purification and subsequent studies involve more effort and technical resources. Usually protein purification must be performed in a series of steps, using different techniques. Obtaining a highly purified protein is usually a formidable task. However, modern chromatographic methods and related techniques have made this work easier and faster.

In Amazonia, determination of total nitrogen or protein content in biological material should be a primary preoccupation. Then subsequent or simultaneous studies could be performed using more sophisticated modern methods. Amino acid composition gives important preliminary information. Edible seeds, fruits, leaves and roots would deserve priority. Ethnobotanical information could give suggestions for food and drug research. The following genera should be considered : *Aptandra, Amaranthus, Bertholletia, Carpotroche, Couroupita, Couralia, Dioscorea, Dipteryx, Enterolobium, Guarea, Guilielma, Glycydendron, Hymenolobium, Ipomoea, Joannesia, Lecythis, Mouriria, Pachylecythis, Protium, Salacia, Stryphnodendron, Theobroma* and *Urospatta*.

A protein or peptide is recognized primarily by its biological function. However, current modern methods allow isolation and characterization of these molecules based only on chemical and physico-chemical properties. In many cases, the use of analytical methods such as electrophoresis and chromatography may be sufficient, independently of the substance's bioactivity. Wild species of palms, for example, should be studied by these methods, to classify and select germoplasm. In the palm group, selection of *Guilielma* ("pupunha") species or varieties could be made by the application of these methods to seeds and complemented, if necessary, with protein sequencing. While snake venoms and similar products, could be studied by chemical and physico-chemical methods only, biochemical and pharmacological monitoring woud be advisable.

From the knowledge of the amino acid sequences of a single homologous protein present in contemporary species, it is possible to analyze evolutionary relationships among these species and construct a phylogenetic tree (2). More accurate phylogenetic trees can be generated using folding conformations, since homologous residues are believed to occupy homologous conformational positions (3). These studies could be made on Amazonian species (plants, animals and microorganisms), complementing classical evolutionary methods.

Systematic screening for bioactive and interesting natural proteins and peptides would be a first step in research. Advances in biotechnology make screening relatively inexpensive and quick (4). After amino acid sequence determination, bioactive peptides could then be synthesized, rather than produced by destruction of natural resources. It is interesting to observe that the range of diversity generated by peptide synthesis has been extended in a major way well beyond natural products in daily usage.

Enterolobin. A large number of substances with the ability to cause cell lysis is known. The chemical nature of these compounds is very diverse, including low molecular weight substances such as saponins, lysophospholipids, ionic and non-ionic detergents, antibiotics, and also many cytolytic peptides and proteins produced by living organisms (5). To date, only one protein, enterolobin, has been purified from plants, and shown to posses cytolytic activity (6). Enterolobin is a large protein (55 kDa) purified from seeds of the forest tree *Enterolobium contortisiliquum* Vell. (Morong) (Leguminoseae-Mimosoideae). Pharmacological studies demonstrated that enterolobin is a very potent inflammatory agent. It induces paw oedema partially dependent on lipoxygenase metabolites and histamine, while PAF-acether and prostaglandins do not seem to be important in this reaction. Enterolobin also causes pleural exudation and cellular infiltration, with the remarkable ability to attract polymorphonuclear neutrophils and eosinophils(7,8). Enterolobin has insecticidal activity for insect larvae as determined in bioassays (9). It is also cytotoxic for cancer cells in culture (10). There are amino acid sequence similarities among enterolobin and the bacterial cytolysins called aerolysins from *Aeromonas hydrophilla* and *Aeromonas sobria* (Sousa, M.V.; Fontes, W.; Richardson, M.; Morhy, L. *J. Prot. Chem.*, in press), as shown in Figure 1.

Sulfur Rich Protein. A sulfur rich albumin purified from *Lecythis usitata* presented two polypeptide chains. The small chain has the following sequence (*11*):

GPRQQCEPREQMQQQMLSHCRMYMRQQMEES

This sequence showed close amino acid composition, hydrophobicity profile and great homology with the small subunit of a 2S sulfur-rich albumin found in *Bertholletia excelsa* seeds (Brazil-nut). Infrared spectra (deuterium oxide solution, dry film) and circular dichroism studies of the small protein subunit from *L usitata*, indicated a great amount of ordered structure (*12*).

In the 2S albumin from *B. excelsa*, all of the 8 cysteine residues are involved in the formation of disulfide bridges. Sequence homology studies showed that *Ricinus comunis* and *Helianthus annuus* albumin have the highest identity score within a super-family of seed storage proteins (*13*).

Acknowledgments. The author thanks Drs.Marcelo Valle de Sousa and Carlos Bloch Jr. for reviewing the English manuscript, and M.D.Wagner Fontes for technical help.

```
Enterolobin       TQRDT.LTNG.AQ
A.hydrophilla     MQK.IKLT.GLSLIISGLLMAQAQAAEPVYPDQLRLFSLGQGVCGDKYRPVN
A.sobria          M.KALKIT.GLSLIISATLAAQTNAAEPIYPDQLRLFSLGEDVCGTDYRPIN
------------------------------------------------------------------
                  WTQNHGLAVN...IDTMATGV.ARINR      RCCY
REEAQSVKSNIVGMMGQW.QISGLA.NGWVI..MGPGYNGEIKPGTASNTWCYPTNPVTGEIPTLSALD
REEAQSVRNNIVAMMGQW.QISGLA.NNWVI..LGPGYNGEIKPGKASTTWCYPTRPATAEIPVLPAFN
------------------------------------------------------------------
                  RLLDAH        TQNHS.W.GFA         RNLGNNN.F
IPDGDEVDVQWRLVHDSANFIKPTSYL.AHYLGYAWVGGNHSQYVGEDMDVTRDGDGWVIR..GNNDGG
IPDGDAVDVQWRMVHDSANFIKPVSYL.AHYLGYAWVGGDHSQFVGDDMDVIQEGDDWVLR..GNDGGK
------------------------------------------------------------------
C          ACIHOYLQFAWN....SF..GDPTV..R        WA..DSDTTNNNS..D.TL.F
CDGYRCGDKTA.IK.VSNFAYNLDPDSFKHGDVTQSDRQLVKTVVGWAVNDSDTP..QSGYDVTLRYDT
CDGYRCNEKSS.IR.VSNFAYTLDPGSFSHGDVTQSERTLVHTVVGWATNISDTP..QSGYDVTLNYTT
------------------------------------------------------------------
           DWFN.FKYETKQE                           S...TV.SR
ATNWSKTNTYGLSEKVTTKNKFKWPLVGETQLSIEIRANQ.SWASQNGGSTTTSLSQSVRPTVPARSKI
MSNWSKTNTYGLSEKVSTKNKFKWPLVGETEVSIEIAANQ.SWASQNGGAVTTALSQSVRPVVPARSRV
------------------------------------------------------------------
                                  KSNYNHK..I..Y.N      IRYLQE
PVKIELYKADISYPIEFKADVSYDLTLSGFLRWGGNAWYTHPDNRPNWNHTFVIGPYKDKASSIRY.QW
PVKIELYKANISYPYEFKADMSYDLTFNGFLRWGGNAWHTHPEDRPTLSHTFAIGPFFKDKASSIRYPQW
------------------------------------------------------------------
DVQVH     WW.W.WSF      LDTMTG.VA..LNR FKASGINGQYLSAR QFSGG.ETVSPYRLAAP
DKRYIPGEVKWWDWNWTIQQNGLSTMQNNLARVL.RPVRAG.ITGDF.SAESQFAGNIEIGAPVPLAA.
DKRYLPGEMKWWDWNWAIQQNGLATMQDSLARVL.RPVRAS.ITGDF.RAESQFAGNIEIGTPVPLGS.
------------------------------------------------------------------
FDSCLWRR..SPNLGTD
 DSKV.RRARSVDGAGQGLRLEIPLDREELSGLGFNK...SAS.A
.DSKV.RRTRSVDGANTGLKLDIPLDAQELAELGFENVTLSVTPARN
```

Figure 1. Alignement of the sequences of some enterolobin peptides with the sequenced aerolysins. The boxed region represents the proposed cytolytic site.

Literature Cited

1. Cheetham, P.S.J. In *Handbook of Enzyme Biotechnology*; Wiseman, A., Ed.; 2nd Ed, Ellis Harwood: Chichester, West Sussex, 1985; pp 274-373.

2. *Methods in Enzymology*; Doolitle, R.F., Ed.; Academic Press: New York, 1990, Vol.183, Section VII.

3. Johnson, M.S.; Sali, A.; Bludell, T.L. In *Methods in Enzymology*; Doolitle, R.F., Ed.; Academic Press: New York, 1990, Vol.183; pp 670-690.

4. Maeji, N.J.; Bray, A.M.; Valerio, R.M.; Seldon, M.A.; Wang, J.-X.; Geysen, H.M. *Pept. Res.* 1991, 4, 142.

5. Sousa, M.V.; Ricart, C.A.; Morhy, L. *Ciênc.Cult.* 1990, 42(7), 495-500.

6. Sousa, M.V.; Morhy, L. *An. Acad. Brasil. Cienc.* 1989, 61, 405-412.

7. Cordeiro, R.S.B.; Castro-Faria-Neto, H.C.; Martins, M.A.; Correia-da-Silva, A.C.V.; Bossa, P.T.; Sousa, M.V.; Morhy, L. *Mem. Inst. Oswaldo Cruz* 1991, 86, 129-131.

8. Castro-Faria-Neto, H.C.; Martins, M.A.; Bozza, P.T.; Perez, S.A.C.; Correa, A.C.V.; Lima, M.R.C.; Cruz, H.N.; Cordeiro, R.S.B.; Sousa, M.V.; Morhy, L. *Toxicon* 1991, 29, 1143-1150.

9.   Sousa, M.V.; Morhy,L.; Richardson, M.; Hilder,V.A.; Gatehousa *Entomol.*
      Exper. et Appl. **1993**, *69*, 31-238.
10.   Sousa, M.V. *Ph.D. Thesis*, University of Durhan, Durhan, England, 1991.
11.   Bloch Jr., C.; Sampaio, M.J.; Morhy, L. *Arq. Biol. Tecnol.* **1988**, *31*, 165.
12.   Bloch Jr., C.; Aragão, J.B.; Morhy, L. *Arq. Biol. Tecnol.* **1988**, *31*, 156.
13.   Da Silva, M.C.M.; Bloch Jr., C.; Morhy, L.; Aragão, J.B.; Neshic, G.
      SBBq-Programa e Resumos da XXII Reunião Anual.; Caxambu, MG,
      Brazil, 1993; p 126.

RECEIVED September 28, 1994

Chapter 10

Terpenoids from Amazonian Icacinaceae

Alaide Braga de Oliveira

Faculdade de Farmácia, Universidade Federal de Minas Gerais, 30180–112, Belo Horizonte, MG, Brazil

Amazonian species of *Emmotum* and *Poraqueiba*, two genera of the Icacinaceae family occuring in the Western Hemisphere, contain a class of rearranged eudesmane sesquiterpenes named emmotins that are of higher frequency in *Emmotum* (27 compounds/4 species) than in *Poraqueiba* (2 compounds/2 species). Chemical transformations were used to interrelate several of the emmotins. While these rearranged eudesmanes are restricted to those two genera, diterpenoids have been found in *Icacina* from tropical Africa and *Humirianthera* from Amazonia. Monoterpene seco-iridoids occur in *Poraqueiba* and in Icacinaceous species from Far Eastern countries.

Icacinaceae is a pantropical family. Its centre of distribution in the New World is the upper Amazon basin. A few representatives reach the southern temperate zone but no genus has been found north of Mexico on the American continent (*1*). According to Engler the family comprises 38 genera (*2*). All twelve genera recognized in the western hemisphere belong to the Icacineae, one of the four tribes of the sub-family Icacinoideae. Eight genera are quoted for Brazil: *Citronella, Dendrobangia, Discophora, Emmotum, Humirianthera, Leretia, Pleurisanthes* and *Poraqueiba*. Species of three of these genera are cultivated as sources of foods in Brazil. *Poraqueiba* species are cultivated in Pará state for the oil and the starchy endosperm of the seed. Fresh fruits are sold in local markets. *Humirianthera* species, that occur also in Brazilian Amazonia, have large tubers which are rich in starch and are used after washing for removal of toxins. *Citronella* is found under cultivation in southern South America, where *C. gongonha* has an extensive use as a substitute for *Ilex paraguayensis* in making "mate" tea (*1*).

The systematic position of the Icacinaceae is controversial. It is placed in different positions in the four most modern classifications of the angiosperms (*3-6*).

NOTE: This chapter is dedicated to the memory of Professor Geovane G. Oliveira.

0097–6156/95/0588–0099$12.00/0
© 1995 American Chemical Society

Table I. Tetralin Sesquiternoids from *Emmotum* and *Poraqueiba*

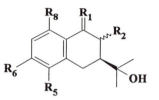

Compound	R_2	R_5	R_6	R_8	R_1	Specie Code
(+)-Rishitinol	βOH	Me	H	Me	H,H	En1
Emmotin-Z	βOH	CHO	OH	Me	H,H	Pg
6-O-Methylemmotin-Z	βOH	CHO	OMe	Me	H,H	Pg, Pp
EO-3	H	Me	H	CHO	H,H	Eo
EO-4	H	Me	H	O—C=O		Eo
EO-5	H	Me	H	CO_2H	H,H	Eo

En1 (*E. nitens* from Diamantina, MG); Eo (*E. orbiculatum* from Manaus, AM);
Pg (*P. guianensis* from Manaus, AM); Pp (*P. paraensis* from Belém, PA)

Table II. Tetralone Sesquiterpenoids from *Emmotum*

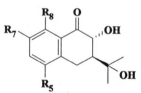

Compound	R_5	R_7	R_8	Species Code
Emmotin-A	Me	H	CH_2OMe	En1, Eg
Emmotin-B	CH_2OH	H	CH_2OMe	En1
Emmotin-F	Me	H	Me	En2
7-Methoxyemmotin-F	Me	OMe	Me	Eg
Emmotin-U	CHO	H	CH_2OMe	En1
Emmotin-V	Me	H	CO_2Me	En1
Emmotin-X	Me	H	CH_2OH	En1

En1 (*E. nitens* from Diamantina, MG); En2 (*E. nitens* from Linhares, ES);
Eg (*E. glabrum* from Manaus, AM).

This divergence is due to difference in importance assigned by each author to morphological, anatomical and embryological characters. Chemical characters can be used as auxiliary criteria of systematic significance (*6-7*). Until Hegnauer's work on chemotaxonomy, chemical data on Icacinaceous species were dispersed in the literature (*8*). The present phytochemical examination of Brazilian Icacinaceae may provide evidence for the placement of this family in the angiosperm system. The interest in this family has increased since the isolation of the antitumoral alkaloid camptothecin from *Nothapodytes foetida* that occurs in India (*9-11*) and *Merriliodendron megacarpum*, a monotypic genus from Malaysia, Melanesia and Micronesia (*12*). Derivatives of camptothecin are either in clinical or preclinical trial stages and are expected to be available for therapeutic use in the near future (*13*).

Of the genera quoted for Brazil three have been investigated: *Emmotum*, *Poraqueiba* and *Humirianthera*. *Emmotum* Desv. and *Poraqueiba* Aubl. are botanically recognized as closely related (*14*).

The thirteen *Emmotum* species occur exclusively in South America. Seven are found in Brazilian Amazonia: *E. nudum, E. acuminatum, E. fagifolium. E. nitens* (extending to central and southern states), *E. glabrum, E. orbiculatum* and *E. holosericeum*. The remaining species occur in Venezuela (*E. conjunctum, E. fulvum, E. argenteum, E.* and *ptarianum*) in Peru (*E. floribundum*). Another species (*E. affine*) is registered only for the states of Bahia and Pernambuco, in northeastern Brazil (*15*).

Only three species of *Poraqueiba* (*P. guianensis, P. sericea* and *P. paraensis*) are recognized. They occur exclusively in Amazonia.. The last two species are cultivated and provide edible fruits known by the vernacular names of "marí" or "umarí" (*16*).

In the course of a systematic investigation of Brazilian Icacinaceae we have examined the chemical composition of trunkwood from *Emmotum nitens, E. fagifolium, E. glabrum, E. orbiculatum, Poraqueiba guianensis* and *P. paraensis*. All of these species, except *E. nitens*, were colleted in the Brazilian states of Amazonas and Pará. Besides our own work on *Emmotum* and *Poraqueiba* species the only additional information on the chemistry of Brazilian Icacinaceae are the reports on *Humirianthera rupestris* and *H. ampla* from Amazonas state (*17-18*).

Rearranged eudesmane sesquiterpenoids with a 1,4-dimethyl-7-isopropyl-decalin skeleton, which we named emmotins, were consistently found in all the *Emmotum* and *Poraqueiba* species that we have analysed. Other isoprenoids, mainly monoterpenoids and triterpenoids, were also found.

The constitution and configuration of the new sesquiterpenoid emmotins have been established by spectroscopic and chemical methodologies. Extensive work on chemical transformations was carried out raising to approximately fifty the total number of compounds obtained as a result of our effort on the chemistry of the Brazilian Icacinaceae, which will be described without going into details on strutuctural determinations but emphasizing the chemical transformations. Tables I-IV and Figure 2 show the structures of the natural sesquiterpenes and Figures 1, 3-6 ilustrate the chemical transformations used to interrelate those compounds.

Table III. Naphthalene Sesquiterpenoids from *Emmotum*

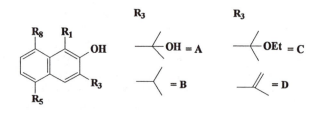

Compound	R₃	R₅	R₈	R₁	Species Code
Emmotin-C	B	Me	CHO	OH	En1
Emmotin-D	A	Me	O—C=O		En1
Emmotin-G	A	Me	Me	H	En2
Emmotin-I	A	Me	CHO	OH	En1
Emmotin-J	A	Me	CO₂H	OH	En1
Emmotin-L	D	Me	O—C=O		En1
Emmotin-M	B	Me	O—C=O		En1
Emmotin-O	A	Me	CHO	H	En1
Emmotin-Q	B	Me	CO₂H	OH	En1
Emmotin-R	A	CHO	CH₂OMe	H	En1
Emmotin-S	A	Me	CH₂OMe	H	En1
Emmotin-D1	C	Me	O—C=O		En1
Emmotin-S1	C	Me	CH₂OMe	H	En1

En1 (*E. nitens* from Diamantina, MG);
En2 (*E. nitens* from Linhares, ES).

Table IV. Naphthoquinone Sesquiterpenoids from *Emmotum*

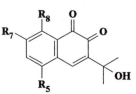

Compound	R₅	R₇	R₈	Species Code
Emmotin-H	Me	H	Me	En2, Ef, Eg
7-Methoxyemmotin-H	Me	OMe	Me	Eg
Emmotin-T	Me	H	CH₂OMe	En1

En1 (*E. nitens* from Diamantina, MG); En2 (*E. nitens* from Linhares, ES);
Ef (*E. fagifolium* from Pará); Eo (*E. orbiculatum* from Manaus, AM);
Eg (*E. glabrum* from Manaus, AM).

Sesquiterpenes from *Emmotum* Species

Emmotum nitens (Benth.) Miers   is an arboreous Icacinaceae with a wide geographical distribution extending from the Amazon to central and southern states. A trunkwood sample from a specimen growing in the vicinity of Diamantina, Minas Gerais state, afforded two tetralone sesquiterpenoids, designated emmotin-A and -B (Table II), and two naphthalene sesquiterpenoids, emmotin-C and -D (Table III). Allocation of the substituents to C-5 and C-8, based on ^{13}C NMR evidence, in the case of emmotin-A and -B, was considered a reasonable postulate also for emmotin-C and -D, on the light of a probable biogenetic relationship (*19*).

The chemical investigation of another trunkwood sample, this time collected from a specimen growing in the Atlantic Forest, at the Linhares Reserve, Rio Doce, Espírito Santo state, led to three new emmotins: a tetralone, emmotin-F (Table II), a naphthalene, emmotin-G (Table III) and a naphthoquinone, emmotin-H (Table IV). Hydrogenolysis of both emmotin-A and -F afforded 2-*epi*-rishitinol whose ^{1}H NMR data for H-2 and H-3 are consistent with a *trans*-diaxial relationship by comparison with data for (+)-rishitinol. The identity of the carbon skeleton of these emmotins, including emmotin-A, was proved by their conversion into the same quinoxaline derivative. A decisive proof of this skeleton relied on the transformation of emmotin-F into (+)-occidol (Figure 1). Additionally this correlation was useful to define the absolute configuration (2R,3S) of emmotin-F. The absolute stereochemistry at C-3 of (+)-occidol must prevail whereas configuration at C-2 was assigned on the basis of the *trans*-diaxial relationship of H-2 and H-3, as evidenced by ^{1}H NMR data. The ORD curves for emmotin-A, -B and -F are superimposable and, thus, they all possess identical absolute configurations (*20*).

Upon catalytic hydrogenation (H_2/Pd-C/AcOEt), emmotin-A was converted quantitatively into emmotin-F (Figure 1), confirming the identity of their skeleton. Finally, X-ray diffractometry of these tetralones confirmed the allocation of substituents at C-5 and C-8 for emmotin-A and -B as well as the relative *trans* configuration at C-2 and C-3 for all the three compounds (*22, 23*).

A further investigation of the trunkwood of the first specimen of *E. nitens* (En1) led to the isolation of three authentic eudesmane derivatives, the new emmotinol-A and -B besides the previously known pterocarptriol (*24*) (Figure 2) and rearranged eudesmane sesquiterpenes, comprising (+)-rishitinol (Table I), a stress compound from Solanaceae (*25*), and 14 new emmotins represented by 3 tetralones (Table II), 10 naphthalenes (Table III) and 1 naphthoquinone (Table IV) (*21*).

The tetralones, emmotin-U, -V and -X, showed spectral data (IR, UV, ^{1}H NMR) very close to those described for emmotin-A, -B (*19*) and -F (*20*). Allocation of C-5 and C-8 substituents relied on the ^{1}H NMR chemical shifts for the oxymethylene groups (emmotin-U and -X), the methylene benzylic protons at

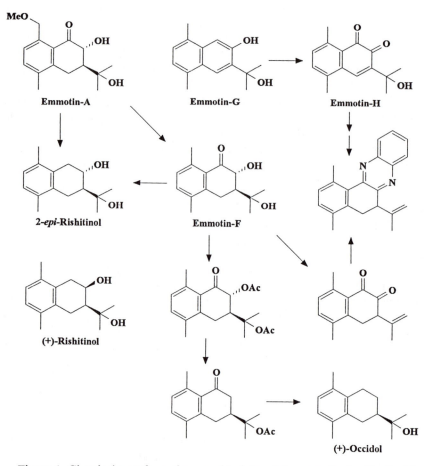

Figure 1. Chemical transformations used to interrelate emmotin-A, -F, -G, -H and (+) - occidol (Reproduced with permission from ref. 46).

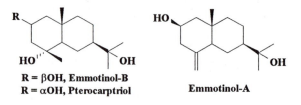

R = βOH, Emmotinol-B
R = αOH, Pterocarptriol

Emmotinol-A

Figure 2. Eudesmanes from *Emmotum nitens.*

C-4 and the aromatic methyl groups, in comparison with those for emmotin-A, -B and -F (*21*).

The new naphthalene emmotins (Table III) can easily be distinguished by ^{1}H NMR, in two groups: those oxygenated at C-1, showing a singlet for a *peri*-naphthalene proton (H-4), and those non-oxygenated at C-1, with two singlets for the distinctly protected *peri*-protons (H-1 and H-4). The isopropyl unit at C-3 is represented either by the alkane, alcohol or alkene function as usually observed in eudesmane sesquiterpenoids. The O-ethyl-derivatives of emmotin-D and -S could even be considered artifacts, but ethoxylated naturally occurring terpenoids are known (*26, 27*). The isolation of the non-lactonized forms of emmotin-D and -M, which have been named emmotin-J and -Q, respectively, is quite surprising. They indeed formed the corresponding methyl esters on reaction with diazomethane in methanol. Chemical transformations (*21*) (Figures 3, 4) allowed the interrellation of several of the naphthalene emmotins and led also to new representatives besides O-methylemmotin-G, previously obtained by total synthesis (*28*).

***Emmotum glabrum* Benth. ex Miers.** The trunkwood of this tree was collected at the Reserva Florestal Ducke, at km 27 of the Manaus-Itacoatiara road. It contains three tetralones, the novel 7-methoxyemmotin-F and the previously known emmotin-A and -F (Table II) along with a new naphthoquinone, 7-methoxyemmotin-H, and emmotin-H (*29*) (Table IV).

The new tetralone was converted to (+)-7-methoxyoccidol by hydrogenation followed by hydrogenolysis at C-1 and deoxygenation at C-2 by radicalar cleavage of the corresponding S-methyl-dithiocarbonate with Bu$_3$SnH (Figure 5). The ORD curve shows a positive Cotton effect, as expected. Allocation of the methoxy group to C-7 was inferred from comparison of the ^{1}H and ^{13}C NMR data for the natural emmotin and the hydrogenolysis product, 7-methoxy-2-*epi*-rishitinol. The ORD curve for this new emmotin and for emmotin-F are very similar, showing a positive Cotton effect, confirming the absolute configuration at C-2 and C-3 (*29*).

The new naphthoquinone was shown to be 7-methoxyemmotin-H by dehydrogenation of the preceding tetralone with Pd/C in toluene. This oxidation occurs also spontaneously when the tetralone is applied to a silica gel TLC plate and left aside for 3-4 days before eluting. The conversion amounts approximately to 60%. The quinoxaline of this naphthoquinone was prepared and characterized (*29*) (Figure 5).

***Emmotum orbiculatum* (Benth.) Miers.** The occurence of this small tree is restricted to Amazonas state and the specimen was collected at km 130 of the Manaus - Caracarí road, on sandy soil. The trunkwood afforded three new sesquiterpenoids along with the previously known emmotins-D and -H. The new compounds were characterized as tetralins, two of which, EO-3 and EO-1, were related to the emmotin group, while to the third one (EO-4) was assigned another rearranged eudesmane skeleton of the 1,6-dimethyl-3-isopropyl-decalin type. The

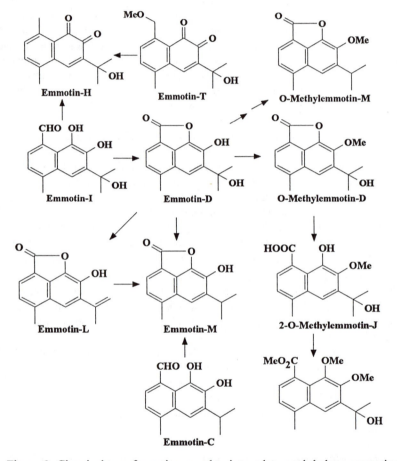

Figure 3. Chemical transformations used to interrelate naphthalene emmotins.

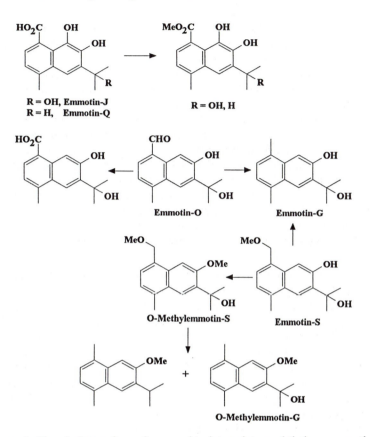

Figure 4. Chemical transformations used to interrelate naphthalene emmotins.

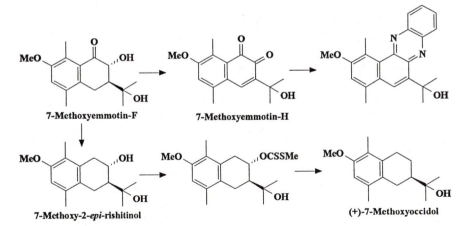

Figure 5. Chemical transformations used to interrelate emmotins from *Emmotun glabrum*.

structural proposals relied on usual spectrometric analysis and comparison with available data for the closely related known emmotins. Chemical transformation (Figure 6) of the aldehyde emmotin into the carboxylic acid confirmed the corresponding location of these groups in both compounds whereas hydrogenolysis of the first one to give (+)-occidol has established the skeleton and configuration of these emmotins. Indeed they gave ORD curves with positive Cotton effects, as expected. Chemical transformation (Figure 6) of emmotin-A into the two new tetralin emmotins confirmed the relative positions of the substituents at the benzenoid ring. The reactions used with this aim could not include hydrogenation since the methoxymethyl group at C-8 had to be preserved for later oxidation to the aldehyde and finally, to the carboxylic acid functions. Deoxygenation was carried out, first at C-2, by reductive cleavage of the acetoxy group of the diacetate, by the same methodology already exploited for emmotin-F (20). The deoxytetralone was converted to the tosylhydrazone that was reduced with NaBH$_3$CN to give the desired tetralin whose oxidative cleavage by DDQ led to the aldehyde which, by further oxidation with Ag$_2$O, gave the corresponding carboxylic acid. Both aldehyde and acid were shown to be identical to the natural emmotins. Exploiting the suitability of DDQ for oxidizing benzylic ether bonds and causing aromatization (dehydrogenation), emmotin-A was converted to emmotin-D and the 2-deoxytetralone derived from emmotin-A gave a new naphthalene lactone, 2-deoxyemmotin-D (29) (Figure 6). The third new tetralin from *E. glabrum* (EO-4, Table I) was, initially, postulated as having the emmotin skeleton. However this structure was revised in favour of a new skeleton on the basis of the ^{1}H NMR chemical shifts for one aromatic proton and one methine hydrogen. They are, respectively, more protected and less protected than expected for the first structural proposal, in disagreement with the observed chemical shifts for related emmotins, but in accordance with the revised structure (29).

***Emmotum fagifolium* Desv. ex Hamilton** is a small tree known by the vernacular name of "marachimbé" (15). Its trunkwood collected in Pará state, afforded emmotin-H (Table IV) which was characterized as its quinoxaline derivative. O-Methylemmotin-G was detected, by ^{1}H NMR, in a mixture with emmotin-H (30).

Sesquiterpenes from *Poraqueiba* Species

***Poraqueiba guianensis* Aublet and *P. paraensis* Ducke.** The trees of this genus are among the tallest of the American Icacinaceae, reaching 90 feet in height (1). Trunkwood samples of these species were collected near Manaus, Amazon state, in the case of *P. guianensis*, and near Belém, as Pará state, in the case of *P. paraensis*. *P. guianensis* afforded two new tetralins, emmotin-Z and its 6-O-methyl ether (Table I). A combination of one and two-dimensional ^{1}H and ^{13}C NMR spectra led to the proposed constitution and relative configuration for emmotin-Z (31) which was confirmed by X-ray diffractometry (32). The absolute configuration at C-2 and C-3 is considered as 2R,3S based the *cis*-relationship of

H-2 and H-3, as evidenced by ^{1}H NMR, and by attributing to C-3 the configuration of all other tetralin emmotins.

Emmotin-Z was the only sesquiterpenoid isolated from *P. paraensis* (*33*).

Mono-, Di- and Triterpenoids from Icacinaceae

The major constituent of *Poraqueiba guianensis* is a monoterpenoid seco-iridoid, secologanoside (0.1% in the trunkwood). It is accompanied by minor ammounts of its methyl ester and by a new highly oxygenated triterpene named icacinic acid (Goulart, M.O.F.; Sant'Ana, A.E.G.; Alves, R.J.; Souza Filho, J.D.; Maia, J.G.S.; Oliveira, G.G.; Oliveira, A.B. *Phytochemistry*, in press.). Iridoids have not been isolated from *P. paraensis* and from the analysed species of *Emmotum*. However they do occur as carbocylic or *seco* derivatives in *Apodytes dimidiata* (*34*), *Cantleya corniculata* (*35, 36*) and *Lasianthera austrocaledonica* (*37*), Icacinaceous species from the Far East (Figure 7). In *Merriliodendron megacarpum*, also from Eastern countries (*12*), *Nothapodytes foetida*, from India (*9-11*), and *Cassinopsis ilicifolia*, from South Africa (*38*), secologanin is the precursor of the monoterpenoid moiety of complex tryptamine derived alkaloids (Figure 8).

Diterpenoids have been reported in *Icacina* species from tropical Africa and *Humirianthera* from tropical South America, i.e. Amazonia. Both genera produce pimarane diterpenoids with similar AB-ring systems bearing a lactone moiety. This group is represented by alkaloids from *I. elaessensis* and *I. guesfeldtii* as well as the non alkaloidal icacinol, from the later species (*39-42*) the humiriantholides A-F from *H. rupestris* (*17*), and the humiriantholide-A, -C and -D from *H. ampla* (*18*) (Figure 9).

Biosynthesis of Sesquiterpenoid Emmotins

Formation of the hydrocarbon precursors of the sesquiterpenes from *Emmotum* and *Poraqueiba* may be rationalized by assuming a cyclization process that is initiated by ionization of 2E,6E-farnesyl pyrophosphate to yield the germacrane cation (*43*). A double chair conformation of this cation, under a eletrophilic attack, explains the formation of (+)-eudesmanes with a *trans*-decalin skeleton, as those of emmotinol-A, -B, pterocarptriol and most of the naturally occurring (+)-eudesmanes showing a typical 7S-configuration as observed in the corresponding C-3 position of the tetralin and tetralone emmotins. It is proposed that the biosynthesis of the emmotins, by analogy with that of (+)-occidol (*44*), should involve a one carbon shift by a dienol-benzene rearrangement of an eudesmane precursor (*20*) (Figure 10).

Concerning the oxygenation of the alicyclic ring of (+)-occidol, it is reasonable to suppose that it occurs, first at C-2 to give compounds of the 2β configuration, in a *cis*-relationship with the substituent at C-3, as in (+)-rishitinol, emmotin-Z and 6-O-methylemmotin-Z. Further oxidation of rishitinol type

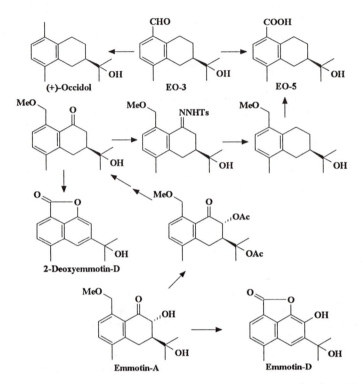

Figure 6. Chemical transformations used to interrelate tetralins from *E. orbiculatum* and Emmotin-A. Conversion of tetralones into naphthalene lactones.

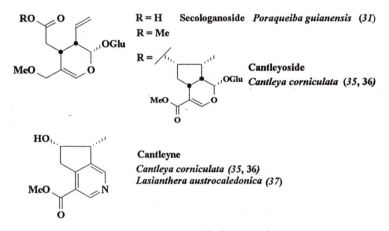

Figure 7. Monoterpenoids from Icacinaceae.

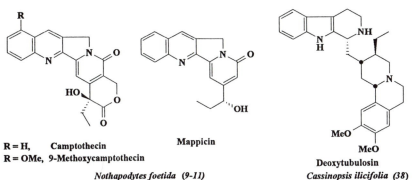

R = H,   Camptothecin
R = OMe, 9-Methoxycamptothecin

Mappicin

Nothapodytes foetida (9-11)
Merriliodendron megacarpum (12)

Deoxytubulosin
Cassinopsis ilicifolia (38)

Figure 8. Monoterpenoids in Icacinaceae: secologanin-tryptamine derived alkaloids.

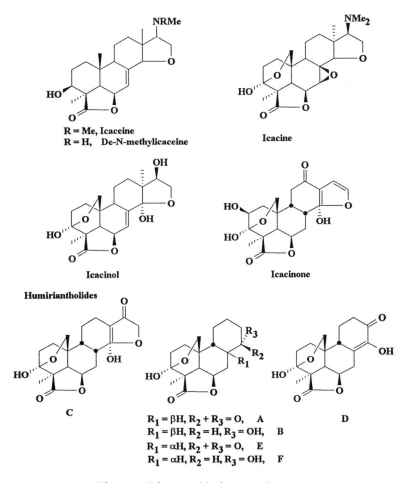

R = Me, Icaceine
R = H,   De-N-methylicaceine

Icacine

Icacinol

Icacinone

Humiriantholides

C

$R_1 = \beta H, R_2 + R_3 = O,$ A
$R_1 = \beta H, R_2 = H, R_3 = OH,$ B
$R_1 = \alpha H, R_2 + R_3 = O,$ E
$R_1 = \alpha H, R_2 = H, R_3 = OH,$ F

D

Figure 9. Diterpenoids from Icacinaceae.

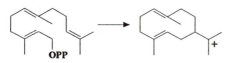

2E,6E-Farnesyl-pirophosphate   **Germacrane cation**

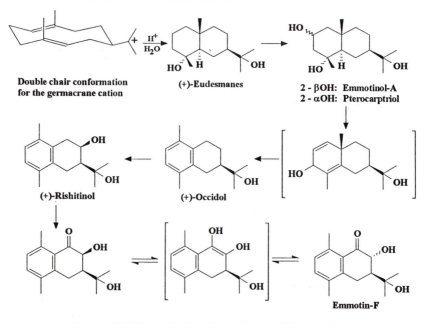

Figure 10. Biosynthesis of sesquiterpenoid emmotins.

compounds should give tetralones keeping, initially, the 2β-configuration, and then, changing, via an enediol, to the 2α-configuration resulting in the more stable *trans*-relationship for C-2 and C-3 observed for all tetralone emmotins, as represented for emmotin-F (Figure 10). Aromatization of rishitinol and tetralone type emmotins would lead to the two groups of naphthalene representatives, non-oxygenated and oxygenated at C-1. Oxidation of one or both aromatic methyl groups could take place at any stage of the biosynthetic route.

Conclusions

The emmotins, a class of sesquiterpenes based on a rearranged eudesmane skeleton of the 1,4-dimethyl-7-isopropyl-decalin type are restricted to *Emmotum* and *Poraqueiba* genera, from the Icacinaceae. (+)-Occidol (Figure 1) and (+)-rishitinol (Figure 1) are representatives of this skeleton that occur in families other than Icacinaceae. (+)-Occidol is supposed to be an intermediate in the biosynthesis of the emmotins (Figure 10). Thus, the close recognized botanical affinity of *Emmotum* and *Poraqueiba* has been shown to extend to chemical characters.

As pointed out by Kaplan *et al.* (*45*), although the chemistry of Icacinaceae does seem rather heterogeneous, a strong chemical relationship linking the genera of this family exists and can be summarized as follows:

1. A pronounced trend towards diversification of mevalonate derived metabolites (mono, sesqui, di and triterpenoids).
2. A uniformly high level of oxidation of the terpenoids, jointly with the presence of *seco*-iridoid derivatives, are recognized as advanced characters supporting the placement of this family in the superorder Cornifloreae, order Cornales, as proposed by Dahlgren (*6*).

Acknowledgments

The work on *Emmotum* and *Poraqueiba* species presented was carried out in collaboration with the late Prof. G.G. Oliveira. The results reported here were taken in part from theses submitted to the Departamento de Química, Instituto de Ciências Exatas, Universidade Federal de Minas Gerais, Belo Horizonte, M.G., Brazil, by M.L.M. Fernandes (M. Sc.), C.T.M. Liberalli (D. Sc.), N.H. Silva (D. Sc.), A.E.G. Sant'Ana (D. Sc.), J.G.S. Maia (D. Sc.), M.O.F. Goulart (D. Sc.), A. Trainotti (D. Sc.).All have been supported by fellowships from CAPES and CNPq.

Financial support by FINEP and CNPq is gratefully acknowledged. I want also to express my warmest thanks to Mrs. C.R. Castro, Ms. A. C. S. Assunção and Mr. J. E. Guimarães for their excellent technical assistance and to Ms. E. G. Machado for her kind help in the preparation of this manuscript.

Literature Cited

1. Howard, R.A. *Contribution from the Gray Herbarium of Harvard University.* CXLII. I. *Studies of the Icacinaceae* IV: *Consideration of the New World Genera*, 1942, 3-60.
2. Engler, A. In *Die Natürlichen Pflanzenfamilien*; Engler, A.; Prantl, K., Eds.; Engelmann, Leipzig, III (5) 233.
3. Thorne, R.I. *Nord. J. Botany.* **1983**, *3*, 85.
4. Cronquist, A. *An Integrated System of Classification of Higher Plants;* Columbia University Press, New York, 1981.
5. Takhtajan, A. *Bot. Rev.* **1980**, *46*, 226.
6. Dahlgren, R. J. *Lin. Soc. (Bot.)* **1980**, *80*, 91.
7. Gottlieb, O.R. *Micromolecular Evolution, Systematics and Ecology. An Essay into a Novel Botanical Discipline*; Springer: Berlin, 1982, 170p.
8. Hegnauer, R. *Chemotaxonomie der Pflanzen*, Birkhäuser Verlag: Berlin, 1966 Vol. IV, 275-277.
9. Agarwal, J.S.; Rastogi, R.P. *Indian J. Chem.* **1975**, *13*, 758.
10. Govindachari, T.R.; Ravindranath, K.R.; Viswanathan, N.J. *Chem. Soc. Perkin Trans I.* **1974**, 1215.
11. Govindachari, T.R.; Viswanathan, N. *Phytochemistry* **1972**, *11*, 3529.
12. Arisawa, M.; Gunasekera, S.P.; Cordell, G.A.; Farnsworth, N.R. *Planta Med.* **1981**, *43*, 404.
13. Chang, C.J.; Ashendel, C.L.; Chan, T.; Geahlen, R.L.; McLaughlin, J.L. In *Advances in New Drug Development*; Kim, B-K.; Lee, E.B.; Kim, C-K.; Han Y.N., Eds.; The Pharmaceutical Society of Korea: Seoul, 1991, 448-457.
14. Carvalho, M.J.C. *O Pólen em Plantas da Amazônia - Gêneros **Poraqueiba** Aubl. e **Emmotum** Desv. (Icacinaceae), Boletim do Museu Paraense Emílio Goeldi*, série Botânica, 42, Belém, 1971, 4pp.
15. Carvalho, M.J.C.; van den Berg, M.E.; Cavalcante, P.B. *O Gênero **Emmotum** Desv. (Icacinaceae) na Amazônia Brasileira*, in O Museu Goeldi no Ano do Sesquicentário, Publicações Avulsas, 20, Belém, 1973, 203-220.
16. Carvalho, M.J.C.; Cavalcante, P.B. *O Gênero **Poraqueiba** Aubl. (Icacinaceae) na Amazônia*, Boletim do Museu Paraense Emílio Goeldi, série Botânica, 39, Belém, 1971.
17. Zoghbi, M. das G.B.; Roque, N.F.; Gottlieb, H.E. *Phytochemistry* **1981**, *20*, 1669.
18. Zoghbi, M. das G.B.; Roque, N.F.; Cabral, J.A. da S. *Acta Amaz.* **1983**, *13*, 215.
19. Oliveira, A.B. de; Fernandes, M. de L.M.; Gottlieb, O.R.; Hagaman, E.W.; Wenkert, E. *Phytochemistry* **1974**, *13*, 1199.
20. Oliveira, A.B. de; Oliveira, G.G.; de Liberalli, C.T.M.; Gottlieb, O.R.; Magalhães, M.T. *Phytochemistry* **1976**, *15*, 1267.

21. Sant'Ana, A.E.G., Ph.D. Thesis, Universidade Federal de Minas Gerais, Belo Horizonte, 1984.
22. Pulcinelli, S.H., M. Sc. Dissertation, Instituto de Física e Química de São Carlos, USP, São Carlos, 1982.
23. Gambardella, M.T.P.; Schpector, J.Z.; Mascarenhas, Y.P. *Ciênc. e Cult.* **1981**, *33* (Supl.), 459.
24. Kumar, N.; Ravindranath, B.; Seshadri, T.R. *Phytochemistry* **1974**, *13*, 633.
25. Katsui, N.; Matsunaga, A.; Imaizumi, K.; Masamune, T. Tomiyama, K. *Tetrahedron Lett.* **1971**, 83.
26. Herz, W.; Aota, K.; Holub, M.; Samek, Z. *J. Org. Chem.* **1970**, *35*, 2111.
27. Tomio, E.; Shimchi, N.; Yanosuke, I. *Yukagaku*, **1970**, *19*, 298. *Apud* Chem Abstr. 73:45627.
28. Reddy, P.A.; Rao, G.S.K. *Ind. J. Chem.*, **1980**, *19*B, 578.
29. Maia, J.G.S., Ph.D. Thesis, Universidade Federal de Minas Gerais, Belo Horizonte, 1983.
30. Oliveira, A.B. de; Magalhães, M.T.; Silva, N.H. da; Gottlieb, O.R. *Ciênc. e Cult.* **1979**, *28* (Supl.), 187.
31. Goulart, M.O.F. Ph.D. Thesis, Universidade Federal de Minas Gerais, Belo Horizonte, 1983.
32. Fernandes, M. de L.M. Ph.D. Thesis, Universidade Federal de Minas Gerais, Belo Horizonte, 1986.
33. Trainotti, A. Ph.D. Thesis, Universidade Federal de Minas Gerais, Belo Horizonte, 1979.
34. Kooiman, P. *Österr. Bot. Z.* **1971**, *119*, 395.
35. Sèvenet, T.; Thal, C.; Potier, P. *Tetrahedron* **1971**, *27*, 663.
36. Sèvenet, T.; Das, B.C.; Parello, J.; Potier, P. *Bull. Soc. Chim. Fr.* **1970**, 3120.
37. Sèvenet, T.; Husson, A.; Husson, H.P. *Phytochemistry* **1976**, *15*, 576.
38. Monteiro, H.; Budzikiewicz, H.; Djerassi, C.; Arndt, R.R.; Baarschers W.H. *J. Chem. Soc. Chem. Commun.* **1965**, 317.
39. On'koko, P.; Hans, M.; Colau, B.; Hootele, C.; Declercq, J.P.; Germain, G.; Van Meerssche, M. *Bull. Soc. Chim. Belg.* **1977**, *86*, 655.
40. On'koko, P.; Vanhaelen, M. *Phytochemistry* **1980**, *19*, 303.
41. On'koko, P.; Vanhaelen, M.; Vanhaelen-Fastré, R.; Declercq, J.P.; Van Meerssche, M. *Tetrahedron* **1985**, *41*, 745.
42. On'koko, P.; Vanhaelen, M.; Vanhaelen-Fastré, R.; Declercq, J.P.; Meerssche, M. *Phytochemistry* **1985**, *24*, 2452.
43. Rucker, G. *Angew. Chem. Int. Ed.* **1973**, *12*, 793.
44. Nakazaki, M. *Chem. and Ind.* **1962**, 413.
45. Kaplan, M.A.C.; Ribeiro, J.; Gottlieb, O.R. *Phytochemistry* **1991**, *30*, 2671.
46. Gottlieb, O.R.; Mors, W.B. *J. Agric. Food Chem.* **1980**, *28*, 215.

RECEIVED October 10, 1994

Chapter 11

Chemical Studies of Myristicaceae Species of the Colombian Amazon

Juan C. Martinez V. and Luis E. Cuca S.

Departmento de Quimica, Universidad Nacional de Colombia, A.A. 14490, Santafé de Bogotá, Colombia

Amazonian societies of Brazil and Colombia make ample use of species belonging to the Myristicaceae family as medicines, hallucinogens and arrow poisons. Analysis of seven such species afford a large number of novel compounds, chiefly neolignans (coupling products of propenylphenols and allylphenols) and iryantherins (coupling products of dehydrochalcones and neolignans).

Since early times, plants have been used by man to cure his diseases. Our natives and Indians knew and manipulated, with surprising effectiveness, the healing powers of plants. This healing power, without doubt, is due to the substances that the plant produces during its development, as a result of its metabolism.

Colombia is a fortunate country in terms of its rich variety of plants, most of which are found in the Amazon Region; but, unfortunately, some species have disappeared and others will follow the same path, in the name of progress, due to the indiscriminate felling of our primary forests, wich will keep us from knowing the metabolites of these species. Besides, the effects of erosion and soil degradation will convert our jungles into deserts.

Our phytochemical research group has centered its interests on studies of the Myristicaceae. Papers that raised most interest in the study of plants of this family were published by Schultes (*1-3*) and Gottlieb (*4*). Table I, adapted from the literature (*4*), gives an idea of the use of Myristicaceae by Amazon Indians. The contents of the Table provide a serious motivation for the necessity of submitting species of this family to phytochemical and pharmacological studies.

The Myristicaceae family belongs to the order Laurales and includes 19 genera (Table II), six of which are native to the American Continent. According to the vouchers that are kept in the Colombian National Herbarium, 50 species of these seven genera have been collected in our country, as shown in Table III.

0097–6156/95/0588–0116$12.00/0

Table I. Uses of Myristicaceae by Indians of the Amazon

Genera	Plant part	Preparations	Uses or treatments
Compsoneura	leaves	tea	mental disorder, trembling
	bark	decoction	infected wounds, ulcers
Dialyanthera	seeds	butter	skin infections of domestic animals, erysipelas, hemorrhoids
Iryanthera	leaves	plaster	infected wounds, infections by fungi, itching of skin
	bark	decoction	infections by fungi, erysipelas
	latex		stomach poisoning
	fruit		fish bait
Virola	leaves	tea	tea substitute
	bark	decoction	vitiligo, tooth ache
		decoction with ashes	hallucinogenic powder and tablets, arrow poison
	seeds	butter	rheumatism, intestinal worms, asthma, bad breath
	kino		colics, bleeding, ulcerated wounds
	sap		hemorrhoids

Adapted from Gottlieb, ref. (*4*).

Table II. Genera of the Myristicaceae family

Brochoneura	Haematodendron	Osteophloeum*
Cephalosphaera	Horsfieldia	Otoba*
Coelocaryon	Iryanthera*	Pycnanthus
Compsoneura*	Knema	Scyphocephalium
Dialyanthera*	Maloutchia	Staudtia
Endocomia	Myristica	Virola*
Gymnacranthera		

* Genera native of the American continent.

Table III. Species of Myristicaceae found in Colombia

Genera	Number of species	Genera	Number of species
Compsoneura	5	Osteophloeum	2
Dialyanthera	5	Otoba	3
Iryanthera	11	Virola	24

Our investigations on species of this family have, as a first, short term objective, the isolation, purification and identification of secondary metabolites of different parts of the plant. We studied ten species (*Virola calophylla, V. calophylloidea, V. carinata, V. elongata, V. sebifera, Iryanthera laevis, I. pavonis, I. tricornis, Compsoneura atopa* and *Osteophloeum sulcatum*). Some of the results of these studies are given below.

Virola calophylla

From bark (*5*) and leaves (*6*) lignans of several types were isolated and identified. The former was shown to contain four neolignans of the dibenzylbutane type **1a**, **1b**, **2a** and **2b**, one of the benzoylbenzylbutane type **3a**, two aryltetralins **4a**, **4b**, and three of the arylnaphthalones **5**, **6a** and **6b**. In the latter three neolignans of the dibenzyl type **1a**, **1c** and **2a**, and one neolignan of the tetrahydrofurane type **7** were identified.

Virola calophylloidea

Benzene extracts of leaves, bark and wood were examined. Metabolites identified were of the neolignan, flavonoid and steroid type. The leaves possess a steroidal fraction. GC, HPLC and GC-MS studies indicated that it was composed of a mixture of colesterol, sitosterol, stigmasterol and three other unidentified sterols (*7*). Four neolignans **8**, **9**, **10** and **2c** were also isolated in pure form (*8,9*). It was shown that compound **10** is produced by dehydration of **9** in acid medium. The substances **8**, **9** and **10** were also identified in bark. Wood afforded three flavonoids (*10*), a flavone **11**, an α-hydroxydehydrochalcone **12**, and a 1,3-diarylpropane **13a**, as well as an alkaloid, 3-indolcarboxaldehyde (*11*).

Virola carinata

Wood of this species was studied in Brazil and bark in Japan. From the benzene extract of the leaves (*12*) we isolated five compounds **14**, **15a**, **15b**, **15c** and **16**, all of the lignoid type. Substances **15a**, **15b**, **15c** and **16** are part of an exceptional series of compounds which we called pseudoneolignans, since the sole oxygenation of C-9 and lack of oxygenation at C-9' does not allow us to include them either

1a Ar=Gu, Ar¹=Pi
1b Ar=Ve, Ar¹=Pi
1c Ar=Ar¹=Gu

2a Ar=Gu, Ar¹=Pi
2b Ar=Ve, Ar¹=Pi
2c Ar=Ar¹=Gu

3a Ar=Gu, Ar¹=Pi
3b Ar=Pi, Ar¹=Ve

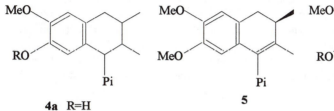

4a R=H
4b R=Me

5

6a R=H
6b R=Me

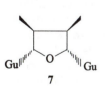

7

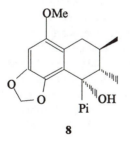

8

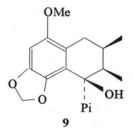

9

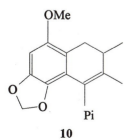

10

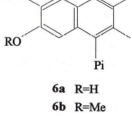

11

among the lignans or the neolignans, the former with oxygenation at both positions and the latter without oxygenation at either position, as defined by Gottlieb (*13*).

Virola elongata

Three neolignans were isolated and identified from the benzene extract of the bark of this species (*14*). Two, **17** and **18**, correspond to the tetralone type and the other, **3b**, to the benzoylbenzylbutane type.

Virola sebifera

V. sebifera is the most widely studied species of the Myristicaceae family. The fruit was studied in Brazil and a large number of metabolites (*15-19*) which correspond to four polyketides, six lignans and 19 neolignans were identified. The bark was studied in Venezuela (*20*) and Japan (*21, 22*), fatty acid derivatives, six alkaloids, one terpene and two lignans having been identified. Leaves, bark and wood were studied by our group. Three furofuran lignans, (+)-sesamine **19a**, (+)-kobusine **19b**, and (+)-eudesmine **19c**, were identified in the leaves (*23*). Two lignans, (+)-sesamine and (+)-kobusine, were identified in the bark (*24*) and three 1,3-diarylpropanes, **13b**, virolanol B **13c**, and virolanol C **13d** were found in the wood (*11*).

Iryanthera laevis

We studied fruit and bark of this species which is known by several common names such as "otoba" or "kimo" in Caquetá, and "mamita" in Meta (region of San Martín). The inhabitants of this last region eat the aryl of the fruit as a sugary preparation. Fruits were separated into mesocarp, aryl and kernel and each of the parts was studied sparately (*9, 25*) revealing dihydrochalcone type compounds in all of them. Compound **20a** was isolated from the three parts of the fruit, compounds **20b**, **21** and **22a** were obtained from the aryl, compounds **21**, **22a** and **22b** from the mesocarp and compound **23** was obtained from the kernel. Four dihydrochalcones, **20a**, **20c**, **20d** and **24** were identified in the bark (*26,27*). In the iryantherins **21**, **22a**, **22b**, **23** and **24** the dihydrochalcone unit bears a lignoid substituent.

Iryanthera tricornis

Dihydrochalcones **20a** and **24** were found in the bark while 1,3-diarylpropanes **13e**, **13f** and **13g** were isolated from the wood (*28*).

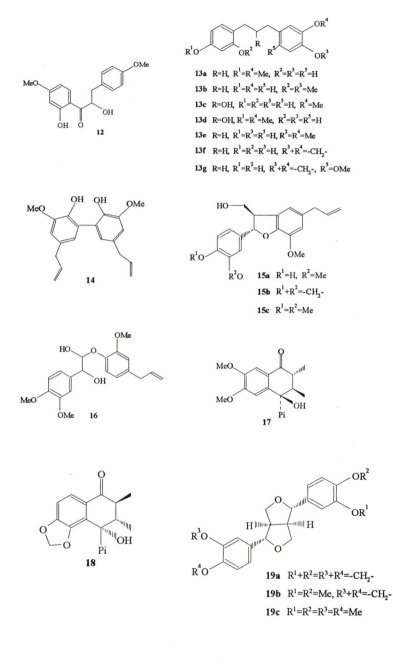

13a R=H, R¹=R⁴=Me, R²=R³=R⁵=H
13b R=H, R¹=R⁴=R⁵=H, R²=R³=Me
13c R=OH, R¹=R²=R³=R⁵=H, R⁴=Me
13d R=OH, R¹=R⁴=Me, R²=R³=R⁵=H
13e R=H, R¹=R³=R⁵=H, R²=R⁴=Me
13f R=H, R¹=R²=R⁵=H, R³+R⁴=-CH₂-
13g R=H, R¹=R²=H, R³+R⁴=-CH₂-, R⁵=OMe

15a R¹=H, R²=Me
15b R¹+R²=-CH₂-
15c R¹=R²=Me

19a R¹+R²=R³+R⁴=-CH₂-
19b R¹=R²=Me, R³+R⁴=-CH₂-
19c R¹=R²=R³=R⁴=Me

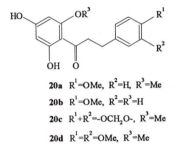

20a R^1=OMe, R^2=H, R^3=Me
20b R^1=OMe, R^2=R^3=H
20c R^1+R^2=-OCH$_2$O-, R^3=Me
20d R^1=R^2=OMe, R^3=Me

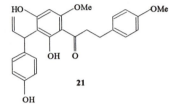

21

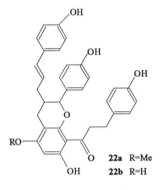

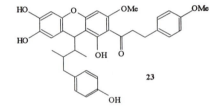

22a R=Me
22b R=H

23

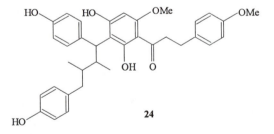

24

Literature cited

1. Schultes, R. E., Botanical Museum Leaflets, Harvard University **1954**, 16, 241-260.
2. Schultes, R. E., Botanical Museum Leaflets, Harvard University **1969**, 22, 229-240.
3. Schultes, R. E. and Holmstedt, B. *Lloydia* **1971**, *34*, 61-78.
4. Gottlieb, O. R. *J. Ethnopharm.* **1979**, *1*, 309-323.
5. Martínez V., J. C.; Yoshida, M.; Gottlieb, O. R. *Phytochemistry* **1990**, *29*, 2655-2657.
6. Alvarez, E.; Cuca, L. E.; Martinez V., J. C. *Rev. Colombiana Quím.* **1985**, *14*, 31.
7. Cuca, L.E.; Martínez V., J. C. *Rev. Colombiana Quím.* **1984**, *13*, 109-118.
8. Martínez V., J. C.; Cuca, L. E.; Yoshida, M.; Gottlieb, O. R. *Phytochemistry* **1985**, *24*, 1867-1868.
9. Garzón, L., Cuca; L. E., Martínez V., J. C.; Yoshida, M.; Gottlieb, O. R. *Phytochemistry* **1987**, *26*, 2835-2837.
10. Martínez V., J.C.; Cuca, L.E. *J. Nat. Prod.* **1987**, *50*, 1045.
11. von Rotz, R.; Cuca, L. E.; Martínez V., J. C. *Rev. Colombiana Quím.* **1990**, *19*, 97-100.
12. Guarin, C.; Cuca, L. E.; Martínez V., J. C. *Spectros. Int. J.* **1988**, *6*, 107-112.
13. Gottlieb, O. R. *Fortsch. Chem. Org. Naturst.* **1978**, *35*, 1-71.
14. Martínez V., J. C.; Cuca, L. E.; Santana, R.; Pombo-Villar, E.; Golding, B. *Phytochemistry* **1985**, *24*, 1612-1614.
15. Lopes, L. M. X.; Yoshida, M.; Gottlieb, O. R. *Phytochemistry* **1982**, *21*, 751-755.
16. Lopes, L. M. X.; Yoshida, M.; Gottlieb, O. R. *Phytochemistry* **1983**, *22*, 1516-1518.
17. Lopes, L. M. X.; Yoshida, M.; Gottlieb, O. R. *Phytochemistry* **1984**, *23*, 2021-2024.
18. Lopes, L. M. X.; Yoshida, M.; Gottlieb, O. R. *Phytochemistry* **1984**, *23*, 2647-2652.
19. Kato, M. J.; Lopes, L. M. X.; Paulino Filho, H. F.; Yoshida, M.; Gottlieb, O. R. *Phytochemistry* **1985**, *24*, 533-536.
20. Corothie, E.;. Nakano, T. *Planta Medica* **1969**, *17*, 184-188.
21. Kawanishi, K.; Uhara, Y.; Hashimoto, Y. *Phytochemistry* **1985**, *24*, 1773-1775.
22. Kawanishi, K.; Hashimoto, Y. *Phytochemistry* **1987**, *26*, 749-752.
23. von Rotz, R.; Cuca, L. E.; Martínez V., J. C. *Rev. Colombiana Quím.* **1987**, *16*, 51-55.
24. Martínez V., J. C.; Cuca, L. E.; Martínez, P. *Rev. Colombiana Quím.* **1985**, *14*, 117-125.
25. Conserva, L. M.; Yoshida, M.; Gottlieb, O. R.; Martínez V., J. C.; Gottlieb, H. E. *Phytochemistry* **1990**, *29*, 3911-3918.
26. Villamil, E.; Cuca, L. E.; Martínez V., J. C. *Spectros. Int. J.* **1988**, *6*, 157-165.
27. Martinez V., J. C.; Cuca, L. E. *Rev. Colombiana Quím.* **1989**, *18*, 37-46.
28. Salazar, L. M.; Cuca, L. E.; Martínez V., J. C. *Rev. Colombiana Quím.* **1988**, *17*, 33-37.

RECEIVED December 16, 1994

NATURAL PRODUCTS:
ECOLOGY AND EVOLUTION

Chapter 12

Plant Chemistry of Amazonia in an Ecological Context

Klaus Kubitzki

Universität Hamburg, Institut für Allgemeine Botanik und Herbarium, Ohnhorststrasse 18, 22605 Hamburg, Germany

Secondary plant chemistry of Amazonia is intimately related to the oligotrophy of the available substrate, which is shown to be the result of the geological history of Amazonia and adjacent Guyana. Patterns of plant defense against herbivores and microorganisms are discussed in the framework of various hypotheses, including the concepts of scleromorphy, of "mobile" vs. "immobile" defense, of nutrient resources, and of defense based on carbon vs. nitrogen. Edaphic processes are mentioned in which the accumulation of acidic mor leads to an increase in soil acidity, followed by selective lixiviation, and an enhancement of polyphenol production in the vegetation.

In view of the vast extension and biotic richness of the "Hiléia Amazônica" the state of its chemical exploration is still unsatisfactory. This is not just due to the enormous size of the task. The difficult access to the region and, last but not least, political sensibilities are additional reasons for this situation.

It is true there exist very valuable contributions made by natural product chemists referring to a selection of Amazonian plant species, mostly focusing on substances with striking pharmaco-dynamic properties, that have been or continue to be employed by indigenous people, such as timbó, curare, and hallucinogenic substances (1,2). There are also numerous studies referring to wood and other plant constituents most valuable in a chemical or sometimes chemosystematic context listed in the "Cadastro Fitoquímico Brasileiro" (3) but we are far from having a coherent picture in this respect, not to speak of the ecologic significance of such compounds. Therefore, in the following, when considering Amazonian plant chemistry on the community level I will have to resort also to contributions to environmental chemistry from other tropical regions of the World.

0097–6156/95/0588–0126$12.00/0
© 1995 American Chemical Society

When trying to deal with ecologic implications of secondary metabolites of Amazonian plants, two aspects appear of foremost importance: firstly, the reaction of the vegetation to the poverty of nutrients that characterizes the major part of Amazonia (which we are beginning to understand); and secondly, the enormous degree of plant diversity which certainly implies an equally large degree of chemical diversity (of which we know next to nothing).

With the exception of the region of the crystalline shields north and south of the Amazon valley, Amazonia is often considered geologically young, and so its flora is supposed to be as well (an aspect not to be discussed here further). This is apparently in contrast to the Guayana Highland which represents one of the oldest regions of South America that certainly has not been inundated since the Middle Cretaceous or, most likely, not even since the Paleozoic. The geologic history of the Guayana highland is closely interrelated with that of its foreland and has strong repercussions on the edaphic conditions beyond the Guayana region. The flora of the Guayana lowland, often considered as representing a separate floristic region, differs mostly at the specific level from that of Amazonia. For these biogeographic and historic reasons it seems necessary to include the Guayana region into our consideration.

Geological History of Amazonia and Guayana

The river Amazon follows an ancient rift valley that existed at least since Jurassic times. The continuation of the Amazon rift valley can be traced in the African Bénoué Graben (*4*). Through major periods of time during the Cretaceous and Tertiary the Amazon valley has been a zone of depression, as is documented by series of deposits composed of clay, sand and gravel, which are up to 600 m thick and are known as Solimões clays, Barreiras layers, and so on. These deposits document the limnic-fluviatile history of Amazonia, which was characterized by freshwater lakes and rivers of vast extension. So long as Africa was connected with South America and the Atlantic Ocean did not exist, the drainage from the Amazon region was directed into the Pacific (*5*) or the Caribbean. Following the opening of the Atlantic Ocean in the Jurassic, a land-locked watershed between the Atlantic and Pacific Ocean was built up, which Grabert (*6*) has identified as being along a line connected by Mount Roraima, the Serra do Divisor and Serra dos Parecis. With the uplift of the Andes, backwaters accumulated in western Amazonia which eventually overflowed and eroded the watershed, and the whole drainage was directed eastwards.

Geo-ecologically, the uplift of the Andes was the event with the greatest impact on Amazonia. Now white water rivers, carrying a load of silt and clay stemming from their source area in the Andes and its foreland crossed Amazonia and began "polluting" it. Nutrient-rich white water influenced also the flood plains which due to the annual deposit of the sediment-load became the most fertile region of Amazonia. In contrast, the clastic sediments that had been deposited in the Amazon valley before the upheaval of the Andes had originated from the erosion of crystalline rocks and sandstones. Only occasionally, limited amounts of calcareous material may have been included in these sediments which was derived from calcareous rocks of Paleozoic age, which fringe the valley of the Lower Amazon. Thus one can conclude that Amazonian rivers previous to the uplift of the Andes had been poor in nutrients, as are present-day "clear" and "black" water rivers (7). Since the sediment load of black and clear water rivers of Amazonia is extremely low, one might conclude that major periods of the limnic history of Amazonia are not at all documented by sediments.

A major portion of the surface of Amazonia is formed by Tertiary sediments, while in its northern and southeastern fringe the products of the decomposition of igneous rocks prevail. However, patches of white sand are found here and there everywhere in Amazonia and in the northwestern sector, the Rio Negro region, areas of white sand are of vast extension. Some of these sand areas have originated from selective lixiviation of soils derived from the weathering of igneous rocks (8). The selective loss of clay minerals from Barreiras layers is also known to have led to podzolic soils (9), but the main source of white sands has been the decomposition of the sandstones of the Roraima formation. These arenitic rocks at present reach a thickness of about 3000 m, but originally must have been much thicker. Based on the presence of low grade metamorphic minerals in the upper part of the Roraima formation it has been argued that originally there had been several thousand meters of rock above the present surface (10). It has also been claimed that originally the Roraima sandstones covered a much larger area than today, though this is not completely known. Whatever their original distribution, horizontal and vertical erosion of the sandstone formation that has led to the present table mountains with deeply incised valleys and steep escarpments must have liberated immense masses of gravel and sand, which were deposited in the foreland of what today is the Guayana highland (Figure 1). Thus the northern savannas of French Guiana, Surinam and Guyana, the Rupununi savannas of Guyana, and the vast white sand areas in the region of the upper Orinoco and Rio Negro largely owe their existence to this process. Also the proportion of sand contained in sediments deposited during the Miocene and Pliocene at a greater distance from the highland, such as in eastern Venezuela and off Trinidad, is certainly due to the weathering of the Roraima sandstone (11).

Thus it can be said the prevailing latosolic forest soils of Amazonia are derived from Tertiary sediments and from weathering of igneous and metamorphic rocks. Interspersed in them are islands of white sand, which prevail in the northwestern sector of the hylaea. The poverty in nutrients even in the more favourable latosols is so pronounced that the well-known short-cut nutrient cycle of humid tropical forests has originated. Jordan (12) has demonstrated that an Amazonian forest on latosol maintained itself on nutrients derived from the atmosphere. Nutrient stress must be even more pronounced on sand-derived "arenosols", which are critical also in terms of their low water-holding capacity. In the major part of Amazonia these precarious edaphic conditions may have prevailed for an extended period at least during the Tertiary, an exception being the richer soils that exist in the southwestern sector of the hylaea. Only with the

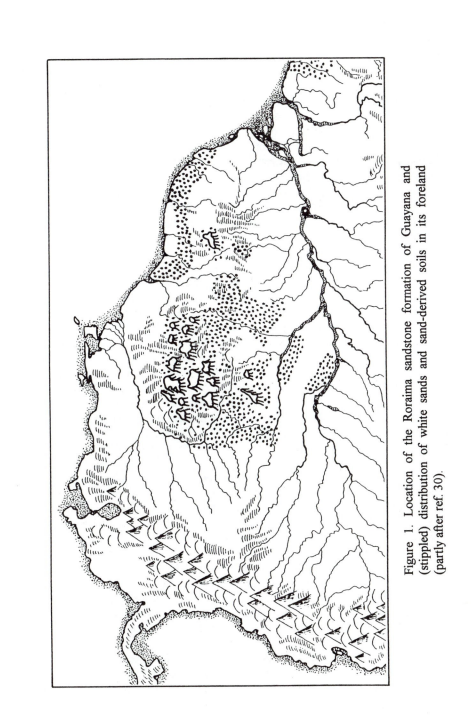

Figure 1. Location of the Roraima sandstone formation of Guayana and (stippled) distribution of white sands and sand-derived soils in its foreland (partly after ref. 30).

upheaval of the Andes did trophic conditions along white water rivers become very favourable.

Nutrient Stress and Plant Chemistry

Now we are prepared to analyse the interrelationships between vegetation and substrate, a topic that has been much discussed during the past decades. Arens (*13,14*) was the first to apply the concept of nutrient stress in an attempt to explain the "sclerophytic" features of Central Brazilian cerrado vegetation. The plants composing this vegetation present features such as leaves rich in sclerenchyma and provided with thick cuticles. These traits would suggest adaptations for diminishing their rate of transpiration. In fact, however, transpiration is not restricted and the scleromorphic features are a reaction to the poverty of nutrients, above all nitrogen and phosphorus. Today we know that this scarcity leads to an elevated rate of cinnamate-derived compounds at the cost of the rate of nucleotides and proteins, and finally results in an increased content in phenolic substances such as flavonoids, condensed tannins, and above all lignin, which directly implies sclerophytism. In numerous experiments plant physiologists were able to modify the phenotype, particularly the rate of sclerophylly, through the variation of nutrient supply. Here, however, we are interested in cases in which sclerophytic traits due to the chronic poverty in nutrients selectively have been incorporated into the genotype. Nutrient stress-induced sclerophylly is noticeable both in specific lineages of plants and at the level of community. Thus the family Ericaceae shows many sclerophytic traits and rich nutrient supply is lethal for most of its members. The sclerophyllous heath forests of Borneo composed the first biome in the humid tropics to which the concept of nutrient stress-induced sclerophytism was applied (*15*). McKey et al. (*16*), working in tropical Africa, found the contents in phenolic substances from tree species of nutrient-poor forests significantly higher than that of nutrient-rich forests. Thus a general chemical dichotomy was recognized: plants which contain alkaloids contain low levels of tannins and other phenolics, while plant species rich in phenolic compounds generally do not contain alkaloids. This has been the experience of various phytochemists since the time of Greshoff (*17*), who thought of an interaction between tannins and alkaloids forming insoluble complexes in which the activity of both were annihilated. Today chemists do not expect to find substances with a striking physiological effect in tanniferous plants.

Also other chemical differences were found to accompany the tannin/alkaloid dichotomy. Plant material rich in tannins normally contains more fibrous material, i.e. lignin, than alkaloid containing leaves. Thus a general dichotomy was envisaged between plants growing in nutrient rich and nutrient poor habitats, the first category having alkaloids, the second tannins and sclerophyllous structures.

Later on it was verified by Coley working in Panamá, and other workers as well (see Coley et al.) (*18*), that leaves rich in tannins have a longer lifetime and are formed by slow-growing plants in comparison with alkaloid-containing plants which are fast-growing and whose leaves are replaced in shorter intervals. By this observation earlier deductions that had been made by Janzen (*19*) were confirmed.

Since it is generally accepted that secondary plant constituents basically owe their existence to the selection by herbivores, microorganisms, etc., the dichotomy between alkaloids and tannins called for an explanation. To this end, various hypotheses were offered, which have been reviewed by Feeny (*20*). One was the "apparency hypothesis", which claimed that the more "apparent" or "predictable and available" plants should depend more on "quantitative" defense, i.e. tannins and other polyphenols, which were considered to be in principle non-detoxifiable. "Qualitative defense" involving smaller quantities of toxins such as alkaloids would then be the viable option for "unapparent" plants. However, these concepts have not been fully supported by observed patterns of predation. It seems that also the postulated effects of tannins can be circumvented by various metabolic properties of the herbivore, and some tannins seem to be directly toxic to some consumers, in contrast to earlier postulates. Nonetheless, many recent studies have strengthened the view that tannins can act as strong deterrents for herbivores, perhaps because of their association with foliage of poor quality.

In the 1980s, and mostly based on work by Coley et al. (*18*) (see also Waterman and McKey) (*21*), the nutrient availability of the habitat again came into focus. The idea of nutrient-induced sclerophytism was taken up and amplified by taking into account differences in growth rates of plants and the lifetimes of their leaves. Investment in polyphenols including lignins in the foliage was classified as "immobile" defense, because these compounds are relatively inactive metabolically (*18*). In more fertile habitats plant growth is more rapid, and "mobile" defense is based on substances such as alkaloids, cyanogenic glycosides, and terpenoids which, though being present in small quantities, continually are turned over (Table I).

Table I. Defense-Oriented, Metabolism-Oriented, and Resource-Oriented Classifications of Secondary Metabolites

alkaloids, cyanogenic glucosides, thiocyanates, etc.	terpenoids, phenolic glucosides	condensed tannins, lignins, hydrolysable tannins, latex ?
micromolecules		**macromolecules**
qualitative defense detoxifiable		**quantitative defense** non-detoxifiable
mobile defense high turnover rate		**immobile defense** low turnover rate
nitrogen-based		**carbon-based**

Thus interpretations have undergone a shift (*22*). Initially defense patterns were interpreted from the point of view of the herbivore, which led to the distinction of qualititative and quantitative defense. Emphasis on the nutrient resources and plant metabolism then led to an amplification of Arens' ideas in the "resource allocation hypothesis" (*18*), from which the distinction between immobile and mobile defense was derived. Somewhat different is the distinction between nitrogen-based and carbon-based defense (*23*), which exclusively is based on trophic conditions and may prove fruitful in the context of forest ecology.

Ecological Plant Chemistry in the Amazon

Since the time of Ducke and Black (*24*) the vegetation on oligotrophic white sand habitats of Amazonia has raised the interest of botanists. There exist close floristic ties between the flora of this so-called "campina-" or "caatinga-"vegetation and the flood plain vegetation of nutrient-poor black and clear water rivers of Amazonia (*25*). The flora of white sand communities - both inundatable and non-inundatable - obviously derives from an old floristic stratum of northern South America that for a long time, perhaps since the Cretaceous, had its development restricted to arenosols, which are highly critical both as to their scarcity of nutrients and unfavourable water balance (*26*). The elements of this vegetation show characteristic physiognomic traits in their sclerophyllous foliage, often tortuous growth, and prominent association with ectotrophic mycorrhizal fungi. The latter were recorded first by Singer and Araujo (*27*) in the floodplain forest of the Rio Negro and more recently have been shown to occur on white sand vegetation in Venezuela. These fungi seem to competitively prevent the growth of litter decomposing basidiomycetes, so that raw humus accumulates in vegetation rich in ectotrophic mycorrhiza. Medina et al. (*28*) studied the sclerophyllous species of the Upper Rio Negro region and found that they structurally and physiologically largely agree with sclerophylls from semiarid regions of the world except for having a larger leaf size. It is noteworthy that the sclerophylls are not more pronouncedly drought-resistant than other less strongly sclerophyllous species. Thus Medina et al. (*28*) agree in considering deficiency in nitrogen and especially phosphorus as the decisive factor for the selection of sclerophylly.

The impact of plant metabolites on edaphic processes

From senescent leaves some secondary compounds and nutrients such as nitrogen and phosphorus are withdrawn, while other compounds, and certainly most polyphenols, are lost with the leaves and other plant litter. Thus together with dead plant biomass, large quantities of polyphenols are deposited on the ground of the forest. In temperate regions, Rice and Pancholy (*29*) estimated the "tannin-fall" at about 85 kg ha^{-1} yr^{-1}, while in tropical humid forests an input of 200 kg ha^{-1} yr^{-1} of phenolics seems reasonable (*21*). It is certain that these compounds will have some favourable effect on the nutrient balance leading to the retention of nitrogen in the soil, either by diminishing its rate of release from decomposing tissue or by preventing its conversion to the NO_3^- form in which it is easily leached out of the soil. The accumulation of tannins and the formation of various tannin/protein complexes enhances the quantity of humic matter in the soil and increases soil acidity. Since in oligotrophic habitats very large quantities of polyphenols will be deposited, the extreme condition of their soils will be aggravated, with the consequence of a further reduction of the growth rate of the vegetation and increasing polyphenol production. These processes would explain the extreme sclerophylly in the campina/caatinga vegetation of the northwestern sector of the hylaea.

The increase in soil acidity also has an effect on the solubility of the components of the soil. At medium values of soil reaction at about pH 5-7 silica is soluble and consequently soils of the humid tropics in very long periods of time can lose all their quartz content and become latosols which are eventually built up exclusively by oxides of aluminum and iron. With increased acidity at values below pH 4 the solubility of silica decreases while ferric and aluminum oxides, now soluble, are lost by lixiviation. Thus there is a feedback between increase in soil acidity, accumulation of acidic mor soils, selective lixiviation of soils ending up in pure quartz sand soils, and an enhancement of polyphenol production in the vegetation.

Acknowledgments

I am most grateful to the Associação Brasileira de Química and Drs. Peter Seidl and Otto R. Gottlieb for inviting me to participate in the First Congress on Chemistry of the Amazon.

Literature Cited

1.   Gottlieb, O. R.; Mors, W. B. *Interciencia* **1978**, *3*, 252-263.
2.   Hofmann, A.; Schultes, R. E. *Plants of the Gods.* Maidenhead, MacGraw Hill: England; 1979.
3.   Sousa Ribeiro, M. N. de; Bichara Zoghbi, M. das G.; Silva, M. L. da; Gottlieb, O. R.; Mata Rezende, C. M.; *Cadastro Fitoquímico Brasileiro*, 2ª Ed., INPA: Manaus, 1987.
4.   Grabert, H.. *Geol. Rundschau* **1983**, *72*, 671-683.
5.   Katzer, F. *Grundzüge der Geologie des unteren Amazonasgebietes.* Leipzig, 1903.
6.   Grabert, H. *At. Simp. Biota Amaz.* Rio de Janeiro **1967**, *1*,. 209-214
7.   Kubitzki, K. *Rev. Acad. Colomb. Cienc.* **1989**, *17*,: 271-276.
8.   Schnütgen, A.; Bremer, H. *Z. Geomorph. N. F. Suppl. Bd.* **1985**, *56*, 55-67.
9.   Chauvel, A.; Lucas, Y.; Boulet, R. *Experientia* **1987**, *43*, 234-241.
10. Urbani, F. *5° Congr. Geol. Venez.* **1977**, *2*, 623-642.
11. Gansser, A. *Verh. Naturf. Ges. Basel* **1974**, *84*, 80-100.
12. Jordan, C. F. *Ecology* **1982**, *63*, 647-654.
13. Arens, K. *Bol. Fac. Fil. Ciênc. Letr. Univ. São Paulo, Botanica* **1958**, *15*: 25-56.

14.   Arens, K. *Bol. Fac. Fil. Ciênc. Letr. Univ. São Paulo, Botanica* **1958**, *15*, 59-77.
15.   Bruenig, E. F. *Ecological Studies in the Kerangas Forests of Sarawak and Brunei.* Kuching: Borneo Lit. Bureau for Sarawak Forest Dept., 1974.
16.   McKey, D.; Waterman, P. G.; Mbi, C. N.; Gartlan, J. S.; Struhsaker, T. T. *Science* **1978**, *202*, 61-64.
17.   Greshoff, M. *Mededel. Plantent. Batavia* **1890**, *7*, 45-69.
18.   Coley, P. D.; Bryant, J. P.; Stuart Chapin, III, F. *Science* **1985**, *230*, 895-899.
19.   Janzen, D.H. *Biotropica* **1974**, *6*, 69-103.
20.   Feeny, P. In *Herbivores: their Interactions with Secondary Plant Metabolites,* second ed., Vol. 2, Rosenthal, G. A.; Berenbaum, M. R., Eds., San Diego; 1992, pp 1-44.
21.   Waterman, P. G.; McKey, D. In *Tropical Rain Forest Ecosystems;* Lieth, H.; Werger, M.J.A., Eds.; Ecosystems of the World 14B. Elsevier: Amsterdam; 1989, pp 513-536.
22.   Salatino, A. *An. Acad. Bras. Ci.* **1993**, *65*, 1-13.
23.   Pastor, J.; Naiman, R. J. *Amer. Natural* **1992**,. *139*, 690-705.
24.   Ducke, A.; Black, G. A. *An. Acad. Bras. Ci.* **1953**, *25*, 1-46.
25.   Kubitzki, K. *Plant Syst. Evol.* **1989**, *162*, 285-304.
26.   Kubitzki, K. *Mem. New York Bot. Gard.* **1990**, *64*, 248-253.
27.   Singer, R.; Araujo Aguiar, I. *Plant Syst. Evol.* **1986**, *153*, 107-117.
28.   Medina, E.; García, V.; Cuevas, E. *Biotropica* **1990**, *22*, 51-64.
29.   Rice, E. L.; Pancholy, S. K. *Amer. J. Bot.* **1974**, *60*, 691-702.
30.   Salgado Vieira, L. T. C.; dos Santos, P. C. *Amazônia seus Solos e Outros Recursos Naturais;* Editora Agronômica Ceres: São Paulo, SP; 1987.

RECEIVED October 4, 1994

Chapter 13

Lignans: Diversity, Biosynthesis, and Function

Norman G. Lewis, Massuo J. Kato, Norberto Lopes, and Laurence B. Davin

Institute of Biological Chemistry, Washington State University, Pullman, WA 99164-6340

The Amazonian region contains one of the world's most important repositories of plant biodiversity, whose careful and lasting stewardship is a most pressing issue facing this and future generations. Remarkably, only a tiny fraction of its plant life has been systematically and comprehensively examined for its (bio)chemical constituents and ultimate benefit to humanity. Consequently preservation of this resource is of paramount importance.

This chapter addresses the occurrence and distribution of a large class of phenylpropanoid metabolites, collectively known as lignans, and their increasing significance as physiologically and pharmacologically active substances e.g., as antiviral, antitumor, biocidal and other bioactive agents.

Based upon emerging chemotaxonomic data and recently discovered biochemical pathways, it appears that a high level of probability can be exercised in predicting lignan structural variants likely to be present in representatives of specific plant family superorders. Using the lignans as an example, it is proposed that further compilation of such data (at the chemical, biochemical, and gene expression levels) will be of great significance in "cataloging" plant biodiversity.

The *ca* 350,000 or so different species in the plant kingdom are all thought to have derived from a common ancestral, aquatic, algal precursor *(1)*. But this transition to a terrestrial environment both necessitated further elaboration of existing biochemical pathways and introduction of new ones. This was required in order that plants could: provide improved means for structural support, water and nutrient conduction; form protective layers to minimize the effects of changes in temperature, humidity, UV-B irradiation and

0097–6156/95/0588–0135$15.25/0

encroachment by pathogens; form layers preventing uncontrolled water diffusion (losses), and sustain a high water potential for active, highly regulated metabolism.

Plant species are fairly readily distinguishable from one another, not just in terms of gross morphology and structure, but also from a chemotaxonomic sense. This is because, as evolution progressed, various organisms established different emphases in their overall biochemical pathways. The competitive advantages gained by specific metabolic changes are most readily observed as different abilities of particular species to thrive under specific climatic/environmental conditions, e.g., in plants favoring rain forest or desert climates, etc. Such differences in habitat adaptation capability are often achieved by exquisitely orchestrated alterations in cell wall and membrane synthesis.

Another notable feature regarding the overall evolution of plant species lies in their different capabilities to accumulate specific low molecular weight metabolites [or micromolecular compounds, (2)], e.g., specific phenylpropanoids, alkaloids, terpenes, etc. Although no structural roles are contemplated, these compounds do nevertheless play important roles in competitive survival. Some exhibit profound effects on growth and development, whereas others are important in plant-host interactions, in allelopathy, and in protective functions (e.g., as antioxidants, biocides, UV-screens, etc.). Many also find important application in pharmacological and medicinal cures/treatments as well as sources of flavors, fragrances and intermediate chemicals. However, it must be emphasized that only a very small number of compounds and plants have been systematically screened for possible use e.g., as medicinal agents.

There is growing concern that many of the plant species which evolved over time will be irretrievably lost, as civilization and its associated technologies encroach upon the remaining virgin (rain) forest stands. Nowhere is this more evident than in the Amazonian (rain) forests, which harbor a great number of angiosperms. The projected disappearance of many of these organisms (and others dependent upon same) is a matter of great concern for the following reasons. They may (1) contain as yet undiscovered cures for existing and/or future diseases; (2) possess unique biological traits which can be subsequently introduced into commercially important species, e.g., to enhance productivity or to impart novel defense functions (antioxidants, biocidal properties, and the like); (3) provide new biopolymers suitable as valuable replacements for existing petroleum-derived polymers; and (4) provide new foodstuffs and/or sources of valuable chemicals for industrial application (e.g., fragrances, intermediate chemicals), and alternate (improved) sources of lumber and paper.

The threat of extinction (of some species and/or the rain forests themselves) has resulted in world-wide concern. One manifestation of this is a desire to catalogue, in some meaningful and lasting manner, the current plant inventory. The mechanisms envisaged to collect such data are varied and range from gathering metabolic/chemotaxonomic profiles and associated biochemical markers to molecular mapping (e.g., via DNA fingerprinting), and so on.

This chapter addresses how, for example, the phenylpropanoid pathway and associated lignan (neolignan) metabolic profiles can be used as a classification aid. It describes the proposed evolution of the pathway and current chemotaxonomic interpretations, the progress in defining biochemical pathways and appropriate markers, and the growing physiological and pharmacological significance.

Nomenclature

The lignans are a widespread and structurally diverse class of phenylpropanoid metabolites whose biosynthesis appears to be restricted to vascular plants *(1)*. Of those isolated and characterized to date, the majority are dimeric *(3)* although a growing number of higher oligomers are being characterized and identified *(4-6)*. There are significant difficulties in usage of current nomenclature, since lignans are currently defined as only those metabolites linked via 8,8' carbon-carbon bonds, with neolignans embracing all other coupling modes [see Ref. *(3)*]. Another school of thought classifies all types of coupling as affording lignans, with neolignans being restricted to those metabolites apparently derived from allyl or propenyl phenols *(7)*.

As a further refinement, we would propose that the term lignans be used to describe all products obtained via direct coupling of at least two moieties *exclusively derived* from phenylalanine **1** or tyrosine **2**, but still containing the requisite C_6C_3 skeleta (see Figure 1). A satisfactory distinction between the terms, lignins and lignans, should follow the widely adopted "rules" in polymer chemistry, i.e., where lignins (cf. cellulose, polyethylene) are polymeric moieties having considerably different physical properties (e.g., viscosity, glass transition temperatures, etc.) and functions, to that of dimeric and oligomeric lignans (cf. glucans, ethylene, etc.).

This refinement in lignan definition would satisfy the following: (1) dimeric and oligomeric products obtained, for example, via O_2-laccase or H_2O_2-dependent peroxidase catalysis of coniferyl alcohol **13** would now all be described as lignans, i.e., the dimers, (±)-pinoresinols **20a/20b**, (±)-guaiacylglycerol-8-*O*-coniferyl alcohol ethers **21**, (±)-dehydrodiconiferyl alcohols **22a/22b**, rather than lignans and neolignans as currently exists; (2) that trimers, etc., containing various "lignan and neolignan" linkages such as buddlenol D **23**, can now be classified simply as lignans; (3) that lignans be systematically classified into families or subgroups based upon basic structural parameters, e.g., tetrahydrofuran, cyclobutane, phenylcoumarin lignans, etc., with appropriate biosynthetic considerations taken into account as needed. An arbitrary classification is given below in Figure 2, indicative of common structural types.

Proposed Evolution of the Phenylpropanoid/Lignan Pathways and Perceived Chemotaxonomic Trends

The envisaged transition from an aquatic to a terrestrial environment was accompanied by, or was dependent upon, evolution of new biochemical pathways, particularly those leading to the phenylpropanoids. The first committed step in the phenylpropanoid pathway involves deamination of phenylalanine, Phe **1** (or, in a few instances, tyrosine, Tyr **2**) to afford the corresponding cinnamic **3** or *p*-coumaric **4** acids, respectively [reviewed in Ref. *(8)*]. Elaboration of the pathway, shown in Figure 1, thus provided a

NOTE: structures appear in Appendix at the end of this chapter

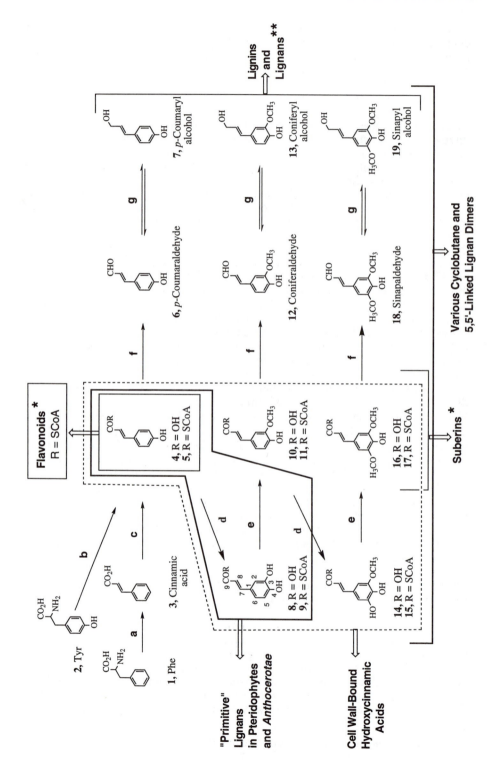

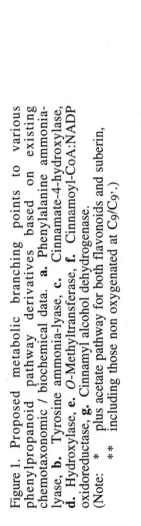

Figure 1. Proposed metabolic branching points to various phenylpropanoid pathway derivatives based on existing chemotaxonomic / biochemical data. **a.** Phenylalanine ammonia-lyase, **b.** Tyrosine ammonia-lyase, **c.** Cinnamate-4-hydroxylase, **d.** Hydroxylase, **e.** *O*-Methyltransferase, **f.** Cinnamoyl-CoA:NADP oxidoreductase, **g.** Cinnamyl alcohol dehydrogenase. (Note: * plus acetate pathway for both flavonoids and suberin, ** including those non oxygenated at C9/C9'.)

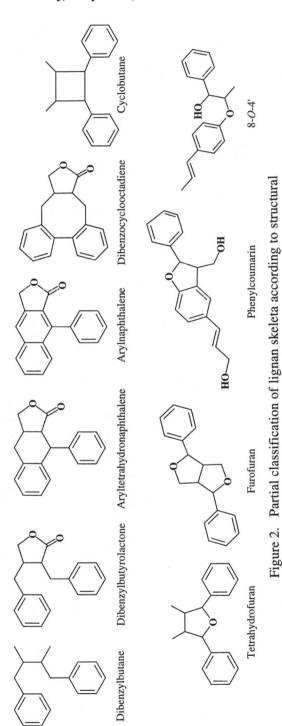

Cyclobutane

8-*O*-4'

Dibenzocyclooctadiene

Phenylcoumarin

Arylnaphthalene

Aryltetrahydronaphthalene

Furofuran

Dibenzylbutyrolactone

Dibenzylbutane

Tetrahydrofuran

Figure 2. Partial classification of lignan skeleta according to structural type.

means to synthesize various metabolites, including those affording lignins *(9)* (for structural support, water/nutrient conduction, and as barriers to opportunistic pathogens), hydroxycinnamic acids [putatively involved in cross-linking primary walls (reviewed in Ref. *(10)*)], lignans/neolignans [which have roles as antioxidants *(11-13)*, putative cytokinins *(14, 15)*, growth inhibitors *(16)*, intermediates in lignin synthesis *(17)*, as well as antiviral *(18)*, fungicidal *(19)* and other biocidal properties *(20)*] and other miscellaneous metabolites, such as vanillin **24**. Offshoots of both the phenylpropanoid and acetate pathways lead to the suberins (providing water diffusion resistance barriers and wound-healing layers) *(8)*, and to the flavonoids (these putatively having roles in protection against UV-irradiation *(21)*, nitrogen fixation *(22)*, defense against phytopathogens *(23)*, pollen germination *(24)*, etc. Although the overall pathway is well established, there is some debate as to whether the free acids or their corresponding CoA derivatives, or both, serves as substrates for *O*-methylation (Ye *et al.*, *Plant Cell*, in press).

Unfortunately, the existing fossil record does not permit delineation of an orderly progression of events from the putative algal precursors to those of the extant plant forms [reviewed in *(1)*]. Consequently, much of our knowledge of presumed evolutionary progression derives from chemotaxonomic considerations.

Algae and Bryophytes. There is no evidence of the biogenetic capability of algae to form either lignin polymers or lignans *(1)*. In a somewhat analogous manner, they are also apparently absent in both the fungi and various classes of bryophytes [i.e., Bryophyta (mosses), Hepatophyta (liverworts), and Takakiophyta (takakiophytes)]. These organisms did, however, partially elaborate the phenylpropanoid pathway to afford *p*-coumaric **4** and caffeic **8** acids and derivatives thereof. This inability to engender formation of lignins may be closely correlated with an apparent inability to regiospecifically *O*-methylate caffeic acid **8** (or its corresponding CoA derivative **9**) at C_3. By contrast, the hornworts (*Anthocerotae*), which are currently classified within the bryophytes do synthesize/accumulate the simple 8,8' linked lignans, (+)-megacerotonic acid **25** (i.e., in *Dendroceros japonicus, Megaceros flagellaris, Notothylas temperata,* and *Phaeoceros laevis*), as well as (+)-anthocerotonic acid **26** in *Anthoceros punctatus (25, 26)*. Significantly, both are apparently formed via coupling of *p*-coumaric **4** and caffeic **8** acid moieties, and it is noteworthy that regiospecific *O*-methylation at C_3 is again lacking. The evolutionary significance of these findings is, however, yet unclear since the hornworts are believed to have evolved during the Cretaceous period *(27)* after evolution of the pteridophytes, gymnosperms, and angiosperms had occurred. Nevertheless testing for the presence of these and related compounds, as well as for the corresponding coupling enzymes, mRNA and DNA, etc. may be excellent phylogenetic markers for hornworts.

Pteridophytes. Evolution of the biochemical pathway to the lignins proper apparently first occurred with the appearance of the pteridophytes, the earliest remains of which to date come from the Devonian period *(27)*. In extant plants, lignins are seemingly present in the ferns (Filicopsida), clubmosses (Lycopsida), horsetails (Equisetopsida), and the Psilotopsida [see

Ref. *(9)]*. Surprisingly, there are few detailed analyses of lignins from these organisms, and perhaps even more surprising there are, to our knowledge, only two reports of pteridophyte lignans *(28, 29)*. The first report was that of glucosides of dihydrodehydrodiconiferyl alcohol **27** (a phenylcoumarin lignan) and lariciresinol **28** (a furanolignan) in *Pteris vittata (28)*, and the second of blechnic **29** and 7-epiblechnic **30** acids (both phenylcoumarin lignans) in six Blechnaceous tree-ferns (e.g., *Blechnum orientale) (29)*.

The tree-fern metabolites are presumably derived via oxidative coupling of two caffeic acid **8** moieties, and perhaps represent the earliest *bona fide* examples of phenolic coupling to give the lignan skeleta. It may also be highly significant that the corresponding *O*-methylated dimers from either ferulic **10** or sinapic **16** acids are apparently absent, again suggestive of some evolutionary clue regarding the importance of timing of *O*-methylation. The optical rotations of both blechnic acid **29**, [α]$_D$=-28°, and epiblechnic acid **30**, [α]$_D$=-145°, are noteworthy since neither are expected to result from typical O_2-laccase or H_2O_2-dependent peroxidase catalyzed transformations which engender racemic product formation. Again, the presence of these metabolites (or analogues thereof) and the identification of their appropriate biochemical/ molecular machinery may be extremely useful in systematic classification of the tree-ferns.

On the other hand, the lignans of the fern, *Pteris vittata*, reveal that *O*-specific methylation has now occurred (as required for lignin synthesis), as well as various reductive modifications (discussed later). However, the aglycones of dihydrodehydrodiconiferyl alcohol glucoside **27**, and lariciresinol glucoside **28** are weakly optically active ([α]$_D$=-8.5° and +15.7°, respectively), suggesting derivation from a racemic precursor and, hence, presumably originate via a typical O_2-laccase or H_2O_2-dependent peroxidase catalyzed conversion. Clearly, the underlying reasons for the differences, noted this far between the ferns and tree-ferns need to be determined to establish evolutionary significance.

Gymnosperms. Although there are no reports of their occurrence in the Cycadales and only one in the Ginkgophytes [i.e., sesamin **31** in *Ginkgo biloba (30)*], lignans are fairly widespread throughout the gymnosperms *(1)*. They are found within the Podocarpaceae, Taxodiaceae, Taxaceae, Cupressaceae, Pinaceae, Ephedraceae, and Araucariaceae families *(1, 7)*. Although various structural trends have surfaced which are described below, these should only be viewed at this point as predictive chemotaxonomic markers until more extensive verification has been obtained. Nevertheless, in the Podocarpaceae, the first appearance of simple aryltetrahydronaphthalene lignans occurs, such as (+)-*O*-methyl-α-conidendral **32** in *Dacrydium intermedium (31, 32)* and (-)-α-conidendrin **33** in *Podocarpus spicatus (33)*. In the Taxodiaceae and Cupressaceae, a new structural feature has been added, namely introduction of a methylenedioxy group, as in taiwanin C **34** in *Taiwania cryptomerioides (34)*, as well as new oxidative patterns such as shown by dihydroxythujaplicatin **35** from *Thuja plicata (35)*. The Pinaceae appear to have a much more restricted structural range, with typical examples being dehydrodiconiferyl alcohol **22** (Nose *et al., Phytochemistry*, in press), pinoresinol **20** *(36, 37)*, (-)-α-conidendrin **33** and (-)-matairesinol **36b** *(38)*, whereas the Taxaceae have lignans apparently derived from monomers containing coniferyl **13** and 3,4-dihydroxycinnamyl **37** alcohols, respectively, e.g., taxiresinol **38** from *Taxus baccata (39)*. Lastly, in the Araucariaceae, such as that of Parana pine (*Araucaria angustifolia*), a new structural modification is evident, as revealed by (-)-galbulin **39** *(40)*. Formation of this metabolite requires either post-coupling reduction of the hydroxymethyl functionality at C_9, or it is derived from allylphenols such as eugenol **40** or isoeugenol **41**. Taken together, characteristic metabolic profiles appear to be emerging for the different gymnosperm families. This in turn means that there is a strong likelihood that assaying for unique enzymatic steps (or genes encoding their products) for particular lignan subgroups will provide a powerful method for their rapid classification.

Angiosperms. The last group are the angiosperms, these being the most diverse form of plant life with *ca* 200,000 different species. Although only a limited number have been examined in detail, they are nevertheless a rich and varied source of lignans, these being found to date in dicots (i.e., the Magnoliiflorae, Nymphaeiflorae, Rosiflorae, Malviflorae, Myrtiflorae, Rutiflorae, Ranunculiflorae, Violiflorae, Primuliflorae, Santaliflorae, Araliiflorae, Asteriflorae, Corniflorae, Gentianiflorae, and Lamiiflorae) and monocots (i.e., the Ariflorae and Commeliniflorae), respectively. As for the gymnosperms, there are some apparent structural progressions within the different superorders that may be of immense predictive value for classification purposes.

Briefly, the angiosperm lignans can tentatively be broken down into three major categories. These are: (1) lignan skeleta apparently derived from pinoresinol **20** (i.e., containing oxygenated functionalities of some type at $C_9/C_{9'}$), (2) lignan skeleta possessing methyl or methylenic groups at $C_9/C_{9'}$, and which are apparently allyl or propenyl phenol derived; and (3) an assortment of aryl cyclobutane dimers, principally found in the monocotyledons.

Of these, the most widely encountered mode of coupling is that affording the 8,8'-linked lignans, which contain oxygenated functionalties at $C_9/C_{9'}$ [Table 1, taken from Refs. *(7, 41, 42)*]. These are *assumed* to be pinoresinol **20** derived based upon biosynthetic findings from this laboratory [reviewed in *(1)*]. As for the gymnosperms, the presence of particular structural groupings seems closely correlated with specific superorders, e.g., the oxydiarylbutanes in the Asterales (Asteriflorae) and the Euphorbiales (Myrtiflorae) *(1, 7)*, and the $C_9/C_{9'}$ oxygenated dibenzylcyclooctadienes in the Araliaceae (Araliiflorae) *(43)*. It should, therefore, be evident that rapid, sensitive assaying for such metabolites (or key branching enzymes and/or corresponding mRNA/DNA therefrom) may be extremely useful for systematic classification.

There are also a relatively large number of lignans possessing structures which seem to be derived from eugenol **40** or isoeugenol **41**. [This assertion is made since a number of organisms accumulate the corresponding monomers (e.g., *Ocotea cymbarum*, Lauraceae *(44)*), indicating that at least in these species the biochemical "machinery" to afford the allyl/propenyl side chains is operative without the need for coupling]. Such lignans are found in the Magnoliiflorae, Lamiiflorae, Myrtiflorae, and Rutiflorae, respectively, and are further distinguished by substantial modifications in the modes of coupling encountered, e.g., gomisin A **42**, an 8,8'; 2,2'-linked dibenzylcyclooctadiene lignan from *Schizandra chinensis* (Magnoliiflorae) *(45)*, with related structures found in the Verbenaceae (Lamiiflorae) (see ref. *(1)*). Other modified 8,8' coupling modes are revealed by (+)-aristolignin **43** from *Aristolochia chilensis (46)*, the 8,8'; 2,7'; 8:8'-7' linked lignans **44** and **45** from *Virola sebifera*, and its seco-analog **46** *(47, 48)*, and carpanone **47** from *Cinnamomum* species *(49)*. Other distinctive coupling modes include those displayed by megaphone **48** (8,1') from *Aniba megaphylla (50)*, mirandin A **49** (8,3') from *Nectandra miranda (49)*, surinamensin **50** and virolin **51** (8-*O*-4') from *Virola surinamensis (51)*, dehydrodieugenol **52** (5,5') from *Litsea turfosa (52)*, isomagnolol **53** (3-*O*-4') from *Sassafras randaiense (53)*, lancilin **54** (2-*O*-3') from *Aniba lancifolia (54, 55)*, chrysophyllon IA **55** (7,1') from *Licaria chrysophylla (56)*, isoasatone **56** (1,5') from *Asarum taitonense (49)*, conocarpin **57** from *Conocarpus erectus (57)*, and ratanhiaphenol I **58** (8,5') from *Krameria cystisoides (58)*. It is proposed that these different metabolites are derived via distinct enzymatic coupling modes. If correct, assaying for such enzymes should provide a rapid mechanism for classifying the corresponding plant types.

Representation of lignans in the monocotyledons is apparently quite restricted, with essentially all reported to date being cyclobutane and 5,5' or 3,3'-linked dimers. Examples of the former include acoradin **59** from *Acorus calamus* (Ariflorae) *(59)* and various dihydroxy truxillic **60** and truxinic **61** acids present in the Poaceae (Commeliniflorae) *(60, 61)*. [Similar cyclobutane dimers have been reported in the quite closely related Magnoliiflorae (dicots) *(62-64)* suggesting some evolutionary "overlap".] The more abundant dihydroxytruxillic and dihydroxytruxinic acids are found in cell wall fractions of various cereals and grasses, and putatively cross-link adjacent hemicellulosic chains in primary walls *(60, 61)*. All of these cyclobutane dimers are considered to be photochemical 2+2 addition adducts rather than formed by enzymatic catalysis. Interestingly, a series of mixed cyclobutane dimers between ferulic acid **10** and coniferyl alcohol **13** have also been detected *(60, 61)*, suggesting that there may be a whole series of undiscovered lignans in the Commeliniflorae. The only other known lignans include the 5,5'-linked and 3,3'-linked diferulic and di-*p*-coumaric acids, respectively, these purportedly being formed via H_2O_2-dependent peroxidase catalysis.

Biosynthesis

Our interest in the biosynthesis of lignans from Amazonian plant species (e.g., the Myristicaceae such as in *Virola* species) stemmed from four considerations: (1) the growing interest in the physiological and pharmacological properties of (neo)lignans. For example, various *Virola* sp. have useful activities against diseases such as schistosomiasis, leishmaniasis and other tropical diseases *(51, 65)*; (2) in aiding a facile systematic classification of rain forest plants, via chemotaxonomic or other markers based on either specific enzymatic activities or the presence of specific mRNAs, etc.; (3) several lignans have structures previously noted as substructures of the lignin polymer, leading to the notion that the formation of each could be

Table I. Distribution of 8.8'-linked lignans, presumably derived from pinoresinol **20**, and classified according to structural types and plant superorders. (Lignan occurrence was compiled from Ref. 7, and plant classification is from Dahlgren *41,42*).

Superorders	Oxydiaryl-butanes	Dibenzyl-butanes	Dibenzyl-butyro-lactones	Dibenzyl-furans	Aryl-naphthalenes	2.7'-Cyclolignan-9'-olides	Aryltetra-hydro-naphthalenes	Dibenzyl-cyclo-octadienes
Araliiflorae			Araliales			Araliales		Araliales
Asteriflorae	Asterales		Asterales			Asterales		
Balanophori-florae								
Caryophylli-florae								
Celastriflorae								
Corniflorae			Cornales			Eucommiales	Eucommiales Ericales	
Fabiflorae								
Gentiani-florae		Gentianales	Gentianales					
Lamiiflorae			Scrophulariales		Scrophulariales	Scrophulariales	Oleales	
Loasiflorae								
Magnolii-florae		Aristolochiales	Aristolachiales Annonales Laurales	Laurales		Annonales Laurales	Annonales Laurales	

Table I. (Continued)

Superorders	Oxydiaryl-butanes	Dibenzyl-butanes	Dibenzyl-butyro-lactones	Dibenzyl-furans	Aryl-naphthalenes	2.7'-Cyclolignan-9'-olides	Aryltetra-hydro-naphthalenes	Dibenzyl-cyclo-octadienes
Malviflorae			Thymelaeales				Urticales	
Myrtiflorae	Euphorbiales				Euphorbiales			
Nymphaei-florae		Piperales	Piperales					
Podostemoni-florae								
Polygoni-florae								
Primuliflorae				Ebenales				
Proteiflorae								
Ranunculi-florae				Ranunculales	Ranunculales	Ranunculales		
Rosiflorae							Rosales Fagales	
Rutiflorae			Rutales		Rutales Polygalales	Rutales Polygalales		
Solaniflorae								
Theiflorae								
Violiflorae								

individually studied; and (4) several species accumulate lignans possessing oxygenated and methyl/methylene functionalities at C9/C9'.

Additionally, *Virola* sp. were a good candidate for studying lignan biosynthesis since various structural forms are present, such as furofurans, dibenzylbutanes, dibenzylbutyrolactones/lactols, aryltetrahydronaphthalenes, phenylcoumarins, and 8-*O*-aryl ethers e.g., (+)-sesamin **31**, (+)-asarinin **62**, (-)-kusunokinin **63**, (-)-kusunokinol **64** and (-)-dihydrocubebin **65** from *Virola venosa (66)*, (-)-cubebin **66**, (-)-hinokinin **67** and (+)-asarinin **62** from fruits of *Virola carinata (67)*, and dihydrocarinatidin **68**, carinatidin **69** *(68)* and dihydrocarinatinol **70** *(69)* from the corresponding bark as well as surinamensin **50** and virolin **51** from *Virola surinamensis (51)* and sesartenin **71** and dihydrosesartemin **72** from *Virola elongata* bark *(70)*.

At the onset of our studies, the only processes known engendering phenolic coupling were catalyzed by H_2O_2-dependent peroxidase(s), O_2-requiring laccases, or phenol oxidases. But such products were racemic, e.g., the coupling of coniferyl alcohol **13** afforded (±)-pinoresinols **20a/20b**, (±)-dehydrodiconiferyl alcohols **22a/22b**, and (±)-guaiacylglycerol-8-*O*-coniferyl alcohol ethers **21**.

We reasoned that racemic coupling was unlikely to be the general case in lignan synthesis, since most are found optically pure although the sign of optical rotation can vary with plant species. There are, however, exceptions such as syringaresinol **73** which can be found either optically pure or racemic *(71)*. Moreover, even though the 8,8' linkage prevailed throughout the plant kingdom, specific plant species were capable of engendering formation and accumulation of other lignans possessing other coupling modes. This suggested that perhaps a whole series of stereoselective oxidases and/or peroxidases awaited discovery.

The plant systems adopted as biological partners were the Amazonian species, *Virola surinamensis* and the temperate plant, *Forsythia intermedia*, respectively. The latter accumulates the 8,8'-linked, optically pure lignans (-)-matairesinol **36b** and (-)-arctigenin **74**, whereas *V. surinamensis* affords surinamensin **50**. Preliminary studies using either plants or cell-free extracts of *V. surinamensis* from *ca* one- to two-year-old plants and eugenol **40**/ isoeugenol **41**/[8-^{14}C]coniferyl alcohol **13** as precursors did not result in formation of either surinamensin **50** or virolin **51** under the conditions employed.

But the investigations with *Forsythia intermedia*, which accumulate lignans similar to those found in *Virola venosa* (e.g., (+)-sesamin **31** vs. (+)-pinoresinol **20a**) were more successful. First, incubation of cell-free *Forsythia* extracts with [8-^{14}C]coniferyl alcohol **13** yielded the desired lignans, but the products were racemic i.e., (±)-pinoresinols **20a/20b** and (±)-dehydrodiconiferyl alcohols **22a/22b**; these were only formed when H_2O_2 was added as cofactor *(72)*. However, when these experiments were repeated, but now using the corresponding *Forsythia* plant residues (i.e., cell wall enriched fraction), a different observation was made. In this instance, [8-^{14}C]coniferyl alcohol **13** was converted into (±)-pinoresinols **20a/20b** as before, but now with the corresponding (+)-antipode **20a** predominating *(73)*. No exogenous cofactors were needed. Subsequent time-course analyses of pinoresinol **20** formation revealed that two distinct modes of coupling were operative. The first was a stereoselective coupling which engendered formation of (+)-pinoresinol **20a**, whereas the other afforded both (±)-pinoresinols **20a/20b** (as well as racemic (±)-dehydrodiconiferyl alcohols **22a/22b**).

Both enzymes have since been dissolved out of *Forsythia* stem residues; the non-stereoselective O_2-requiring enzyme has been purified to apparent homogeneity using a combination of ionic, affinity and gel filtration chromatographic separations. The resulting purified enzyme (Mw ~ 100,000) "is tentatively" a Cu-containing laccase, whereas the other O_2-requiring enzyme catalyzed conversion of *E*-coniferyl alcohol **13** into (+)-pinoresinol **20a** *(74)*. This latter enzyme is the first example of stereoselective coupling of plant phenols, and we tentatively propose that it represents the entry point into the various lignan skeleta containing oxygenated functionalities at $C_9/C_{9'}$. This enzyme, trivially described as (+)-pinoresinol synthase, has been partially purified with some of its properties described elsewhere *(74)*. Its proposed mode of action is shown in Figure 3, where the two bound oxidized forms derived from *E*-coniferyl alcohol **13** approach each other from their *si-si* faces, with intramolecular ring closure affording the product, (+)-pinoresinol **20a** as shown. No other evidence for any other stereoselective conversion of any other precursor was obtained.

Given the discovery of the long sought after, novel, stereoselective coupling enzyme, attention was next given to formation of the dibenzylbutyrolactones, (-)-matairesinol **36b** and (-)-arctigenin **74**. Returning to the *F. intermedia* cell-free extracts, we next discovered that (+)-[8,8-^{14}C]pinoresinol **20a** was converted, in the presence of NADPH, into two new metabolites, subsequently shown to be (+)-[8,8'-^{14}C]lariciresinol **75a** and (-)-[8,8-^{14}C]secoisolariciresinol **76b**, respectively *(75, 76)*. [The corresponding (-)-[8,8'-^{14}C]pinoresinol **20b** did not serve as substrate for formation of either (±)-lariciresinols **75a/75b** or (±)-secoisolariciresinols **76a/76b**.] Confirmation of this highly unusual benzylic ether reduction, which to our knowledge is the first report in plants of such a conversion, was obtained using (+)-[9,9'-^{2}H$_2$, OC^{2}H$_3$]pinoresinol **20a** as substrate, and isolating and identifying the corresponding (+)-[9,9'-^{2}H$_2$, OC^{2}H$_3$]lariciresinol **75a** and (-)-[9,9'-^{2}H$_2$, OC^{2}H$_3$]secoisolariciresinol **76b**, respectively *(77)*.

The pinoresinol/lariciresinol reductase(s) has since been purified to homogeneity via a combination of affinity, hydrophobic and gel filtration chromatographies (Dinkova *et al.*, in press). The enzyme has a M.W. ~ 40,000, and only utilizes the pro-*R* hydride of NADPH, i.e., it is a type A reductase *(78)*.

The next conversion to be demonstrated was the stereo- and enantiospecific dehydrogenation of (-)-secoisolariciresinol **76b** into (-)-matairesinol **36b**, in a conversion requiring NADP as cofactor *(79, 80)*; the final step leading to (-)-arctigenin **74** involves regiospecific methylation *(81)*.

Taken together, the results reveal that *E*-coniferyl alcohol **13** undergoes novel stereoselective coupling to give the furofuran lignan, (+)-pinoresinol **20a**, followed by sequential reduction to yield the corresponding furano derivative, (+)-lariciresinol **75a** and the dibenzylbutanol, (-)-secoisolariciresinol **76b**, respectively. Subsequent dehydrogenation affords the corresponding dibenzylbutyrolactone, (-)-matairesinol **36b** (Figure 4). It can, therefore, be proposed that formation of (+)-pinoresinol **20a** serves as an entry point into the various lignan skeleta. In the case of the Amazonian *Virola* species, a possible biosynthetic route is shown in Figure 4. Most importantly, these findings, now, allow us to begin to systematically classify plants based both on their mode of stereospecific coupling, and the presence/absence of various other modifications leading to formation of the various skeleta observed. This should be of immense value in the future for systematic classification of Amazonian species.

Biological Activities

The lignans are becoming increasingly important for their physiological and pharmaceutical properties.

Physiological Properties in Plants. From a physiological perspective, various roles have been demonstrated. For example, fungicidal and bactericidal properties have been reported for magnolol **77**, isolated from *Sassafras randaiense (53)* and *Magnolia virginiana (19)*, respectively; dehydrodiisoeugenol **78** and 5'-methoxydehydrodiisoeugenol **79**, from *Myristica fragrans (20)* and obovatol **80** and obovatal **81**, from *Magnolia obovata (82)*, inhibit the growth of *Streptococcus mutans*.

Haedoxan A **82**, isolated from the herbaceous perennial plant *Phryma leptostachya* exhibits insecticidal activity *(83)*, whereas, the lignans, sesamin **31** and asarinin **62** show synergistic properties with pyrethrum insecticides *(84, 85)*. (+)-Epimagnolin A **83**, from the flower buds of *Magnolia fargesii*, exhibits growth inhibitory activity against the larvae of *Drosophila melanogaster (86)*.

Nordihydroguaiaretic acid **84**, from the creosote bush (*Larrea tridentata*) reduces the growth of the seedling roots of grasses (e.g., barnyard, green foxtail and perennial rye grasses) *(87)*, whereas, the lignan **85** from *Aegilops ovata*, inhibits germination of *Lactuca sativa* achenes, in white light; it is, however, inactive in the dark *(16, 88)*.

Finally, the 8,5'-linked lignan dehydrodiconiferyl alcohol glucoside **86** has cytokinin-like properties: It induces plant cell division and can replace cytokinin in pith and callus cell cultures of tobacco (*Nicotiana tabacum*) *(14, 15, 89)*.

Medicinal Properties. In terms of medicinal properties, plants containing lignans have been used as folk medicines by many different cultures. For instance, both the North American Indians and the Native of the Himalayas have utilized, for about 400-600 years, alcoholic extracts of *Podophyllum* rhizomes (rich in the lignan podophyllotoxin **87**) as a cathartic and a poison *(90)*.

Podophyllotoxin **87**, isolated from *Podophyllum* species shows antitumor properties; it inhibits microtubule assembly, *in vivo*, the result of which is the destruction of the cytoskeleton in the cytoplasm. As a consequence, the cell division is stopped at the mitotic stage of the cell cycle *(3)*. Semi-synthetic derivatives of podophyllotoxin, i.e., Etoposide **88** and Teniposide **89** have been developed. They do not possess the toxicity of podophyllotoxin **87** and are now being used (alone or in conjunction with other drugs) in treatment for germinal testicular cancer, small cell lung cancer, and certain form of leukemia *(3)*. They have been shown to induce, both, *in*

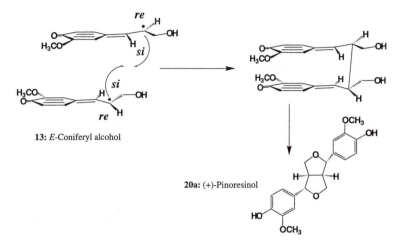

Figure 3. Proposed mechanism of (+)-pinoresinol **20a** formation in *Forsythia* sp. from *E*-coniferyl alcohol **13**. [Adapted from Ref. *(74)*]

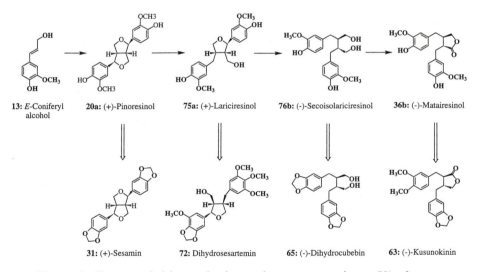

Figure 4. Proposed biosynthetic pathway to various *Virola* sp. lignans.

vivo and *in vitro*, single and double-stranded DNA breaks through interaction with topoisomerase II *(91)*.

Other lignans with known antitumor properties are (+)-wikstromol **90** isolated from *Wikstroemia foetida* var. *oahuensis* Gray (Thymelaeaceae) *(92)* and (-)-burseran **91** from *Bursera microphylla* (Burseraceae) *(93)*. (-)-Steganacin **92** and (-)-steganangin **93**, isolated from *Steganotaenia araliacea* stem bark and wood, exhibit antileukemic activity in the *in vivo* murine P-388 lymphocytic leukemia test system *(43)*.

Antiviral properties have also been described i.e., (-)-arctigenin **74** and (-)-trachelogenin **94**, isolated from the tropical climbing shrub *Ipomoea cairica* (Convolvulaceae) *(94)* and the lignans **95** and **96**, from *Anogeissus acuminata* var. *lanceolata* (Combretaceae) *(95)* exhibit strong anti-HIV properties, *in vitro*: they inhibit the HIV-1 reverse transcriptase.

Kadsurenone **97** isolated from *Piper futokadsura* (Piperaceae) is a platelet-activating factor (a potent mediator of inflammation and asthma) antagonist *(96)*. Antagonism with the platelet-activating factor has also been observed with nectandrin A **98** and B **99**, isolated from the Brazilian *Nectandra rigida (97)*, and fargesin **100** and dimethylpinoresinol **101** from *Magnolia biondii (98)*.

The lignans (+)-pinoresinol **20a** and (-)-matairesinol **36b**, isolated from *Forsythia* species inhibit cAMP phosphodiesterase which catalyses the breakdown of the second messenger, cAMP, inside the cells *(99)*. *Cis*-hinokiresinol **102** and oxy-*cis*-hinokiresinol **103** from *Anemarrhena asphodeloides (100)*, and (+)-syringaresinol-di-O-β-D-glucopyranoside **104** and (+)-hydroxypinoresinol 4',4''-di-O-β-D-glucopyranoside **105** from *Eucommia ulmoides* (Eucommiaceae) bark *(101)* are also inhibitors of this phosphodiesterase.

Additional pharmacological effects include the antidepressant activity of prostalidins A, B and C **106-108**, from *Justicia prostata* (Acanthaceae), a plant native to the Western Himalayas *(102)*. Magnoshinin **109**, isolated from *Magnolia salicifolia* buds, shows anti-inflammatory effect comparable to hydrocortisone acetate *(103)*. Gomisin A **42** *(104, 105)*, isolated from *Schizandra* fruits, has been shown to improve liver injuries caused by hepatotoxic chemicals (e.g., CCl$_4$) and other biphenyl lignans, such as schizandrin **110** and isoschizandrin **111**, exhibit inhibitory effects on stress induced gastric-ulceration *(106)*.

The cardiovascular effects of lignans are also worthy of mention. It has been known for some time that certain plants, such as Siberian ginseng (*Acanthopomax senticosus*), have an ability to sustain cardiovascular activity during prolonged exercise (see Ref. *(107)*), and this effect was recently attributed to (+)-syringaresinol di-O-β-D-glucoside **104** *(108)*.

Lastly, about 15 years ago, lignans have also been isolated from human and animals, i.e., they have been found in serum, urine and seminal fluids *(109-111)*. Enterolactone **112** and enterodiol **113** are the major mammalian lignans; it is believed they are derived from the metabolic action of bacterial flora on matairesinol **36** and secoisolariciresinol **76**, respectively *(112)*. They are excreted in large amounts in urine by persons having a diet rich in berries, grain, seeds, vegetables and whole grain products (e.g., rye products) *(113)*. They are thought to be a significant factor in reducing prostate and breast cancer risks by modulating the synthesis of sex hormones *(113)*.

Concluding Remarks

Lignans are a widespread class of phenylpropanoid metabolites, which according to our definition, can be linked via a large number of different coupling modes (e.g., 8,8'; 8,5'; etc.).

They are becoming an increasingly important class of compounds, for their physiological and pharmacological properties which are only now emerging. Based on existing chemotaxonomic data, various organisms ranging from algae, fungi, mosses, pteridophytes, gymnosperms and angiosperms (monocotyledons and dicotyledons) can be readily distinguished by differences in the degree/extent of phenylpropanoid metabolism. But there appears to be a significant progression in the (bio)chemical complexity of the lignans in the different superorders of the pteridophytes, gymnosperms and angiosperms, i.e., in terms of degree of oxidative substitution, structural variant, potential monomeric precursor, etc.. It is proposed that these differences (i.e., in terms of rapidly assaying for specific enzymes or genes encoding these enzymes) will be of considerable importance in cataloging the biodiversity of plant life, particularly these in the Amazonian rain forest which contains so many distinctive structural types.

Acknowledgments. The authors wish to thank the National Science Foundation (MCB 9219586) and the U.S. Department of Agriculture (91371036638) for financial assistance.

Literature Cited

1.   Lewis, N. G.; Davin, L. B. In *Isopentenoids and Other Natural Products: Evolution and Function*; Nes, W. D. Ed.; ACS Symposium Series: Washington, DC, 1994; Vol. 562; pp 202-246.
2.   Gottlieb, O. R. *Micromolecular Evolution, Systematics and Ecology*; Springer-Verlag: Berlin, Germany, 1982; pp 170.
3.   Ayres, D. C.; Loike, J. D. *Chemistry and Pharmacology of Natural Products. Lignans. Chemical, Biological and Clinical Properties*; Cambridge University Press: Cambridge, England, 1990; pp 402.
4.   Abe, F.; Yamauchi, T.; Wan, A. S. C. *Phytochemistry.* **1989**, *28*, 3473-3476.
5.   Abe, F.; Yamauchi, T.; Wan, A. S. C. *Phytochemistry.* **1988**, *27*, 3627-3631.
6.   Sakakibara, A.; Sasaya, T.; Miki, K.; Takahashi, H. *Holzforschung.* **1987**, *41*, 1-11.
7.   Gottlieb, O. R.; Yoshida, M. In *Natural Products of Woody Plants. Chemicals Extraneous to the Lignocellulosic Cell Wall*; Rowe, J. W., Kirk, C. H. Eds.; Springer Verlag: Berlin, 1989; pp 439-511.
8.   Davin, L. B.; Lewis, N. G. In *Rec. Adv. Phytochemistry*; Stafford, H. A., Ibrahim, R. K. Eds.; Plenum Press: New York, NY, 1992; Vol. 26; pp 325-375.
9.   Lewis, N. G.; Yamamoto, E. *Annu. Rev. Plant Physiol. Plant Mol. Biol.* **1990**, *41*, 455-496.

10.   Yamamoto, E.; Bokelman, G. H.; Lewis, N. G. In *Plant Cell Wall Polymers. Biogenesis and Biodegradation*; Lewis, N. G., Paice, M. G. Eds.; ACS Symposium Series: Washington, DC, 1989; Vol. 399; pp 68-88.

11.   Oliveto, E. P. *Chem. Ind.* **1972**, 677-679.

12.   Osawa, T.; Nagata, M.; Namiki, M.; Fukuda, Y. *Agric. Biol. Chem.* **1985**, *49*, 3351-3352.

13.   Fukuda, Y.; Osawa, T.; Namiki, M.; Ozaki, T. *Agric. Biol. Chem.* **1985**, *49*, 301-306.

14.   Binns, A. N.; Chen, R. H.; Wood, H. N.; Lynn, D. G. *Proc. Natl. Acad. Sci. USA.* **1987**, *84*, 980-984.

15.   Lynn, D. G.; Chen, R. H.; Manning, K. S.; Wood, H. N. *Proc. Natl. Acad. Sci. USA.* **1987**, *84*, 615-619.

16.   Gutterman, Y.; Evenari, M.; Cooper, R.; Levy, E. C.; Lavie, D. *Experientia.* **1980**, *36*, 662-663.

17.   Rahman, M. M. A.; Dewick, P. M.; Jackson, D. E.; Lucas, J. A. *Phytochemistry.* **1990**, *29*, 1971-1980.

18.   Markkanen, T.; Makinen, M. L.; Maunuksela, E.; Himanen, P. *Drugs Exptl. Clin. Res.* **1981**, *7*, 711-718.

19.   Nitao, J. K.; Nair, M. G.; Thorogood, D. L.; Johnson, K. S.; Scriber, J. M. *Phytochemistry.* **1991**, *30*, 2193-2195.

20.   Hattori, M.; Hada, S.; Watahiki, A.; Ihara, H.; Shu, Y.-Z.; Kakiuchi, N.; Mizuno, T.; Namba, T. *Chem. Pharm. Bull.* **1986**, *34*, 3885-3893.

21.   Schmelzer, E.; Jahnen, W.; Hahlbrock, K. *Proc. Natl. Acad. Sci. USA.* **1988**, *85*, 2989-2993.

22.   Long, S. *Cell.* **1989**, *56*, 203-214.

23.   Lamb, C. J.; Lawton, M. A.; Dron, M.; Dixon, R. A. *Cell.* **1989**, *56*, 215-224.

24.   Vogt, T.; Pollak, P.; Tarlyn, N.; Taylor, L. P. *Plant Cell.* **1994**, *6*, 11-23.

25.   Takeda, R.; Hasegawa, J.; Shinozaki, M. *Tetrahedron Lett.* **1990**, *31*, 4159-4162.

26.   Takeda, R.; Hasegawa, J.; Sinozaki, K. In *Bryophytes. Their Chemistry and Chemical Taxonomy*; Zinsmeister, H. D., Mues, R. Eds.; Oxford University Press: New York, NY, 1990; Vol. 29; pp 201-207.

27.   Taylor, T. N.; Taylor, E. L. *The Biology and Evolution of Fossil Plants*; Prentice Hall, Inc,: Englewood Cliffs, NJ, 1993; pp 982.

28.   Satake, T.; Murakami, T.; Saiki, Y.; Chen, C.-M. *Chem. Pharm. Bull.* **1978**, *26*, 1619-1622.

29.   Wada, H.; Kido, T.; Tanaka, N.; Murakami, T.; Saiki, Y.; Chen, C.-M. *Chem. Pharm. Bull.* **1992**, *40*, 2099-2101.

30.   Kariyone, T.; Kimura, H.; Nakamura, I. *Yakugaku Zasshi.* **1958**, *7*, 1152-1155.

31.   Cambie, R. C.; Parnell, J. C.; Rodrigo, R. *Tetrahedron Lett.* **1979**, *12*, 1085-1088.

32.   Cambie, R. C.; Clark, G. R.; Craw, P. A.; Jones, T. C.; Rutledge, P. S.; Woodgate, P. D. *Aust. J. Chem.* **1985**, *38*, 1631-1645.

33.   Briggs, L. H.; Cambie, R. C.; Hoare, J. L. *Tetrahedron.* **1959**, *7*, 262-269.

34.   Lee, C. L.; Hirose, Y.; Nakatsuka, T. *Mokuzai Gakkaishi.* **1974**, *20*, 558-563.

35. MacLean, H.; MacDonald, B. F. *Can. J. Chem.* **1967**, *45*, 739-740.
36. Popoff, T.; Theander, O.; Johansson, M. *Physiol. Plant.* **1975**, *34*, 347-356.
37. Weinges, K. *Tetrahedron Lett.* **1960**, *20*, 1-2.
38. Rao, C. B. S. *Chemistry of Lignans*; Andhra University Press: Andhra Pradesh, India, 1978; pp 377.
39. Mujumdar, R. B.; Srinivasan, R.; Venkataraman, K. *Indian J. Chem.* **1972**, *10*, 677-680.
40. Fonseca, S. F.; Nielsen, L. T.; Rúveda, E. A. *Phytochemistry.* **1979**, *18*, 1703-1708.
41. Dahlgren, R. M. T. *Bot. J. Linn. Soc.* **1980**, *80*, 91-124.
42. Dahlgren, R. *Nord. J. Bot.* **1983**, *3*, 119-149.
43. Kupchan, S. M.; Britton, R. W.; Ziegler, M. F.; Gilmore, C. J.; Restivo, R. J.; Bryan, R. F. *J. Amer. Chem. Soc.* **1973**, *95*, 1335-1336.
44. De Diaz, A. M. P.; Gottlieb, H. E.; Gottlieb, O. R. *Phytochemistry.* **1980**, *19*, 681-682.
45. Nakajima, K.; Taguchi, H.; Ikeya, Y.; Endo, T.; Yosioka, I. *Yakugaku Zasshi.* **1983**, *103*, 743-749.
46. Urzúa, A.; Freyer, A. J.; Shamma, M. *Phytochemistry.* **1987**, *26*, 1509-1511.
47. Lopes, L. M. X.; Yoshida, M.; Gottlieb, O. R. *Phytochemistry.* **1984**, *23*, 2647-2652.
48. Lopes, L. M. X.; Yoshida, M.; Gottlieb, O. R. *Phytochemistry.* **1984**, *23*, 2021-2024.
49. Gottlieb, O. R. *Progr. Chem. Org. Nat. Prod.* **1978**, *35*, 1-72.
50. Kupchan, S. M.; Stevens, K. L.; Rohlfing, E. A.; Sickles, B. R.; Sneden, A. T.; Miller, R. W.; Bryan, R. F. *J. Org. Chem.* **1978**, *43*, 586-590.
51. Barata, L. E. S.; Baker, P. M.; Gottlieb, O. R.; Rùveda, E. A. *Phytochemistry.* **1978**, *17*, 783-786.
52. Holloway, D. M.; Scheinmann, F. *Phytochemistry.* **1973**, *12*, 1503-1505.
53. El-Feraly, F. S.; Cheatham, S. F.; Breedlove, R. L. *J. Nat. Prod.* **1983**, *46*, 493-498.
54. Diaz, D. P. P.; Yoshida, M.; Gottlieb, O. R. *Phytochemistry.* **1980**, *19*, 285-288.
55. Whiting, D. A. *Nat. Prod. Rep.* **1985**, *2*, 191-211.
56. Lopes, M. N.; da Silva, M. S.; Barbosa-Filho, J. M.; Ferreira, Z. S.; Yoshida, M.; Gottlieb, O. R. *Phytochemistry.* **1986**, *25*, 2609-2612.
57. Hayashi, T.; Thomson, R. H. *Phytochemistry.* **1975**, *14*, 1085-1087.
58. Achenbach, H.; Groß, J.; Dominguez, X. A.; Cano, G.; Star, J. V.; Brussolo, L. D. C.; Muñoz, G.; Salgado, F.; López, L. *Phytochemistry.* **1987**, *26*, 1159-1166.
59. Patra, A.; Mitra, A. K. *Indian J. Chem.* **1979**, *17B*, 412-414.
60. Hartley, R. D.; Ford, C. W. In *Plant Cell Wall Polymers. Biogenesis and Biodegradation*; Lewis, N. G., Paice, M. G. Eds.; ACS Symposium Series: Washington, DC, 1989; Vol. 399; pp 137-145.
61. Ford, C. W.; Hartley, R. D. *J. Sci. Food Agric.* **1990**, *50*, 29-43.
62. Whiting, D. A. *Nat. Prod. Rep.* **1987**, *4*, 499-525.
63. Yamamura, S.; Niwa, M.; Nonoyama, M.; Terada, Y. *Tetrahedron Lett.* **1978**, *49*, 4891-4894.

64.   Yamamura, S.; Niwa, M.; Terada, Y.; Nonoyama, M. *Bull. Chem. Soc. Jpn.* **1982**, *55*, 3573-3579.
65.   Ferri, P. H.; Barata, L. E. S. *Phytochemistry.* **1991**, *30*, 4204-4205.
66.   Kato, M. J.; Yoshida, M.; Gottlieb, O. R. *Phytochemistry.* **1992**, *31*, 283-287.
67.   Cavalcante, S. H.; Yoshida, M.; Gottlieb, O. R. *Phytochemistry.* **1985**, *24*, 1051-1055.
68.   Kawanishi, K.; Uhara, Y.; Hashimoto, Y. *Phytochemistry.* **1983**, *22*, 2277-2280.
69.   Kawanishi, K.; Uhara, Y.; Hashimoto, Y. *Phytochemistry.* **1982**, *21*, 2725-2728.
70.   MacRae, W. D.; Towers, G. H. N. *Phytochemistry.* **1985**, *24*, 561-566.
71.   Yamaguchi, H.; Nakatsubo, F.; Katsura, Y.; Murakami, K. *Holzforschung.* **1990**, *44*, 381-385.
72.   Umezawa, T.; Davin, L. B.; Yamamoto, E.; Kingston, D. G. I.; Lewis, N. G. *J. Chem. Soc., Chem. Commun.* **1990**, 1405-1408.
73.   Davin, L. B.; Bedgar, D. L.; Katayama, T.; Lewis, N. G. *Phytochemistry.* **1992**, *31*, 3869-3874.
74.   Paré, P. W.; Wang, H. B.; Davin, L. B.; Lewis, N. G. *Tetrahedron Lett.* **1994**, *35*, 4731-4734.
75.   Katayama, T.; Davin, L. B.; Lewis, N. G. *Phytochemistry.* **1992**, *31*, 3875-3881.
76.   Lewis, N. G.; Davin, L. B.; Katayama, T.; Bedgar, D. L. *Bull. Soc. Groupe Polyphénols.* **1992**, *16*, 98-103.
77.   Katayama, T.; Davin, L. B.; Chu, A.; Lewis, N. G. *Phytochemistry.* **1993**, *32*, 591-591.
78.   Chu, A.; Dinkova, A.; Davin, L. B.; Bedgar, D. L.; Lewis, N. G. *J. Biol. Chem.* **1993**, *268*, 27026-27033.
79.   Umezawa, T.; Davin, L. B.; Lewis, N. G. *J. Biol. Chem.* **1991**, *266*, 10210-10217.
80.   Umezawa, T.; Davin, L. B.; Lewis, N. G. *Biochem. Biophys. Res. Commun.* **1990**, *171*, 1008-1014.
81.   Ozawa, S.; Davin, L. B.; Lewis, N. G. *Phytochemistry.* **1993**, *32*, 643-652.
82.   Ito, K.; Iida, T.; Ichino, T.; Tsunezuka, M.; Hattori, M.; Namba, T. *Chem. Pharm. Bull.* **1982**, *30*, 3347-3353.
83.   Taniguchi, E.; Imamura, K.; Ishibashi, F.; Matsui, T.; Nishio, A. *Agric. Biol. Chem.* **1989**, *53*, 631-643.
84.   Haller, H. L.; McGovran, E. R.; Goodhue, L. D.; Sullivan, W. N. *J. Org. Chem.* **1942**, *7*, 183-184.
85.   Haller, H. L.; LaForge, F. B.; Sullivan, W. N. *J. Org. Chem.* **1942**, *7*, 185-188.
86.   Miyazawa, M.; Ishikawa, Y.; Kasahara, H.; Yamanaka, J.-I.; Kameoka, H. *Phytochemistry.* **1994**, *35*, 611-613.
87.   Elakovich, S. D.; Stevens, K. L. *J. Chem. Ecol.* **1985**, *11*, 27-33.
88.   Lavie, D.; Levy, E. C.; Cohen, A.; Evenari, M.; Guttermann, Y. *Nature.* **1974**, *249*, 388-390.
89.   Orr, J. D.; Lynn, D. G. *Plant Physiol.* **1992**, *98*, 343-352.
90.   Kelly, M. G.; Hartwell, J. L. *J. Nat. Cancer Inst.* **1954**, *14*, 967-1010.

91. Ross, W.; Rowe, T.; Glisson, B.; Yalowich, J.; Liu, L. *Cancer Res.* **1984**, *44*, 5857-5860.
92. Torrance, S. J.; Hoffmann, J. J.; Cole, J. R. *J. Pharm. Sci.* **1979**, *68*, 664-665.
93. Cole, J. R.; Bianchi, E.; Trumbull, E. R. *J. Pharm. Sci.* **1969**, *58*, 175-176.
94. Schröder, H. C.; Merz, H.; Steffen, R.; Müller, W. E. G.; Sarin, P. S.; Trumm, S.; Schulz, J.; Eich, E. *Z. Naturforsch.* **1990**, *45c*, 1215-1221.
95. Rimando, A. M.; Pezzuto, J. M.; Farnsworth, N. R.; Santisuk, T.; Reutrakul, T.; Kawanishi, K. *J. Nat. Prod.* **1994**, *57*, 896-904.
96. Shen, T. Y.; Hwang, S.-B.; Chang, M. N.; Doebber, T. W.; Lam, M.-H. T.; Wu, M. S.; Wang, X.; Han, G. Q.; Li, R. Z. *Proc. Natl. Acad. Sci., USA.* **1985**, *82*, 672-676.
97. Braquet, P.; Godfroid, J. J. *Trends in Pharm. Sci.* **1986**, 397-403.
98. Pan, J.-X.; Hensens, O. D.; Zink, D. L.; Chang, M. N.; Hwang, S.-B. *Phytochemistry.* **1987**, *26*, 1377-1379.
99. Nikaido, T.; Ohmoto, T.; Kinoshita, T.; Sankawa, U.; Nishibe, S.; Hisada, S. *Chem. Pharm. Bull.* **1981**, *29*, 3586-3592.
100. Nikaido, T.; Ohmoto, T.; Noguchi, H.; Kinoshita, T.; Saitoh, H.; Sankawa, U. *Planta Med.* **1981**, *43*, 18-23.
101. Deyama, T.; Nishibe, S.; Kitagawa, S.; Ogihara, Y.; Takeda, T.; Ohmoto, T.; Nikaido, T.; Sankawa, U. *Chem. Pharm. Bull.* **1988**, *36*, 435-439.
102. Ghosal, S.; Banerjee, S.; Frahm, A. W. *Chem. Ind.* **1979**, 854-855.
103. Kadota, S.; Tsubono, K.; Makino, K.; Takeshita, M.; Kikuchi, T. *Tetrahedron Lett.* **1987**, *28*, 2857-2860.
104. Takeda, S.; Arai, I.; Kase, Y.; Ohkura, Y.; Hasegawa, M.; Sekiguchi, Y.; Sudo, K.; Aburada, M.; Hosoya, E. *Yakugaku Zasshi.* **1987**, *107*, 517-524.
105. Nomura, M.; Nakachiyama, M.; Hida, T.; Ohtaki, Y.; Sudo, K.; Aizawa, T.; Aburada, M.; Miyamoto, K.-I. *Cancer Lett.* **1994**, *76*, 11-18.
106. Ikeya, Y.; Taguchi, H.; Mitsuhashi, H.; Takeda, S.; Kase, Y.; Aburada, M. *Phytochemistry.* **1988**, *27*, 569-573.
107. Farnsworth, N. R.; Kinghorn, A. D.; Soejarto, D. D.; Waller, D. P. In *Economic and Medicinal Plant Research*; Wagner, H., Hikino, H., Farnsworth, N. R. Eds.; Academic Press: London, 1985; Vol. 1; pp 155-215.
108. Nishibe, S.; Kinoshita, H.; Takeda, H.; Okano, G. *Chem. Pharm. Bull.* **1990**, *38*, 1763-1765.
109. Axelson, M.; Setchell, K. D. R. *FEBS Let.* **1981**, *123*, 337-342.
110. Setchell, K. D. R.; Lawson, A. M.; Mitchell, F. L.; Adlercreutz, H.; Kirk, D. N.; Axelson, M. *Nature.* **1980**, *287*, 740-742.
111. Setchell, K. D. R.; Lawson, A. M.; Conway, E.; Taylor, N. F.; Kirk, D. N.; Cooley, G.; Farrant, R. D.; Wynn, S.; Axelson, M. *Biochem. J.* **1981**, *197*, 447-458.
112. Borriello, S. P.; Setchell, K. D. R.; Axelson, M.; Lawson, A. M. *J. Appl. Bacteriol.* **1985**, *58*, 37-43.
113. Adlercreutz, H. In *Nutrition, Toxicity, and Cancer*; Rowland, I. R. Ed.; CRC Press: Boca Raton, Fl, 1991; pp 137-195.

Appendix

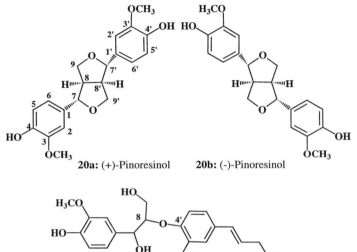

20a: (+)-Pinoresinol   **20b:** (-)-Pinoresinol

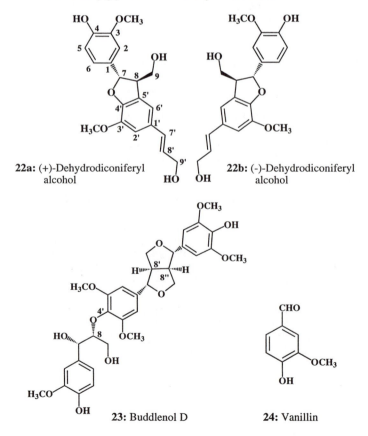

21: Guaiacylglycerol- 8-*O*- coniferyl alcohol ether

22a: (+)-Dehydrodiconiferyl alcohol   **22b:** (-)-Dehydrodiconiferyl alcohol

23: Buddlenol D   **24:** Vanillin

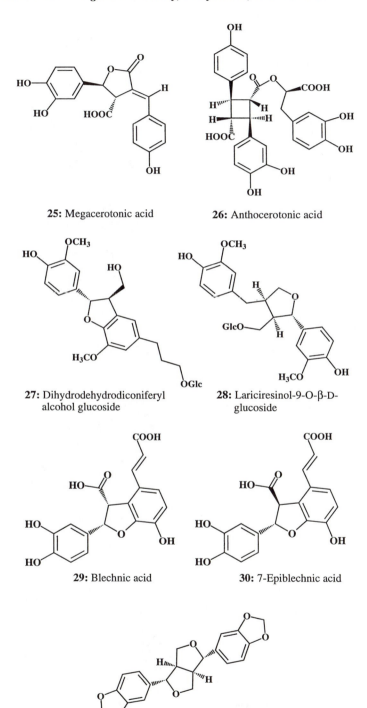

25: Megacerotonic acid

26: Anthocerotonic acid

27: Dihydrodehydrodiconiferyl
alcohol glucoside

28: Lariciresinol-9-O-β-D-
glucoside

29: Blechnic acid

30: 7-Epiblechnic acid

31: (+)-Sesamin

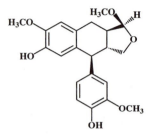

32: (+)-*O*-Methyl-α-conidendral

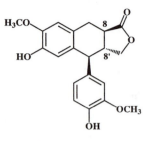

33: (-)-α-Conidendrin

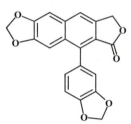

34: Taiwanin C

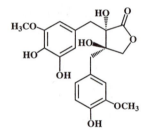

35: Dihydroxythujaplicatin

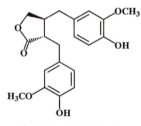

36a: (+)-Matairesinol

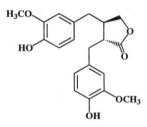

36b: (-)-Matairesinol

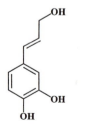

37: 3,4-dihydroxycinnamyl
alcohol

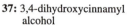

38: Taxiresinol

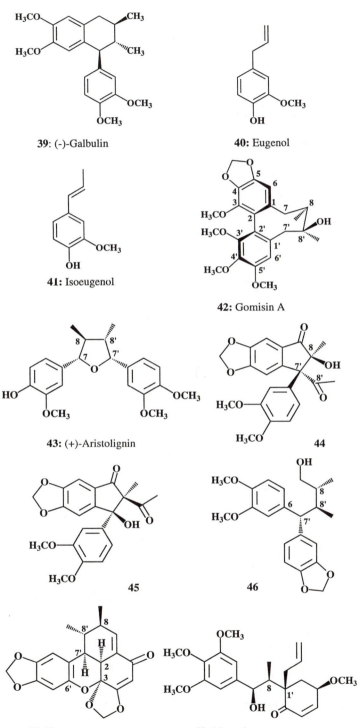

39: (-)-Galbulin

40: Eugenol

41: Isoeugenol

42: Gomisin A

43: (+)-Aristolignin

44

45

46

47: Carpanone

48: Megaphone

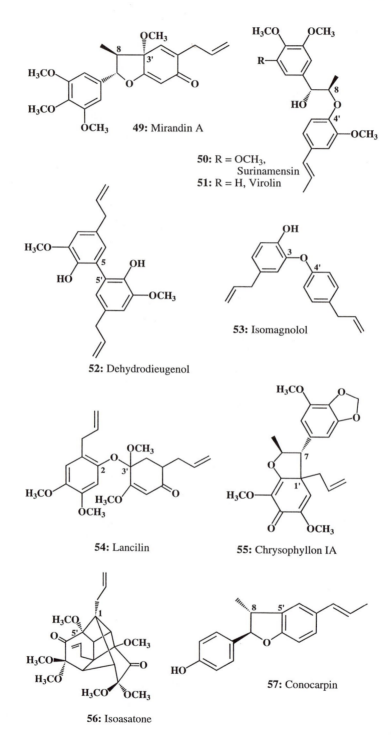

49: Mirandin A

50: R = OCH$_3$,
Surinamensin
51: R = H, Virolin

52: Dehydrodieugenol

53: Isomagnolol

54: Lancilin

55: Chrysophyllon IA

56: Isoasatone

57: Conocarpin

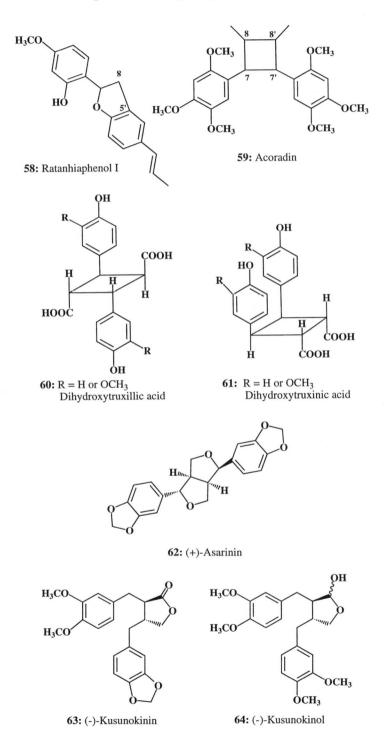

58: Ratanhiaphenol I

59: Acoradin

60: R = H or OCH$_3$
Dihydroxytruxillic acid

61: R = H or OCH$_3$
Dihydroxytruxinic acid

62: (+)-Asarinin

63: (-)-Kusunokinin

64: (-)-Kusunokinol

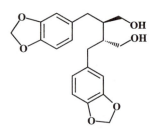

65: (-)-Dihydrocubebin

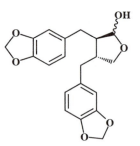

66: (-)-Cubebin

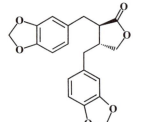

67: (-)-Hinokinin

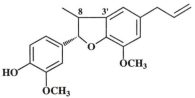

68: Dihydrocarinatidin

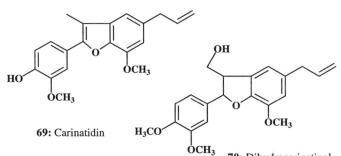

69: Carinatidin

70: Dihydrocarinatinol

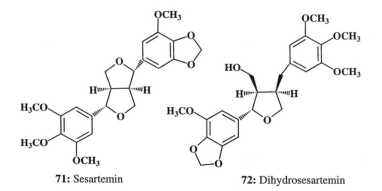

71: Sesartemin

72: Dihydrosesartemin

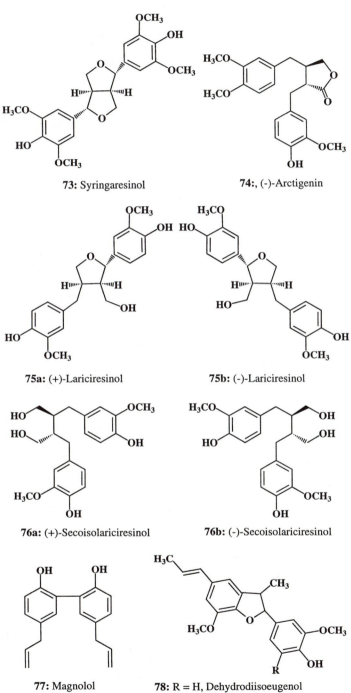

73: Syringaresinol

74:, (-)-Arctigenin

75a: (+)-Lariciresinol

75b: (-)-Lariciresinol

76a: (+)-Secoisolariciresinol

76b: (-)-Secoisolariciresinol

77: Magnolol

78: R = H, Dehydrodiisoeugenol
79: R = OCH₃, 5'-methoxy-
dehydrodiisoeugenol

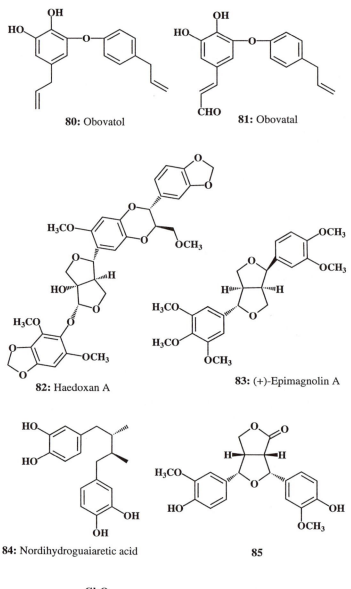

80: Obovatol

81: Obovatal

82: Haedoxan A

83: (+)-Epimagnolin A

84: Nordihydroguaiaretic acid

85

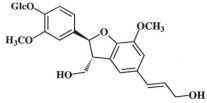

86: Dehydrodiconiferyl alcohol glucoside

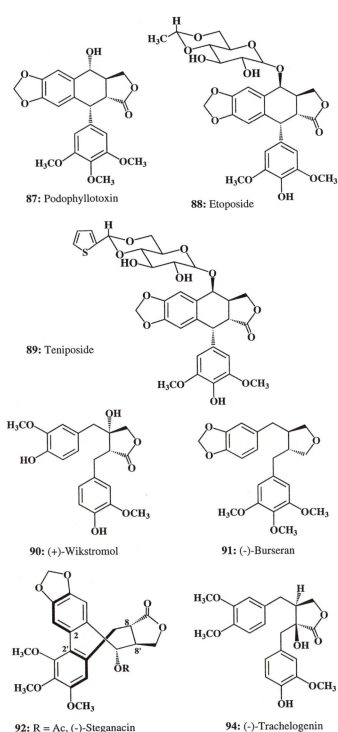

87: Podophyllotoxin

88: Etoposide

89: Teniposide

90: (+)-Wikstromol

91: (-)-Burseran

92: R = Ac, (-)-Steganacin
93: R = Angeloyl, (-)-Steganangin

94: (-)-Trachelogenin

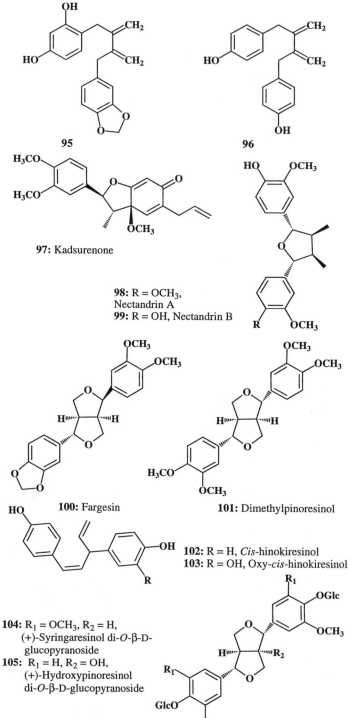

95

96

97: Kadsurenone

98: R = OCH₃,
Nectandrin A
99: R = OH, Nectandrin B

100: Fargesin

101: Dimethylpinoresinol

102: R = H, *Cis*-hinokiresinol
103: R = OH, Oxy-*cis*-hinokiresinol

104: R₁ = OCH₃, R₂ = H,
(+)-Syringaresinol di-*O*-β-D-
glucopyranoside
105: R₁ = H, R₂ = OH,
(+)-Hydroxypinoresinol
di-*O*-β-D-glucopyranoside

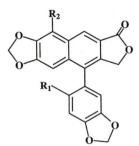

106: R_1 = OH, R_2 = OCH$_3$, Prostalidin A
107: R_1 = R_2 = OCH$_3$, Prostalidin B
108: R_1 = OH, R_2 = H, Prostalidin C

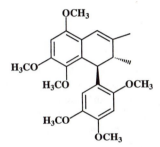

109:, Magnoshinin

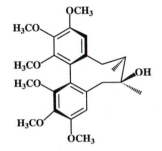

110: Schizandrin

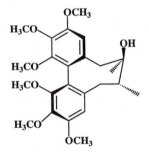

111: Isoschizandrin

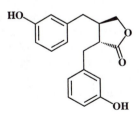

112 : Enterolactone

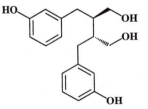

113 : Enterodiol

RECEIVED December 19, 1994

Chapter 14

The Chemistry of Amazonian Myristicaceae
Developmental, Ecological, and Pharmacological Aspects

Massuo J. Kato

Instituto de Quimica, Universidade de São Paulo, Caixa Postal 01498, 20718–970 São Paulo, SP, Brazil

Earlier phytochemical investigations on Amazonian Myristicaceae species were carried out based on ethnopharmacological reports among Indians. A remarkable variation in compositions on reproductive tissue such as flowers, different parts of the fruits and seedlings have been observed in *Virola* species. Different lignans and neolignans occur in high concentration in pericarps, arils, seed coats and seeds. Consideration of the different allelopathic activity of the respective extracts suggests some of the possible ecological roles. Other dynamic aspects include displacement of flavones by lignans during the ripening process and variability of chemical profiles on seedlings. Work of such aspects of reproductive phases reveals a still unexploited source of new bioactive molecules from tropical species.

Species of the Myristicaceae family are widely distributed in the Amazon region, occurring in "terra firme" and inundated forests. Popularly known as "virola" their wood has been extensively exploited in carpentry and cellulose manufacture (*1*). The Indian name "ucuúba" means fat producing tree, suggesting one of the most important features provided by their seeds.

Ethnopharmacological investigations during the sixties among Indian tribes from northwest of Amazon indicated that several *Virola* species are included in their culture (*2, 3*). They prepare hallucinogenic snuffs and arrow poison from resins and in addition, use several products prepared from leaves, seeds, and resins to treat many diseases (*4*). These aspects triggered a series of phytochemical reports on Myristicaceae describing a number of compounds linked to the phenylpropanoid pathway. Nevertheless, few correlations between presence of secondary compounds and different pharmacological activities have been achieved (*4*).

0097–6156/95/0588–0168$12.00/0

The main objective of this chapter is to highlight new aspects on chemical composition of Myristicaceae species jointly with recent data on plant development and ecology.

Plant Development and Ecology

Fruit chemistry. Most of the knowledge available on chemical composition of tropical plants results from investigations on economically or medicinally important species, generally carried out on adult specimens (*5*, *6*). In the case of Myristicaceae, the earlier phytochemical examinations were addressed to species involved in ethnopharmacological uses (*4*, *7*, *8*). In addition to tryptamine and β-carboline alkaloids, isolated from resins as active principles of hallucinogenic snuffs and arrow poisons (*2*, *9*, *10*), other classes of secondary metabolites, isolated from several parts of the Myristicaceae species, include diarylpropanoids (*11-13*), stilbenes (*8*, *14*), lignans, neolignans (*14-25*), arylalkanones (*25*, *26*), tocotrienols (*27*), and γ-lactones (*28*).

Myristicaceae trees reach their adult stage in approximately 10 years, then starting to produce fruits normally during the rainy season. The red arils are ingested by birds, thus effecting seed dispersal. The limited number of predators is supposedly due to the presence of toxins (*29*).

A summary of the phytochemical investigations on fruits of some Myristicaceae species that have been described in the last couple of decades is represented in Figure 1. The widely distributed furofuran (**2.1**) and dibenzylbutyrolactone (**7.1**) lignans accumulate in the three groups of *Virola* species. In groups A and B these types occur in whole fruit (*30*, *31*). In group C they occur only in pericarps and arils but not in kernels in which the pathways producing aryltetralins (**4.1**, **4.2**) and aryltetralones (**5.1-5.6**) and other related compounds are dominant (*18-21*, *25*). It is assumed that chemical homogeneity and simplicity in several parts of fruits, as observed in species in groups A and B, represents the primitive pattern. On the other hand, differentiation in fruit tissue, as observed in group C, should represent an advanced character. An outstanding point is the occurrence of trioxygenated aromatic rings observed in the dibenzylbutirolactone (**6**), lariciresinol (**2.2**) and secoisolariciresinol (**2.3**) types, and also in aryltetralone (**5.3**, R=OH) neolignans and arylalkanones (**8.2**, **9.3**, R=OH, OMe) from the seed coat of *V. elongata* (*25*) in contrast to analogous compounds isolated from other tissues or species.

The seed coat of this species exhibits the highest chemical diversity, including lignans of the magnostelin type (**3**). The aryltetralone lignans bearing methylenedioxyphenyl groups (**5.2**, **5.3**) are the major constituents in the seeds of species belonging to group C, where they occur in higher concentrations than in the bark (*23*, *24*). Finally arylalkanones (**8.2**, **9.2**, **9.3**) have been isolated from all parts of fruits. Considering the diversity of structures in several fruit parts and also the higher oxidation pattern of the lignoids accumulating in several species, the evolutionary sequence indicated in Figure 1 is suggested. This scheme is consistent

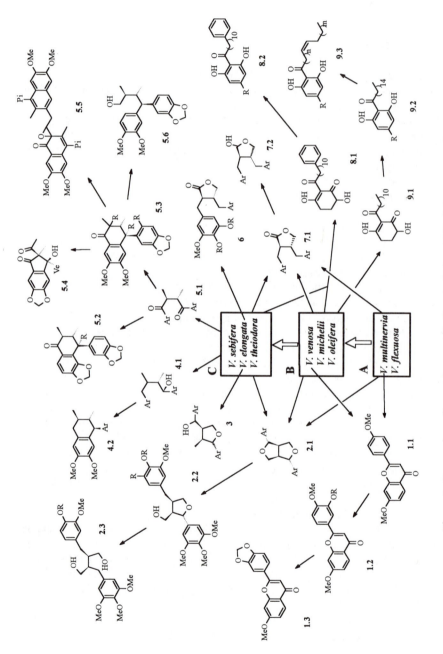

Figure 1: Expression of secondary metabolism in fruits of Myristicaceae species.

with the accumulation of aryltetralins in the seed of *Osteophloeum platyspermum* (*32*), the most advanced genus in the family (*33*). *V. pavonis* accumulates, in addition to furofuran, dibenzylbutyrolactone, and dibenzyl-butyrolactol lignans, benzodioxane type neolignans in pericarps and kernels (*30*).

The examination of *V. venosa* during the development process revealed the occurrence of flavones (**1.1-1.3**) in flowers and in unripe fruits development, followed by substituted later in by lignans, especially dibenzylbutyrolactones (**7.1**) and dibenzylbutyrolactols (**7.2**) (*31*).

All the data obtained so far on the chemistry of Myristicaceae fruits support the prediction that the larger the seeds the higher their content of secondary metabolites (*34*). While this is certainly true for the family as a whole, no such positive correlation between the size of seeds and metabolic content was observed comparing *O. platyspermum* and *Virola* or *Iryanthera* seeds by TLC analysis and ^{1}H-NMR spectra. Several arguments support a defensive role for metabolites isolated from different part of the fruits (*35, 36*). It has been argued that the presence of methylenedioxy functionalities in neolignans causes toxicity of the seeds of *Ocotea veraguensis* against the spiny pocket mouse (*37*). Indeed, compounds containing the methylenedioxyphenyl group are in commercial use due to their powerful synergistic action on pyrethrins (*38*). It has been claimed that seeds of *V. surinamensis* contain an extraordinarily high concentration of soluble tannins, probably with a defensive function (*39*). Yet speculatively, secondary compounds in seeds are considered to be antifeedants (*40*, 41); synergists (*38, 42, 43*); insecticides (*44*); fungicides (*45*); phytoalexins (*46-51*); stress compounds (*52*); microbicides and fungicides (*53*). All these informations are suggestive of the biological importance of lignoids in fruits of Myristicaceae species, but requires further work on the reproductive tissue as a target of phytochemical and biological investigations.

Seed dispersal and allelopathy.   Ecological investigations on the dispersal systems of *Virola surinamensis* have revealed additionally a strong dependence of seed germination on removal of aril by birds or by handling (*29, 39*). The present work confirmed the inhibitory effects of seed germination by the presence of aril of two *Virola* species and also revealed the allelopathic effects on lettuce seeds (*Lactuca sativa* var-grand rapids) by crude extracts of pericarps, arils and seeds of eight Myristicaceae species belonging to three genera. The fruits of the following Myristicaceae species were collected during April to August (1989): *Iryanthera lancifolia, I. paraensis, Osteophloeum platyspermun, Virola theiodora* and *V. venosa* (Manaus, AM); *V. surinamensis* and *V. michelli* (Belém, PA), *V. oleifera* ("Mata Atlântica", SP), *V. sebifera* (Araraquara, SP). The fruits were separated in pericarps, arils, seed coats and kernels, dried, milled and extracted with dichloromethane. Parts of fruits of *V. venosa* were sequentially extracted with methanol. The allelopathic assay was based on effect on lettuce seeds exactly as described (*54*). In case of *O. platyspermum* leachings from fifteen seeds soaked in 50 ml of distilled water during 4 h were used. For *I. lancifolia* only pericarp provided enough material to be evaluated. The results of the assay for all extracts, including methanolic extracts from *V. venosa*, and extracts from whole fruits in early stage of development (WF) are represented in Figure 2.

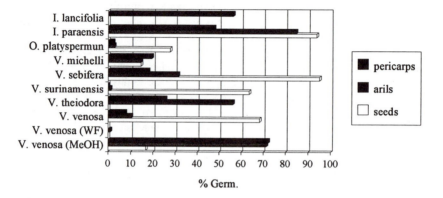

Figure 2: Effect of crude extracts of different part of fruits of Myristicaceae species on lettuce seed germination.

As a general trend higher inhibitory effects were observed for extracts of pericarps and arils than of kernels. Opposite results were obtained with methanolic extracts from *V. venosa*, indicating that apolar compounds inhibit the lettuce seed germinations more effectively. Additionaly, a strong allelopathic effect was observed in the soil underneath *Virola* trees against lettuce seeds (*55*). However, considering the absence of detailed chemical investigations on soil samples, it is difficult to ascertain whether the net effect was due to constituents of pericarps accumulated in the soil. The evidence for allelopathic effect of lignans is only circumstantial considering cases in which inhibitory effects have been detected (*56-58*). These aspects of tropical biology need further investigation in order to confirm the presence of allelopathic compounds in fruits of Myristicaceae and other tropical families.

Seed germinations.   The germination behavior of Myristicaceae species was evaluated on seeds of *V. venosa*, *V. sebifera* and *O. platyspermum* (Figure 3). It was observed that *O. platyspermum* seeds showed the lowest percent of germination, unchanged after scarification procedures. The seeds of the Amazonian species, *V. venosa*, showed a proportion of germination near 76%, with initial emergence of radicle occurring after the first week. Germination experiments with *V. sebifera* seeds were carried out under several conditions: with and without arils, and with different levels of dehydrations (Figure 4). A period of dormancy or quiescency of about five weeks was observed. The presence of arils inhibited almost totally the germination of *V. sebifera* seeds, similarly to *V. surinamensis* in which no germination took place at all in the same condition (*39*). The dehydration treatment caused not only a faster rate in the germination process but also increased the proportion of germinations. This aspect can be associated with the small size and high seed productivity (*59*). The abnormal richness of viable seeds in young plants (Kato, M. J., Universidade de São Paulo, unpublished data) may explain the wide occurrence of *V. sebifera* in South America.

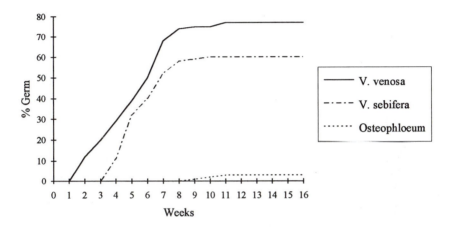

Figure 3. Germinations of Some Myristicaceae Seeds.

Seedlings. The ecology of seeds and seedlings in tropical forests has gained importance in the understanding of forest structure, and in the management and conservation of biodiversity (*60-62*).

The processes of seed dispersal, germination and seedling establishment are crucial to plant reproduction and depend on secondary metabolites. It has been assumed that the higher content of secondary metabolites in large seeds can protect also the seedlings through translocations (*34*). In case of *V. surinamensis* seedling survival depends on the distance from the parent tree (*63*). The seedling predations become lower as distance from the parent tree increases. The adaptation of some weevils causes disproportionate seed and seedling mortality under fruiting trees.

A detailed chemical investigation carried out on each part of *V. venosa* seedlings (*31*) showed that two main lignans, (-)-cubebin and (-)-dihydrokusunokinin (**7.2**), accumulated in mature seeds were not detected in any part of seedlings. On the other hand, the seedling leaves accumulate flavones (**1.1-1.3**) in higher concentrations than those detected in flowers, unripe pericarps or in the leaves from adult plants. The arylalkanone (**8.1**) is the only compound that apparently has been translocated to the whole seedling, but there is no conclusive evidence favoring *de novo* synthesis or translocation. The lignan (+)-sesamin,

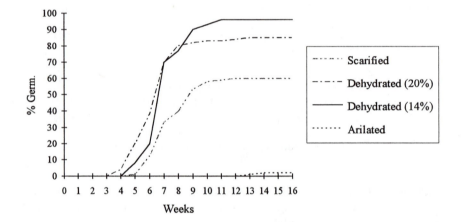

Figure 4: Germinations of *Virola sebifera* seeds under several treatments

isolated as minor compound from the seeds, was detected as the major compound in the roots. As the chemical ecology of this interaction remains unknown it is interesting to note that similar polyketides (**8.1, 9.1**), isolated from *V. elongata* and *V. venosa* seeds, were described to have kairomonal activity in an insect parasite (*64*).

A totallly different chemical composition of adult leaves and seedlings has been observed in *V. surinamensis*. While several neolignans have been isolated from the adult leaves (*65*), γ-lactones were detected as the main metabolites in leaves of germinated seedlings or in micropropagated ones (*66*). Phytochemical investigations on seedlings of some species so far revealed a profile totally different from adult plants with a variety of metabolic events occurring during the reproductive phases. At least in case of *Virola* systems it seems that the translocation (or *de novo* synthesis) is very specific and may include other physiological purposes in addition to defense.

Pharmacological. The ethnopharmacological investigations on hallucinogenic snuffs and arrow poisons from Myristicaceae species have also resulted in the description of lignans as monoamino-oxidase inhibitors (*67*) and as the reducer of induced agression when administered to mice (*7*). Lignans and neolignans represent the most promising classes of secondary compounds described from Myristicaceae, considering the variety of different biological activities exhibited (*68-70*). They have been mentioned as anti-PAF (*71, 72*), antimicrobial (*73*) and antioxidant agents (*74, 75*). Podophyllotoxin, a well known bioactive lignan, was isolated as the main active principle of *Podophyllum peltatum* L. and *P. emodi* Wallich. Its ethylidene (etoposide) and thenylidene (teniposide) derivatives were developed as anticancer drugs and are still clinically in use (*76*).

The possibilities to interconvert some classes of lignans have been exploited in the cases of dibenzylbutyrolactones yielding dibenzocyclootadiene lignans (*77*) or in case of (-)-dihydrocubebin convertible to different lignans (*78*). Interesting possibilities to obtain furofuran lignans with highly oxygenated aromatic rings were demonstrated in the regioselective demethylation at 4-positions followed by additional hydroxylations at 5-positions by *Aspergillus niger* (*79*).

The antioxidant activity, described for lignans such as otobain bearing methylenedioxy groups (*80*), have been exploited in detail only recently. Shimizu *et al.* (*81*, 82) have shown inhibition of Δ5-desaturase by sesamin and similar lignans in microbes and in rat liver microsomes. The antioxidative activity of schizandrin type lignans in aged and ischemic rat brain is due to methylenedioxy and phenolic functionalities (*83*). Ortho-methoxy-phenol systems in lignans proved to be effective as chain-breaking antioxidants (*84*). The inhibition of microsomal lipid peroxidation by podophyllotoxin derivative VP-16, also requires the presence of a free OH at 4'-position (*85*). It also has been demonstrated that sesamin inhibits cholesterol synthesis and absorption in rats (*86*) and that such compounds act as synergists of γ-tocopherol to produce vitamin E activity in rats (*87*).

Many lignans and neolignans isolated from Myristicaceae fruits contain the methylenedioxy group, essential in inseticide synergism, antioxidant and antitumoral activities.

Conclusions

The potential of higher plants as sources for new drugs is still largely unexplored. Only a small percentage has been investigated phytochemically and the fraction submitted to biological screening is even smaller. The ethnopharmacological report on indigenous uses of Myristicaceae species probably represents one of the last discoveries from primitive cultures considering their fast aculturation process. Indeed as it seems that only a minute percentage has been selected for medicinal uses and almost all plant species in tropical regions remain untouched by researchers (88), the modern society cannot depend only on ethnopharmacological approaches to supply the demand for new lead compounds.

The investigation of dynamic aspects of secondary metabolism at reproductive phases in Myristicaceae and other tropical families is important to establish a rational approach in the search for biologically active compounds and also to help understanding biodiversity in the tropics.

Acknowledgments

The author is grateful to Dr. Otto R. Gottlieb for his kind revision of this manuscript. The investigation was supported by Fundação de Amparo à Pesquisa do Estado de São Paulo (FAPESP, 301520/84-4) and Conselho Nacional de Desenvolvimento Científico e Tecnológico (CNPq).

Literature Cited

1.   Rodrigues, W.A. *Acta Amazonica* **1980**, *10*, 1.
2.   Schultes, R. E.; Holmstedt, B. *Rhodora* **1968,** *70*, 113.
3.   Schultes, R. E. *Bot. Mus. Leaflets* **1969**, *22*, 229.
4.   Gottlieb, O.R. *J. Ethnopharm.* **1979**, *1*, 309.
5.   Gottlieb, O.R.; Mors, W. B. *J. Agric. Food Chem.* **1980**, *28*, 196.
6.   Downum, K., Romeo, J. T.; Stafford, H. A., Eds.; *Phytochemical Potential of Tropical Plants*; Rec. Adv. Phytochem.; Plenun Press: New York, NY, **1993**, 27, 299.
7.   McRae, W.D.; Towers, G.H.N. *J. Ethnopharm.* **1984**, *12*, 75.
8.   McRae, W.D.; Towers, G.H.N. *Phytochemistry* **1985**, *24*, 561.
9.   Agurell, S.; Holmstedt, B.; Lindgren, J.E.; Schultes, R.E. *Biochem. Pharmac. 1* **1968**, *17*, 2478.
10.   Agurell, S.; Holmstedt, B.; Lindgren, J.E.; Schultes, R.E. *Acta Chem. Scand.* **1969,** *23*, 903.

11.   Gottlieb, O. R.; Loureiro, A. A.; Carneiro, M. S.; Rocha, A. I. *Phytochemistry* **1973**, *12*, 1830.

12.   Kijjoa, A.; Giesbrecht, A.M.; Gottlieb, O.R.; Gottlieb, H.E. *Phytochemistry* **1981**, *20*,1385.

13.   Conserva, L. M.; Yoshida, M.; Gottlieb, O.R. *Phytochemistry* **1990**, *29*, 1986.

14.   Blair, G. E.; Cassady, J. M.; Robbers, J. E.; Tyler, V. E.; Raffauf, R. F. *Phytochemistry* **1969**, *8*, 497.

15.   Gottlieb, O. R.; Maia, J. G.S.; Ribeiro, M. N. S. *Phytochemistry* **1976**, *15*, 773.

16.   Kawanishi, K.; Uhara, Y.; Hashimoto, Y. *Phytochemistry* **1982**, *21*, 929.

17.   Kawanishi, K.; Uhara, Y.; Hashimoto, Y. *Phytochemistry* **1982**, *21*, 2725.

18.   Lopes, L. M. X.; Yoshida, M.; Gottlieb, O. R. *Phytochemistry* **1982**, *21*, 751.

19.   Lopes, L. M. X.; Yoshida, M.; Gottlieb, O. R. *Phytochemistry* **1983**, *22*, 1516.

20.   Lopes, L. M. X.; Yoshida, M.; Gottlieb, O. R. *Phytochemistry* **1984**, *23*, 2021.

21.   Lopes, L. M. X.; Yoshida, M.; Gottlieb, O. R. *Phytochemistry* **1984**, *23*, 2647.

22.   Cavalcante, S.H.; Yoshida, M.; Gottlieb, O.R. *Phytochemistry* **1985**, *24*, 1051.

23.   Martinez, J.C.; Cuca, S.L.E.; Yoshida, M.; Gottlieb, O.R. *Phytochemistry* **1985**, *24*, 1867.

24.   Martinez, J.C.; Cuca, L.E.; Santana, M.A.J.; Pombo-Villar E.; Golding, B.T. *Phytochemistry* **1985**, *24*, 1612.

25.   Kato, M.J.; Yoshida, M.; Gottlieb, O.R. *Phytochemistry* **1990**, *29*, 1799.

26.   Kato, M.J.; Lopes, L. M. X.; Paulino Filho, H. F.; Yoshida, M.; Gottlieb, O.R. *Phytochemistry* **1985**, *24*, 533.

27.   Vieira, P. C.; Gottlieb, O. R.; Gottlieb, H. E. *Phytochemistry* **1983**, *22*, 2281.

28.   Vieira, P. C.; Yoshida, M.; Gottlieb, O. R.; Paulino Filho, H. F.; Nagem, T. J.; Braz Filho, R. *Phytochemistry* **1983**, *22*, 711.

29.   Howe, H.F.; Smallwood, J. *Ann. Rev. Ecol. Syst.* **1982**, *13*, 201.

30.   Cavalcante, S.H.; Fernandes, D.; Paulino Filho, H.F.; Yoshida M.; Gottlieb, O.R. *Phytochemistry* **1985**, *24*, 1865.

31.   Kato, M.J.; Yoshida, M.; Gottlieb, O.R. *Phytochemistry* **1992**, *31*, 283.

32.   Braz Filho, R.; Carvalho, M. G.; Gottlieb, O. R. *Planta Medica* **1984,** 53.

33.   Morawetz, W. *Pl. Syst. Evol.* **1986**, *152*, 49.

34.   Foster, S.A. *The Bot. Rev.* **1986**, *52*, 260.

35.   Bell, E.A. In *Ann. Proc. Phytochem. Soc. Europe*; *Biochemical Aspects of Plant and Animal Coevolution*; Harborne, J. B., Ed.; 1978, 15, 145.

36. Janzen, D.H. In *Physiological Ecology of Fruits and Their Seeds*; Lange,
     O.L.; Nobel, P. S.; Osmond, C. B.; Ziegler, H., Eds.; Physiol. Plant Ecol.,
     Springer-Verlag:Wurzburg, 1983, III, 625.
37. Dodson, C. D.; Stermitz, F.R.; Castro, O. C.;Janzen, D. H.
     Phytochemistry **1987**, *26*, 2037.
38. Casida, J.E. *J. Agr. Food Chem.* **1970**, *18*, 753.
39. Howe, H.F.; Vande Kerckhove, G.A. *Ecology* **1981**, *62*, 1093.
40. Matsui, K.; Munakata, K. *Tetrahedron Letters* **1975**, 1905.
41. Munakata, K. In *Host Resistance to Pests*; Hedin, P.A., Ed.; Amer. Chem.
     Soc. Symposium Series 1977, *62*, 185.
42. Bowers,W. S. *Science* **1968**, *161*, 895.
43. Kamikado, T.; Chang, C.; Murakoshi, S.; Sakurai, A.; Tamura, S. *Agric.
     Biol. Chem.* **1975**, *39*, 833.
44. Burden, R. S.; Crombie, L.; Whiting, D. A. *J. Chem. Soc.* (C) **1969**, 693.
45. Adjangba, M. S. *Bull. Soc. Chim. France* **1963**, 2344.
46. Kemp M.S.; Burden, R.S. *Phytochemistry* **1986**, *25*, 1261.
47. Hasegawa, M.; Shirato, T. *J. Jap. Forest Soc.* **1959**, *41*, 1.
48. Stoessl, A. *Tetrahedron Lett* **1966**, 2287.
49. Shain, L.; Hillis, W. E. *Phytopathology* **1971**, *61*, 841.
50. Smith, T. A.; Best, G. R. *Phytochemistry* **1978**, *17*, 1093.
51. Takasugi, N.; Katsui, N. *Phytochemistry* **1986**, *25*, 2751.
52. Yoshihara, T.; Yamaguchi, K.; Sakamura, S. *Agric. Biol. Chem.* **1982**,
     43, 853.
53. Clark, A. M.; El-Feraly, F.; Li, W. S. *J. Pharm.Sci.* **1981**, *70*, 951.
54. Gutterman, Y.; Evenary, M.; Cooper, R.; Levy, E.C.; Lavie, D.
     Experientia **1980**, *36*, 662.
55. Campbell, D. G.; Richardson, P. M.; Rosas Jr, A. *Biochem. Syst. Ecol.*
     1989, *17*, 403.
56. Cooper, R.; Levy, E.C.; Lavie, D. *J. Chem. Soc. Chem. Comm.* **1977**,
     749.
57. Roy, S.; Guha, R.; Chakraborty, D. P. *Chem Ind.* **1977**, *19*, 231.
58. Chakraborty, D.P.; Roy, S.; Roy, S.S.P.; Majumber, S. *Chem. Ind.* **1979**,
     6, 667.
59. Howe, H.F. *The Auk* **1981**, *98*, 88.
60. Ng, F. S. P. In *Tropical trees as living systems*; Tomlinson, P.B.;
     Zimmermann, Eds.; Cambridge University Press. **1978**, pp. 129.
61. Garwood, N. C. *Ecol. Monog.* **1983**, *53*, 159.
62. Bawa, K. S.; Hadley, M., Eds.; *Reproductive Ecology of Tropical Forest
     Plants*, Unesco and The Parthenon Publishing Group: Halifax, *1990*, 7.
63. Howe, H.F.; Schupp, E.W.; Westley, L.C *Ecology* **1985**, *66*, 781.
64. Mudd, A. *J. Chem. Soc.Perkin Trans.* I, **1983**, 2161.
65. Barata, L.E.S.; Baker, P. M.; Gottlieb, O.R.; Rúveda, E.A.
     Phytochemistry **1978**, *17*, 783.

66. Lopes, N. P.; França, S. C.; Pereira, A. M. S.; Maia, J. G. S.; Kato, M. J.; Cavalheiro, A. J.; Gottlieb. O. R.; Yoshida, M. *Phytochemistry* **1994**, *35*, 1469.

67. McKeena, D.J.; Towers, G.H.N.; Abbott, F.S. *J. Ethnopharm.* **1984**, *12*, 179.

68. McRae, W.D.; Towers, G.N.N. *Phytochemistry* **1984**, *23*, 1207.

69. Pelter, A. *Rec. Adv. Phytochem.* **1986**, *20*, 201.

70. Ayres, D. C.; Loike, J. D., Eds.; *Lignans: Chemical, Biological and Clinical Properties.* Cambridge University Press, Cambridge, CB, **1990**, pp 402.

71. Biftu, T.; Stevenson, R. *Phytotherapy Research* **1987**, *1*, 97.

72. Janssens, J.; Laekeman, M. G.; Pieters, L. A. C.; Totte, J.; Herman, A. G.; Vlietinck, A. *J. Ethnopharm.* **1990**, *29*, 179.

73. Nitao, J. K.; Nair, M. G.; Thorogood, D.L.; Johnson, K. S.; Scriber, J. M. *Phytochemistry* **1991**, *30*, 2193.

74. Fukuda, Y.; Osawa, T.; Namiki, M.; Ozaki, T. *Agric. Biol. Chem.* **1985**, *49*, 301.

75. Fukuda, Y.; Nagata, M.; Osawa, T.; Namiki, M. *J. Am. Oil Chem. Soc.* **1986**, *63*, 1027.

76. Loicke, J. D.; Horwitz, S. B. *Biochemistry* **1976**, *15*, 5435.

77. Fernandes, J. B.; Fraga, R. L.; Capelato, M. D.; Vieira, P. C.; Yoshida, M.; Kato, M. J. *Synthetic Comm.* **1991**, *21*, 1331.

78. Pelter, A.; Ward, R. S. *Tetrahedron* **1991**, *47*, 1275.

79. Miyazawa, M.; Kasahara, H.; Kameoka, H. *Phytochemistry* **1993**, *34*, 1501.

80. Dugall, S. P.; Kartha, A. R. S. *Indian J. Agric. Sci.* **1956**, *26*, 391.

81. Shimizu, S.; Akimoto, K.; Shinnen, Y.; Kawashima, H.; Sugano, M.; Yamada, H. *Lipids* **1991**, *26*, 512.

82. Shimizu, S.; Kawashima, H.; Akimoto, K.; Shinnen, Y.; Sugano, M.; Yamada, H. *Phytochemistry* **1992**, *31*, 757.

83. Xue, J-Y.; Liu, G-T.; Wei, H-L ; Pan, Y. *Free Radical Biol. & Medicine* **1992**, *12*, 127

84. Fauré, M.; Lissi, E.; Torres, R.; Videla L. A. *Phytochemistry* **1990**, *29*, 3773.

85. Sinha, B. K.; Trush, M. A.; Kalyanaraman, B. *Biochem. Pharm.* **1985**, *34*, 83.

86. Hirose, N.; Inoue, T.; Nishihara, K.; Sugano, M.; Akimoto, K.; Shimizu, S.; Yamada, H. *J. Lip. Res.* **1991**, *32*, 629.

87. Yamashita, K.; Nohara, Y.; Katayama, K.; Namiki, M. *J. Nutr.* **1992**, *122*, 2440.

88. Hamburger, M.; Hostettmann, K. *Phytochemistry* **1991**, *30*, 3864.

RECEIVED October 24, 1994

Chapter 15

Amazonia versus Australia
Geographically Distant, Chemically Close

Maria Auxiliadora Coelho Kaplan

Núcleo de Pesquisas de Produtos Naturais, Centro de Ciêancias de Saúde, Universidade Federal do Rio de Janeiro, 21941–590 Rio de Janeiro, RJ, Brazil

Species belonging to the genera *Alexa* from Amazonia and *Castanospermum* from Australia contain polyhydroxylated piperidines, pyrrolidines, indolizines and pyrrolizidines. The presence of this related series of alkaloids suggests the original derivation of both genera from a common ancestor thriving on the old South Panagean continent.

The legumes can be considered true generalists, comprising plants of different habits, from huge trees, shrubs, climbers, down to tiny herbs, showing an enormous variety of methods of reproduction and growth besides an enormous variety of chemical defences. Although the family can be found in all terrestrial habitats, from the equator to the limits of dry and cold deserts, it shows greater biodiversity in tropical and subtropical regions including mainly the Amazon basin.

The versatility of Leguminosae enhances their economic importance. For a long time their usage was restricted to the supply of wood and fuel, followed by the use in nutrition and traditional medicine. These plants give nitrogenated foods, either through their seeds or through their vegetative parts, and are equally employed as feeds.

The family is considered by many botanists to consist in three sub-families: Caesalpinioideae, Mimosoideae and Papilionoideae (*1*). Nevertheless, more recent proposals aim to raise their status to level of families: the Caesalpiniaceae, Mimosaceae and Fabaceae (*2,3*). The Caesalpinioideae comprise about 150 genera in five tribes distributed in tropical and subtropical regions as well as in areas of temperate climate (*4*). The Mimosoideae with about 60 genera in 5 tribes are distributed over Central and South America as well as Africa and Australia (*5*). The Papilionoideae comprise about 440 genera, the species of which are amplely

0097–6156/95/0588–0180$12.00/0

dispersed and well represented in tropical Asia, Australia and the Amazon region. These plants are classified in 32 tribes (*6a*).

The fossil record also does not clarify evolutionary relationships within the Leguminosae. The three subfamilies most certainly were already well differentiated and established during the Eocene. The considerable differences between the few primitive genera which survived, taken jointly with biogeography and fossil record, securely place the origin of Caesalpinioideae in the late Cretaceous, but no real evidence has so far been produced for the precise localization within this period, and the oldest fossil yet discovered is not older than 70 My (*7*).

In Papilionoideae the tribe Sophoreae is a group of convenience between Caesalpinioideae and the major part of Papilionoideae with no clear connections to either one. The interrelations were exhaustively evaluated, leading to the recognition of a great diversity of flowers, fruits and seeds. With the exception of *Baphia*, *Ormosia* and *Sophora* the majority of genera are small and discrete, sometimes with disjunctions of species, and more closely placed genera. Already the sole genus *Sophora* expands significantly beyond the tropics (*6a*).

Among the 32 tribes of Papilionoideae the tribe Sophoreae merits to be mentioned because it comprises among its 48 genera, *Alexa* and *Castanospermum* which aroused our special interest on account of their similar chemical compositon. A priori this is rather surprising since the genus *Alexa* occurs in the tropical forests of South America. Its 7 or 8 species are native of Guyana, Suriname, French Guyana, Venezuela and the Amazon basin. This genus belongs to the Dussia group. In contradistinction the genus *Castanospermum* is monospecific. Its sole species thrives in Northeast Australia in some coastal forests and on beaches. The genus belongs to the Angylocalyx group and does not keep close relationship with any other known on this continent. Plants of both genera have imparipenated leaves, calyxes divided in small lobes, large red and orange coloured petals, dehiscent fruits and well developed seeds. Preliminary studies of pollen indicated a uniform pattern without specialization; the most conspicuous anomaly consisting in the presence of columellas on the two layers of ectesine on the big flowers of *Alexa* and *Castanospermum* (*6b*).

Chemical Aspects

Discoveries of novel polyhydroxy-alkaloid derivatives in both genera *Alexa* and *Castanospermum* on one hand, and *Swainsonia*, *Astragalus* and *Oxytropis* on the other, also suggest the relationship of these goups (*8*).

In recent years polyhydroxy-alkaloids that structurally and stereochemically mimic sugars, have been found in a variety of organisms including higher plants. Many of them are potent inhibitors of glycosidase activities in insects, mammals and microorganisms, and it is suggested that these properties contribute to natural chemical defences of those plants in which they accummulate (*9*).

Four structural types of such alkaloids can be distinguished, namely polyhydroxy-derivatives of piperidine, pyrrolidine, indolizine and pyrrolizidine.

Such compounds have so far been isolated from species of five plant families Leguminosae, Moraceae, Polygonaceae, Aspidiaceae and Euphorbiaceae (*10*).

. POLYHYDROXYPIPERIDINES

DNJ - deoxynojirimycin, first shown as microbial metabolite, was found in roots of *Morus* spp (Moraceae) (*11*). It is clear that this compound can be seen as an azapyranose analog of glucose. The respective mannose analog DMJ - deoxymannojirimycin, has been isolated from seeds of the tropical legume *Lonchocarpus sericeus* (*12*). DNJ and DMJ are inhibitors of α-glucosidase and mannosidase respectively.

HNJ - homonojirimycin was isolated from leaves of *Omphalea diandra* (Euphorbiaceae) (*13*) and it has similar properties as DNJ. Other alkaloidal glycosidase inhibitors of the azapyranose type are: nojirimycin, nojirimycin B, galactostatin, fagomine, fagomine glucoside XZ-1 and BR-1 (Figure 1) (*14*).

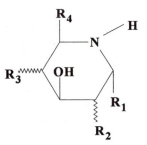

	R_1	R_2	R_3	R_4
DNJ	H	αOH	αOH	CH_2OH
DMJ	H	βOH	αOH	CH_2OH
HNJ	CH_2OH	αOH	αOH	CH_2OH
NOJIRIMYCIN	OH	αOH	αOH	CH_2OH
NOJIRIMYCIN B	OH	βOH	αOH	CH_2OH
GALACTOSTATIN	OH	αOH	βOH	CH_2OH
FAGOMINE	H	H	αOH	CH_2OH
FAGOMINE GLUCOSIDE	H	H	αOGlc	CH_2OH
BR-1	H	αOH	αOH	CO_2H

Figure 1. Examples of natural polyhydroxypiperidines.

. POLYHYDROXYPYRROLIDINES

DMDP -   2R,5R-dihydroxymethyl-3R,4R-dihydroxypirrolidine,   an azafuranose analog, was isolated from leaves of *Derris elliptica* (Leguminosae) (*15*), from seeds of *Lonchocarpus* spp (Leguminosae) (*16*) and from leaves of *Omphalea diandra* (Euphorbiaceae) (*17*). DMDP is a strong inhibitor of α and β-glucosidases. Other alkaloidal glycosidase inhibitors of the azafuranose type are: DAP-1, CYB-3 and FR-900483 (Figure 2) (*14*).

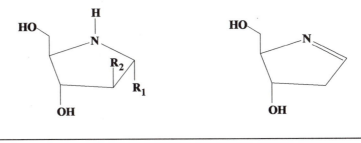

	R₁	R₂	FR-900483
DMDP	CH₂OH	OH	
DAP-1	H	OH	
CYB-3	H	H	

Figure 2. Examples of natural polyhydroxypyrrolidines.

. POLYHYDROXYINDOLIZINES

Swainsonine was isolated from the Australian legume *Swainsonia canescens* (*18*) and in the United States from two species of locoweed *Astragalus lentiginosum* and *Oxytropis sericea* (*19*). Swainsonine is a strong inhibitor of α-mannosidase.

Castanospermine was isolated from seeds of the Australian tree *Castanospermum australe* (*20*) and from the South American genus *Alexa* (*21*). Castanospermine is a potent inhibitor of α and β-glucosidase. 6-Epicastanospermine, an inhibitor of amylglycosidase, has been found in seeds of *C. australe* (*22*). 7-Deoxy-6-epicastanospermine was also found in seeds of *C. australe* (*23*). This compound showed a weak inhibitory effect (Figure 3).

. POLYHYDROXYPYRROLIZIDINES

A series of alexines was found in *C. australe* and/or in *Alexa* spp.

7-Epialexine = australine was isolated from seeds of *C. australe* (*24*) and from pods of A. *leiopetala* (*25*).

1,7a-Diepialexine = 1-epiaustraline as well as 3,7a-diepialexine = 3-epiaustraline were found in *C. australe* (*26*). Alexine was isolated from *Alexa* spp (*25*). All these compounds are good inhibitors of α-glycosidase.

7a-Epialexaflorine isolated from different organs of *Alexa grandiflora* (*27*) is an inhibitor of fungal amylglucosidase. This compound was isolated for the first time by us and it represents the first example of a pyrrolizidine alkaloid which is at the same time an α-aminoacid.

Lentinoside and epilentinoside were isolated from *Astragalus lentiginosum*. The first one is an inhibitor of amylglicosidase and the latter has no inhibitory effect on glycosidases (Figure 4) (*23*).

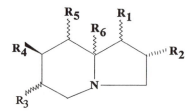

	R_1	R_2	R_3	R_4	R_5	R_6
SWAINSONINE	αOH	OH	H	H	βOH	βH
CASTANOSPERMINE	βOH	H	αOH	OH	αOH	αH
6-EPICASTANOSPERMINE	βOH	H	βOH	OH	αOH	αH
7-DEOXY-6-EPICASTANOSPERMINE	βOH	H	βOH	H	αOH	αH

Figure 3. Examples of natural polyhydroxyindolizines.

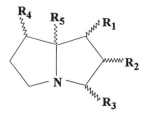

	R_1	R_2	R_3	R_4	R_5
AUSTRALINE	βOH	αOH	βCH₂OH	αOH	βH
ALEXINE	βOH	αOH	βCH₂OH	αOH	αH
3,7a-DIEPIALEXINE	βOH	αOH	αCH₂OH	αOH	βH
7a-EPIALEXAFLORINE	βOH	αOH	βCO₂H	αOH	βH
1,7a-DIEPIALEXINE	αOH	αOH	βCH₂OH	αOH	βH
LENTIGENOSIDE	αOH	βOH	H	H	αH
EPILENTIGENOSIDE	αOH	αOH	H	H	αH

Figure 4. Examples of natural polyhydroxypyrrolizidines.

Conclusion

In view of their extraordinary chemical relationship, it is conjectured that *Alexa* and *Castanospermum* had a common ancestor thriving on the old South Panagean continent, a terrestrial mass which comprised modern South America, Antarctica and Australia (Figure 5). The flora of the regions destined to become South America and Australia was separated by the intrusion of masses of ice and by the rupture of the earth crust due to the movement of tectonic plates. It is possible that the ancestors of *Astragalus* and *Oxytropis* (unknown in Australia) and *Swainsonia* (to be found only in Australia and New Zealand) also were located on this continent. They too possess a number of common morphological characteristics (*28,29*).

A major point of general interest of the present work concerns the correlation of three types of independent evidence: geographical location, peculiar chemical composition and comparable morphological features. Whenever all these informations concur in the case of a plant group, little doubt remains that data collected from presently available specimens can be extrapolated back to reveal the characteristics of a common ancestor.

In the present case the existence of such an ancestor constitutes an important clue revealing evolution and spacial radiation of Leguminosae.

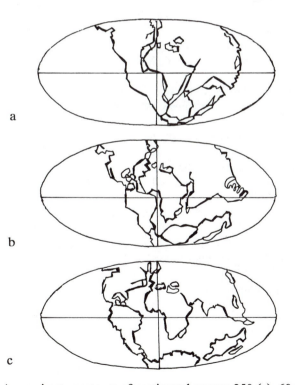

Figure 5. Approximate contours of continental masses 250 (a), 60 (b) and 2 (c) millions of years before present.

Literature Cited

1. Brummitt, R.K. *Vascular Plant Families and Genera*, The Royal Botanic Gardens, Kew, **1992**, pp 595-602.
2. Cronquist, A. *The Evolution and Classification of Flowering Plants*, 2nd. ed., New York Botanic Garden, New York, **1988**.
3. Dahlgren, R.M.T. *J. Linn. Soc. Bot.* **1980**, *80*, 91-124.
4. Cowan, R.S. In *Advances in Legume Systematics*, Part 1, Polhill, R.M.; Raven, P.H., Eds.; The Royal Botanic Gardens, Kew, **1981**, pp 57-63.
5. Elias, T.S. In *Advances in Legume Systematics*, Part 1, Polhill, R.M.; Raven, P.H, Eds.; The Royal Botanic Gardens, Kew, **1981**, pp 143-151.
6. Polhill, R.M. In *Advances in Legume Systematics*, Part 1, Polhill, R.M.; Raven, P.H., Eds.; The Royal Botanic Gardens, Kew, **1981**, a) pp 191-204; b) pp 213-230.
7. Herendeen, P.S.; Crept, W.L.; Dilcher, D.L. In *Advances in Legume Systematics*, Part 4, Herendeen, P.S.; Dilcher, D.L., Eds.; The Royal Botanic Gardens, Kew, **1992**, pp 303-316.
8. Fellows, L.E. *New Scientist* **1989**, 45-48.
9. Elbein, A.D.; Molyneux, R.J. In *Alkaloids: Clinical and Biological Perspectives*, Pelletier, S.W., Ed., Wiley Interscience, New York, **1987**, vol. 5, pp 1-54.
10. Hartmann, T. In *Herbivores: Their Interactions with Secondary Plant Metabolites*, 2nd ed., Rosenthal, G.A.; Berenbaum, M.R., Eds., Academic Press, New York, **1991**, vol. 1, pp 79-121.
11. Yagi, M.; Kouno T.; Aoyagi, Y.; Murai, H. *Nippon Nogei Kagaku Kaishi*, **1976**, *50*, 571-572.
12. Fellows, L.E.; Bell, E.A.; Lynn, D.G.; Pilkiewicz, F.; Miura, I.; Nakanishi, K. *J. Chem. Soc., Chem. Comm.* **1979**, 977-978.
13. Kite, G.C.; Fellows, L.E.; Fleet, G.W.J.; Liu, P.S.; Scofield, A.M.; Smith N.G. *Tetrahedron Letters* **1988**, *29*, 6483-6486.
14. Fellows, L.E.; Lite, G.C.; Nash, R.J.; Simmonds, M.S.J.; Scofield, A.M. *Recent Adv. Phytochem.* **1989**, *23*, 395-427.
15. Welter, A.; Jadot, J.; Dardenne, G.; Marlier, M.; Casimir J. *Phytochemistry* **1976**, *15*, 747-749.
16. Fellows, L.E.; Evans, S.V.; Nash, R.J.; Bell, E.A. *ACS Symposium Ser.* **1986**, *296*, 72-78.
17. Horn, J.M.; Lees, D.C.; Smith, N.G.; Nash, R.J.; Fellows, L.E.; Bell, E.A. In *Proceedings of the 6th International Symposium of Plant-Insect Relationships*, Labeyrie, V.; Fabres G.; Lachaise, D.,Eds.; Dr. W. Junk, Dordrecht, **1986**, p 394.
18. Colegate, S.M.; Dorling, P.R.; Huxtable, C.R. *Aust. J. Chem.* **1979**, *32*, 2257-2264.
19. Molyneux, R.J.; James, L.F. *Science* **1982**, *216*, 190-191.

20   Hohenschutz, L.D.; Bell, E.A.; Jewess, P.T.; Leworthy, D.P.; Pruce, R.J.;
      Arnold, E.; Clardy, J. *Phytochemistry* **1981**, *20*, 811-814.

21.   Nash, R.J.; Fellows, L.E.; Dring, J.V.; Stirton, C.H.; Carter, D.; Hegarty,
      M.P.; Bell, E.A. *Phytochemistry* **1988**, *27*, 1403-1404.

22.   Molyneux, R.J.; Roitman, J.N.; Dunnhein, G.; Szumilo, T.; Elbein, A.D.
      Arch. Biochem. Biophys. **1986**, *251*, 450-457.

23.   Molyneux, R.J.; Tropea, J.E.; Elbein, A.D. *J. Nat. Prod.* **1986**, *53*, 609-
      614.

24.   Molyneux, R.J.; Benson, M.; Wong, R.Y.; Tropea, J.E.; Elbein, A.D. *J.
      Nat. Prod.* **1988**, *51*, 1198-1206.

25.   Nash, R.J.; Fellows, L.E.; Dring, J.V.; Fleet, G.W.J.; Derome, A.E.;
      Hamor, T.A.; Scofield, A.M.; Watkin, D.J. *Tetrahedron Letters* **1988**, *29*,
      2487-2490.

26.   Nash, R.J.; Fellows, L.E.; Plant, A.C.; Fleet, G.W.J.; Derome, A.E.; Baird,
      P.D.; Hegarty, M.P.; Scofield, A.M. *Tetrahedron* **1988**, *44*, 5959-5964.

27.   Pereira, A.C.S.; Kaplan, M.A.C.; Maia, J.G.S.; Gottlieb, O.R.; Nash, R.J.;
      Fleet, G.; Pearce, L.; Watkin, D.J.; Scofield, A.M. *Tetrahedron* **1991**, *47*,
      5637-5640.

28.   Smith, A.G; Briden, J.C.; Drewry, G.E. *Spec. Pap. Paleontol.* **1973**, *12*, 1-
      42.

29.   Wrenn, J.H.; Beckman, S.W. *Science* **1992**, *216*, 187-189.

RECEIVED October 13, 1994

Chapter 16

Chemodiversity of Angiosperms

Latitude- and Herbaceousness-Conditioned Gradients

Maria Renata de M. B. Borin and Otto R. Gottlieb

Departamento de Fisiologia e Farmacodinâmica, Instituto Oswaldo Cruz,
FIOCRUZ, Avenida Brasil 4365, 21045–900 Rio de Janerio, RJ, Brazil

Novel methodology was developed in order to investigate the spatial differentiation of angiospermous phytochemicals. Its application to the biogeography of selected regions, including Amazonia, revealed: 1. chemo-diversity and morpho-diversity to compensate for each other in opposed latitudinal gradients; 2. acetate-derived aliphatics to replace shikimate-derived aromatics upon passing from forest to savanna; 3. morphology and metabolism to be both conditioned by lignin-requirement.

The evolutionary history of organisms is accompanied by increase in biological diversity. The number of species alive on the Earth today is conjectured to be in the range of tens of millions. However, only a small fraction of this number, about 1.4 million species, has been identified (*1, 2*). The contemporary land flora comprises about 300,000 vascular plant species, among which more than 80% are flowering plants. Indeed angiosperms feature among the areas of highest diversity in biology. Special metabolism (i.e. secondary compound chemistry) is another large area, and about 80%, 26,000 (*3*) or 30,000 (*4*), of the identified compounds are of plant origin (*5*). Quite obviously, even these relatively large numbers only represent a small fraction of those thought to exist. The coincidence of these areas of enormous diversity in flowering plants stresses the importance of metabolism in the functioning of their cells with respect to evolution, adaptation and bioactivity. Thus, recognition and measurement of chemical diversification are essential for the understanding of the causes and consequences of biodiversity and can be powerful clues for the localization of biodiversity, a fundamental preliminary objective for efforts towards conservation strategies.

However, the micromolecular trends which accompany the chemical diversification of the flowering plants are still unknown, since very few taxa have been chemically examined in their entire geographical extension (or even in circumscribed areas). This is understandable since pertinent practical work would be a tremendous

0097–6156/95/0588–0188$12.00/0

task to be accomplished, even with the available modern methods of investigation. In order to approach the study of chemo-geographical variation, it was necessary first to introduce adequate methodologies endowed with predictive value.

Methodology

Micromolecular Profiles.   One procedure consists in the inference of the composition of a chosen micromolecular category in a taxon (e.g. a family) through literature data for all its species, wherever they may have been collected. Thus a universal profile for a micromolecular category is generated for each taxon in the hope, based on chemosystematics (*6*), that it may adequately represent its species which occur in the selected regions (*7*).

The micromolecular categories were selected in order to be representative partially of the shikimate pathway and partially of the acetate pathway (Fig. 1).

Floristic Inventories. The floristic inventories selected corresponded to five different regions: a tropical rain forest, a semideciduous mesophytic forest, a "cerrado", a "campo rupestre" and an arctic tundra (Table I). These regions were chosen because they have different types of vegetation.

Sporne Index.   Sporne (*8*) quantified evolutionary, mainly morphology based, features of angiosperm families by a percentage advancement index (SI). This was calculated considering the frequency in which 30 characters (woodiness included), regarded as primitive (within a total of 107), appear in each of the envisaged 291 families. The Sporne index of each region, represented by the mean SIs of the inventoried families, is considered indicative of morphological trends (Table II).

Herbaceousness Index. We previously quantified the herbaceousness of angiosperm families by a percentage index (*7, 9*). This was calculated by input values from 1 to 100, according to indications by Cronquist (*10*) concerning the preponderant habit (Table III). Thus, for instance, the herbaceousness index is 100 for families constituted only by "herbs", and 12.5 for families characterized by "trees predominating over shrubs". The herbaceousness index of each region, represented by the mean HIs of the inventoried families, is considered indicative of morpho-chemical trends (Table II) (Gottlieb, O.R.; Borin, M.R. de M.B.; Kaplan, M.A.C., FIOCRUZ & UFRJ, Rio de Janeiro, unpublished data).

Number of Occurrences. Furthermore, we represented micromolecular features in plant groups by frequency (= number) of occurrences (NO) (*9*). The NO of a selected biosynthetic category known to occur in a family is established by the total number of compounds registered for their species.

In this work the chemical composition of each region is considered to be represented by NO/F, the chemical diversity of a hypothetical family with average characteristics (Table II).

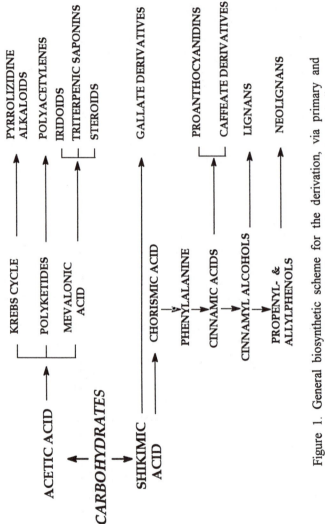

Figure 1. General biosynthetic scheme for the derivation, via primary and intermediate metabolism, of ten selected categories of secondary metabolites.

Table I. Characteristics of Five Floristic Inventories

Regions	Type of Vegetation	Localities	Families	Genera	Spp
			Number of		
Rondônia	Tropical rain forest. "Terra firme" forest with medium height of 15m	Near the Cuiabá-Porto Velho road (BR-364), Rondônia State, BR (*13*)	59	180	277
Serra do Japi	Semideciduous mesophytic forest	Atlantic plateau, 6Km from Jundiaí City, São Paulo State, BR (*14*)	48	90	128
Moji Guaçu Biological Reserve	"campo cerrado" "cerrado"	Municipality of Moji Guaçu, São Paulo State, BR (*15*)	82	274	521
Serra do Cipó	Gallery forests "capões" "cerrado" "campo rupestre"	Espinhaço Range of mountains in the Municipality of Santana do Riacho, Minas Gerais State, BR (*16*)	125	499	1551
Arctic Tundra	Many herbs (grasses and related taxa) and a few shrubs	Near Meade River, Alaska (*17*)	27	63	89

Table II. Latitude, Sporne Indices (SI), Herbaceousness Indices (HI), Numbers
of Occurrences per Family (NO/F) of Neolignans (NLG), Lignans (LGN)
(*18,19*), Caffeoyltannins (CAF), Proanthocyanidins (PRO), Gallo- and
Ellagitannins (GAL) (*20*), Steroids (STE) (*9*), Triterpenic Saponins (SAP)
(*21-27*), Iridoids (IRI) (*28*), Polyacetylenes (POL) (*29*) and Pyrrolizidine
Alkaloids (PYR) (Ferreira, Z.S., Universidade de São Paulo, unpublished data)
and Percent Values of Shikimate Derivatives (NLG+LGN+CAF+PRO+GAL)
(SH%) of Five Different Regions

	Forest	*Mata*	*Cerrado*	*C. Rupestre*	*Tundra*
Latitude	9°20'-40'S	23°11'S	22°15'-16'S	19°12'-20'S	70°28'N
SI	48	50	51	50	56
HI	30	36	48	54	72
NLG	9	8	7	4	-
LGN	4	7	5	4	3
CAF	4	12	12	9	26
PRO	8	15	9	8	17
GAL	11	19	11	9	16
STE	12	25	33	23	42
SAP	9	8	9	7	18
IRI	4	4	6	4	8
POL	1	45	28	19	85
PYR	11	26	16	11	46
ΣNO/F	73	169	136	98	261
SH%	50	36	32	34	24

Table III. Herbaceousness Indices (HI) Assigned to Families
Characterized by Types of Habit

Types of Habit	*HI*
Trees	1.0
Trees predominating over shrubs	12.5
Trees and shrubs	25.0
Shrubs predominating over trees	37.5
Shrubs	50.0
Shrubs predominating over herbs	62.5
Shrubs and herbs	75.0
Herbs predominating over shrubs	87.5
Herbs	100.0

SH/AC. Usually chemotaxonomic work has relied on particular micromolecular markers. The present chemo-geographic effort required the development of more general chemical criteria. The most significant one concerns the evolutionary replacement of shikimate derived aromatics by acetate derived aliphatics. Lignans (LGN), neolignans (NLG), proanthocyanidins (PRO), caffeic acid derivatives (CAF) and gallic acid derivatives (GAL) (particular metabolic categories collectively designated SH) were selected to represent the former group, and steroids (STE), triterpenic saponins (SAP), iridoids (IRI), polyacetylenes (POL), and pyrrolizidine alkaloids (PYR) (other particular metabolic categories collectively designated AC) were selected to represent the latter one. The SH/AC relationship is established by the sum of the NO/F-values of the aromatic categories (NO/F-SH = NO/F-NLG + NO/F-LGN + NO/F-CAF + NO/F-PRO + NO/F-GAL) *versus* the sum of the NO/F-values of the aliphatic categories (NO/F-AC = NO/F-STE + NO/F-SAP + NO/F-IRI + NO/F-POL + NO/F-PYR) with respect to each region:

$$SH\% = \frac{NO / F - SH \times 100}{NO / F - SH + NO / F - AC} \qquad AC\% = \frac{NO / F - AC \times 100}{NO / F - SH + NO / F - AC}$$

Oxidation level of compounds. Another procedure for the evaluation of a family's tendency toward micromolecular diversity involves the determination of the mean oxidation level of constituents pertaining to selected biosynthetic categories. The higher this average, the more diversified the categories (*11*). Positive correlations between increasing metabolic diversification [conditioned by an increase of the product's oxidation level], and decreasing woodiness (i.e., increasing herbaceousness) of the corresponding angiosperm families, have been registered not only for steroids (*9*) and diterpenoids (*12*), but also for flavonoids, benzylisoquinoline alkaloids, iridoids, pyrrolizidine alkaloids and polyacetylenes (*11*, Gottlieb, O.R.; Borin, M.R. de M.B.; Kaplan, M.A.C. - FIOCRUZ & UFRJ, Rio de Janeiro, unpublished data).

Metabolic Differentiation of Angiosperms

According to a concept discussed in the preceding chapter, the phenotypic levels of manifestation of the genotype, morphology and metabolism, are interdependent. This postulate does not by itself solve the question if the correlation is positive or negative. Accepting the supposition that morpho-diversity increases towards the equator, chemical diversity could follow the same direction or increase in the opposite direction.

Chemical diversity can be considered according to two points of view, a quantitative and a qualitative one.

Latitude-Conditioned Gradients. With respect to the quantitative aspect, i.e., to diversity proper, our evidence suggests morphology and metabolism to increase in opposite directions (Σ NO/F *versus* latitude of the regions in Table II and Fig. 2). The corresponding increase of quantitative micromolecular diversity with latitudinal gradient could be due to decrease of light-intensity. According to a concept

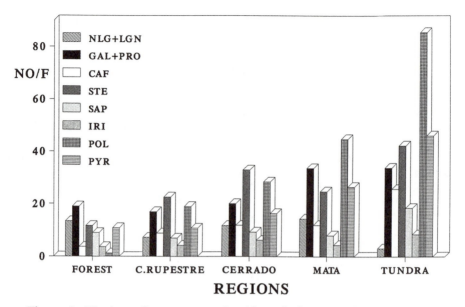

Figure 2. Number of occurrences for biosynthetic categories of chemical characters for five hypothetical families (NO/F), each representative of one of the five regions (see Table II and text), arranged according to increase of latitude.

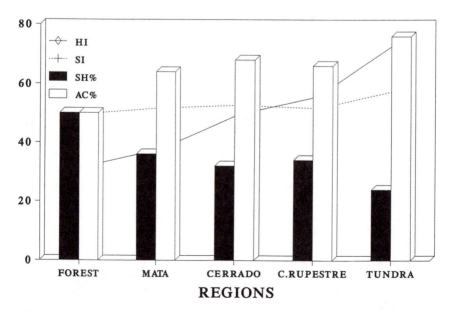

Figure 3. Sporne indices (SI), herbaceousness indices (HI) and percent values of shikimate derivatives (SH%) *versus* acetate derivatives (AC%) of five regions, arranged according to increase of HI.

developed in the precedent chapter: lowering of solar energy results in smaller rates of carbon flow through metabolic cycles, with enhancement of micromolecular biosynthesis.

Herbaceousness-Conditioned Gradients. With respect to the qualitative aspect, i.e. to chemical nature, one observes the replacement of shikimate derived aromatics by acetate derived aliphatics to follow advance in herbaceousness (HI) and morphology (SI) (Table II and Fig. 3).

The fact that herbaceousness is connected with micromolecular composition of plants is demonstrated by the percentage of the ten metabolic categories in each region. The evidence reveals all shikimate derived categories to decrease and all acetate derived categories to increase proportionally with the advance of the respective herbaceousness indices (HI) (Figs. 4a-4e). The observed interdependence of morphology and metabolism in angiosperms may be due to the connection of both phenotypic characters with lignification, a morpho-chemical driving force of plant development.

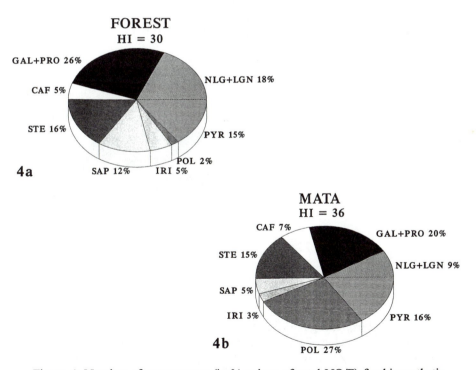

Figure 4. Number of occurrences (in %-values of total NO/F) for biosynthetic categories of chemical characters for forest (Fig. 4a), mata (Fig. 4b), "cerrado" (Fig. 4c), "campo rupestre" (Fig. 4d) and tundra (Fig. 4e). *Continued on next page*

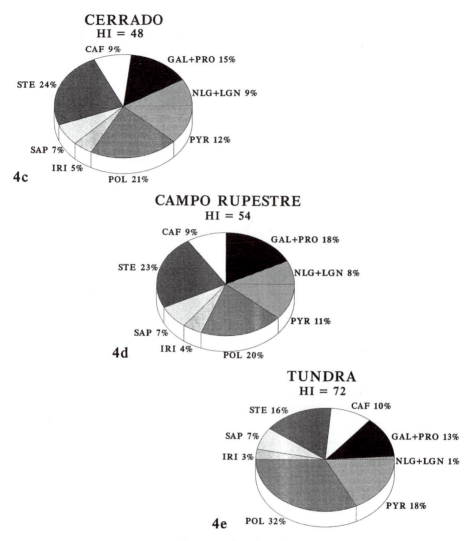

Figure 4. Continued

Conclusions

A practical way of expressing these results is to state that the simple determination of biodiversity via morphology-based inventories is at best an incomplete clue for the evaluation of a region for preservation or other purposes. It is imperative to decide at the outset of efforts toward zonation if the aim includes preservation of morphological or chemical diversity. If later situation is considered of importance, as indeed it should, a further decision will have to concern what kind of chemistry, a shikimate derived one or an acetate derived one, should be preserved preferentially.

If, for instance, it is desired to obtain phenolic compounds, lignified plants would be a better source, whereas, the chances of finding novel, aliphatics becomes brighter, the more herbaceous the investigated plant groups.

Obviously the most rational way to proceed for each nation would be the selection for preservation purposes of representative lots endowed with the widest range of desired characteristics.

Acknowledgments

The authors are grateful to CNPq, Brasil, for financial support.

Literature Cited

1. May, R.M. *Scient. Am.* **1992**, *267*, 18.
2. Wilson, E.O. In *Biodiversity*; Wilson, E.O., Ed.; National Academy Press: Washington, D.C., 1988, pp. 3-18.
3. Luckner, M. *Secondary Metabolism in Microorganisms, Plants, and Animals*; 3rd ed;. VEB Gustav Fischer Verlag: Jena, 1990; pp. 17-20.
4. Harborne, J.B. *Introduction to Ecological Biochemistry*; 3rd ed.; Academic Press: London, 1988; pp. 148.
5. Edwards, P.J.; Wratten, S.D. *Ecologia das Interações entre Insetos e Plantas*; Coleção "Temas de Biologia"; EPU: São Paulo, SP, 1981; Vol. 27, pp. 19.
6. Gottlieb, O.R. *Micromolecular Evolution, Systematics and Ecology*; Springer-Verlag: Berlin, 1982.
7. Borin, M.R. de M.B.; Gottlieb, O.R. *An. Acad. bras. Ci.* **1992,** *64*, 325.
8. Sporne, K.R. *New Phytol.* **1980**, *85*, 419.
9. Borin, M.R. de M.B.; Gottlieb, O.R. *Pl. Syst. Evol.* **1993**, *184*, 41.
10. Cronquist, A. *An Integrated System of Classification of Flowering Plants*; Columbia University Press: New York, 1981.
11. Gottlieb, O.R. *Phytochemistry* **1989**, *28*, 2545.
12. Figueiredo, M.R.; Gottlieb, O.R.; Kaplan, M.A.C. *1st International Symposium on Chemistry of the Amazon*; Abstract nº 23; Associação Brasileira de Química, Rio de Janeiro, RJ, 1993.
13. Absy, M.L.; Prance, G.T.; Barbosa, E.M. *Acta Amazonica* **1986/1987**, *16/17* (Sup.), 85.
14. Rodrigues, R.R.; Morellato, L.P.C.; Joly, C.A.; Leitão Filho, H. de F. *Revta brasil. Bot.* **1989**, *12*, 71.
15. Mantovani, W.; Martins, F.R. *Acta bot. bras.* **1993**, *7*, 33.
16. Giulietti, A.M.; Menezes, N.L. de; Pirani, J.R.; Meguro, M.; Wanderley, M. das G.L. *Bolm Botânica* **1987**, *9*, 1.
17. Jung, H.-J.G.; Batzli, G.O.; Seigler, D.S. *Biochem. Syst. Ecol.* **1979**, *7*, 203.
18. Gottlieb, O.R.; Yoshida, M. In *Natural Products of Woody Plants - Chemicals Extraneous to the Lignocellulosic Cell Wall*; Rowe, J.W., Ed.; Springer-Verlag: Berlin, 1989, Vol. I; pp 439-511.
19. Romoff, P. *Lignóides como Marcadores Quimiossistemáticos* Doctorate thesis, Instituto de Química, Universidade de São Paulo: São Paulo, 1991.

20. Borin, M.R. de M.B. *Polifenóis: Indicadores da Evolução de Plantas Floríferas* Doctorate thesis, Instituto de Química, Universidade de São Paulo: São Paulo, 1993.
21. Basu, N.; Rastogi, R.P. *Phytochemistry* **1967**, *6*, 1249.
22. Agarwal, S.K.; Rastogi, R.P. *Phytochemistry* **1974**, *13*, 2623.
23. Hiller, K.; Voigt, G. *Pharmazie* **1977**, *32*, 365.
24. Chandel, R.S.; Rastogi, R.P. *Phytochemistry* **1980**, *19*, 1889.
25. Hiller, K.; Adler, C. *Pharmazie* **1982**, *37*, 619.
26. Adler, C.; Hiller, K. *Pharmazie* **1985**, *40*, 676.
27. Mahato, S.B.; Sarkar, S.K.; Poddar, G. *Phytochemistry* **1988**, *27*, 3037.
28. Kaplan, M.A.C.; Gottlieb, O.R. *Biochem. Syst. Ecol.* **1982**, *10*, 329.
29. Ferreira, Z.S.; Gottlieb, O.R. *Biochem. Syst. Ecol.* **1982**, *10*, 155.

RECEIVED September 28, 1994

Chapter 17

Future-Oriented Mapping of Biodiversity in Amazonia

Otto R. Gottlieb

Departamento de Fisiologia e Farmacodinâmica, Instituto Oswaldo Cruz, FIOCRUZ, Avenida Brasil 4365, 21045–900 Rio de Janerio, RJ, Brazil

The study of Amazonian plants leads to a comprehensive morpho-chemical concept of biodiversity. The latitudinal gradients from poles to equator for increasing richness run in the opposite direction for metabolic features. This is due to the fact that the production of both major macromolecular plant products, cellulose and lignins, is light-intensity dependent. High solar energy input around the equator conditions rapid metabolic turnover, while low energy input results in smaller rates of carbon flow through metabolic cycles with consequent micromolecular diversification potential. Crash programs for the investigation of Amazonian morpho-chemical biogeography would acquire predictive value if the usual count of species in restricted areas was replaced by surveys across broad geographic transects. Novel methodology, needed for the investigation of such mostly longitudinal trends, should help assign the regions of highest biodiversity, as well as the corridors leading up to them. Further work along these lines will clarify the fascinating question if morphological and chemical evolutionary or spacial gradients consistently point in the same direction, whatever the plant groups under scrutiny.

According to a recent book by Wilson (*1*), "tropical rain forests, though occupying only 6% of the land surface, are believed to contain more than half the species of organisms on Earth. An explicit example is offered by the vascular plants. Of the approximately 250,000 species known, 170,000 (68%) occur in the tropics and subtropics, especially in the rain forests. The peak of global plant diversity is the combined flora of the three Andean countries of Colombia, Ecuador and Peru. There over 40,000 species occur on just 2% of the world's land surface. [This number] is to be compared with 700 native species found in all of the United States

0097–6156/95/0588–0199$12.00/0

and Canada, in every major habitat from the mangrove swamps in Florida to the coniferous forests of Labrador.

To summarize the present global pattern, latitudinal diversity gradients rising [from north and south] toward the tropics are an indisputable general feature of life. And on land biodiversity is heavily concentrated in the tropical rain forests.

The cause of tropical preeminence poses one of the great theoretical problems of evolutionary biology. Many [biologists] have called the problem intractable, suposing its solution to be lost somewhere in an incomprehensible web of causes or else dependent on past geological events that have faded beyond recall. Yet a light glimmers. Enough solid analyses and theory have locked together to suggest a relatively simple solution, or at least one that can be easily understood: the Energy-Stability-Area Theory of Biodiversity. In a nutshell, the more solar energy, the more stable the climate, and finally, the larger the area, the greater the diversity (*1*)."

"There is growing recognition of the need for a crash program to map biodiversity in order to plan its conservation and practical use. With up to a fifth or more of the species of all groups likely to disappear over the next 30 years, as human population doubles in the warmer parts of the world, we are clearly faced with a dilemma. But what is the best way to proceed?".

Raven and Wilson (*2*) provide answers to their own question. "Some systematists have urged the initiation of a global biodiversity survey, aimed at the ultimate full identification and biogeography of all species. Others, noting the shortage of personnel, funds, and above all, time, see the only realistic hope to lie in overall inventories of those groups that are relatively well known now, including flowering plants, vertebrates, butterflies, and a few others. In order to accomplish this second objective as quickly as possible, it would be necessary to survey transects across broad geographic areas and to examine a number of carefully selected sites in great detail. A reasonable number of specialists is available to begin this task, and with adequate funding it could be applied directly to problems of economic development, land use, science, and conservation. Meanwhile, adequate numbers of specialists could be trained and supported to deal with all of the remaining groups of organisms. The aim would be to gain a reasonably accurate idea of the representation of these groups on Earth while attempting complete inventories of all the global biota over the course of the next 50 years. As most of the tropical rain forests of the world are likely to be reduced to less than 10 percent of their original extent during this half-century, adequate planning is of the essence. The results from inventories should be organized in such a way as to apply directly to the development of new crops, sustainable land use, conservation, and the enhancement of allied disciplines of science (*2*)."

These overall inventories are clearly restricted to morphological and anatomical aspects, in spite of the fact that further on in the article one reads that "as networks of expertise and monographing grow, ecologists, population biologists, biochemists and others will be drawn into the enterprise. It is also inevitable that genome descriptions will feed into the data-base. Molecular biology is destined to fuse with systematics. [Furthermore,] chemical prospecting, the search for new natural products, is readily added to the collection of inventories. So is screening for species and gene complexes of special merit in agriculture, forestry, and land reclamation (*2*)."

Although it is probably true that macromolecular data are destined to shortly join morphological and anatomical ones in systematics, if indeed this has not yet already happened, it is to be doubted if this fusion will be of substantial relevance for the purpose of the present program, the mapping of biodiversity. This is due to the paradigm of unit and diversity in evolution formulated by Eigen and Schuster (*3*). "Why do millions of species, plants and animals, exist, while there is only one basic machinery of the cell; one universal genetic code and unique chiralities of the macromolecules?" Hence, a glimpse into biodiversity preferentially requires analysis of the phenotype, and this consists in form and physiology in intimate association. That we do not yet perceive the mechanistic details of this association is of little importance in connection with the present task. The point remains that we must report both characteristics together if we really want to understand development, radiation and relevance of biodiversity. Hence, chemical prospecting must be promoted to one of the most important aims of the survey, although, even in Raven and Wilson's definition (*2*), it should already please the funding agencies in view of its possible direct returns in benefits to mankind.

Morpho-Chemical Biodiversity

Why should morphology and metabolism be linked into an inseparable fabric, or even more fundamentally, before we takle this question, are these features indeed interdependent? A simple, even if circumstantial evidence for the connection of form and chemistry refers to the fact that great numbers of species (defined morphologically) and micromolecular constituents characterize plants and invertebrates, small numbers of species and constituents characterize vertebrates. This does of course not mean that morphological and metabolic biodiversity must necessarily co-occur in plant taxa. Indeed, while in one group many species may possess a rather uniform chemical composition, in another few species may show a quite heterogeneous one.

Morpho-Chemical Biogeography

Latitudinal Radiation of Angiosperms. This caveat does not concern the validity of the visual observation. Travel from both poles to the equator quite obviously demonstrates a prodigious increase in plant diversity. The doubts concern the comprehensiveness of the latitudinal gradient. The concept, as defined so far, involves only the macroscopic features of the phenotype. The chemical aspects, which do not contribute less to plant diversity, and hence to adaptation potential, are usually neglected. Do such cryptic attributes occur invariably according to analogous latitudinal gradients? A generally valid answer to this question has yet to be found.

Nevertheless, all data obtained so far (next chapter) indicate chemical diversity of floras to increase upon travelling from the equatorial forests into regions of lower latitude. On the other hand, qualitative micromolecular gradients seem to be closer related with habit, rather than with habitat. The alternate expression of morphological and metabolic biodiversity, controlled conceivably by an on/off mechanism, suggests the integration of both phenotypic features, only the operation of the switch leading to the alternatives remaining to be discussed.

Both, form and chemistry, are manifestations of metabolism (4). Metabolic differentiation involves initially so called primary (ubiquous) and secondary (special) micromolecules. Biosynthesis and biodegradation of these metabolites occur from and to carbon dioxide along cyclic reaction pathways (Figure 1) with two major outlets towards polymeric (macromolecular) material. In other words, part of the oxygen produced by photosynthesis in green plants is recycled through the atmosphere and used up in the other fundamental activity of life, respiration. The remaining part of the oxygen is retained in the atmosphere (at least during most of the plant's lifetime) in view of the relative stability of the polymers, in plants chiefly cellulose and lignins. Now, it so happens that sunlight intervenes in the formation of both these materials. Energy is required in the reductive process leading to glucose, dehydration of which gives cellulose; and light activation of phenylalanine ammonia-lyase (PAL) is implicated in the deaminative process leading to the phenylpropanoid-precursors of lignins. Thus, also in theory, one would expect that in equatorial regions, characterized by the highest solar energy input, production of ligno-cellulosic biomass should attain the highest rate. In consequence, internally, micromolecular turnover, responsible for the molecular continuity between the two biomass yielding processes, must display a correspondingly high rate. This speedy carbon flow through the cycles would be expected to lessen the opportunities for stabilization of special micromolecules and shorten their half-lives. Externally, however, the faster the production of biomass, the vaster the area covered per unit time and the greater the chances of morphological adaptation (cf. the ESA theory, above).

Accordingly, lowering of energy input must lead to gradual decrease in biomass deposition. The consequent diminishing rate of carbon flow through the cycles would induce scavenging of intermediates by condensation into more difficultly degradable, more complex (often considered bizarre) "natural products". The rationale behind this concept lies in the positive evolutionary relationship of increasing oxidation level of special metabolites and versatility in the elaboration of molecular protection devices, such as Schiff bases, methyl ethers and hydrogenated derivatives during the biosynthesis of alkaloids, polyphenols and terpenoids, respectively (5).

Raven and Wilson's advice that inventories should be limited to groups that are relatively well known now (2) is extremely relevant. Crash programs are incompatible with the slowness inherent in experimental verifications. I doubt that in the next say 5 years the presently available mass of data can be increased significantly in the sense that the new data added to the existing ones will relevantly alter trends which are not already perceptible through the evaluation of

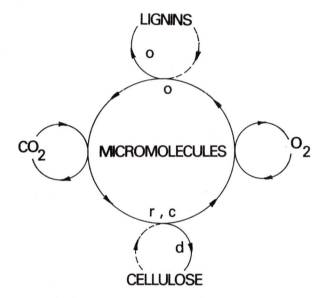

Figure 1. Schematic representation of carbon flow through metabolic pools in plants. Major reaction types: r...reduction, c...condensation, d...dehydration, o...oxidation (broken lines ... post mortem events).

data accumulated in the past 50 years with respect to micromolecular botany and in the last 250 years with respect to morphological botany. Clearly the statement stands on less firm grounds concerning chemical features. However, we ourselves have repeatedly verified its validity. To give only one example: a Master thesis listed the reported distribution of complex steroids (6). Later, in order to publish this work in the primary scientific literature (7), the data were updated by an increase of about 20% in the number of entries. Nevertheless, it was hardly necessary to introduce alterations in the text. The interpretations of the results based on distributional trends had remained valid.

The survey of plant transects across broad geographic areas has in the past referred nearly always to the macroscopic, morphological part of the phenome. Unless the submicroscopic, chemical characteristics are considered additionally, we will never learn the lesson biogeography is able to teach on the subject of biodiversity.

So far examples of integration of morpho-chemical biogeography were elaborated through two different types of approaches. In the first one we selected a botanical group, circumscribed its subgroups on a map, characterized each of these regionalized subgroups by their chemical composition and tried to interpret these compositions by mechanistically or biogenetically acceptable chemical gradients. In the second approach not the plant group, but the class of micromolecular components was selected and its variance, properties and frequencies in ecogeographically characterized plant groups was gauged. The following items summarize some of the results.

Icacinaceae.   Chemical evolution in the Icacinaceae, postulated to involve oxidative sequences within monoterpenoids, sesquiterpenoids and diterpenoids, is accompanied by spacial radiation of genera from Melanesia in a western direction along the tropical belt to Amazonia (8).

Simaroubaceae.   Specialization of quassinoid skeletons is accompanied by a West-East spacial radiation of the simaroubaceous lineage. Indeed the transition from American and West African genera to East African and Asian genera is accompanied by diversification of oxygenation and unsaturation patterns, as well as by increase in oxidation level of the quassinoids (9).

Fabaceae.   Quinolizidine alkaloidal evolution proceeded by skeletal specialization in tropical regions and by variation of oxidation level in temperate regions (10).

Pyrones in Lauraceae, xanthones in Gentianaceae.   The variation of secondary plant constituents in major plant taxa takes place in small steps which allow to trace plant evolution and at the same time also reflect the dispersion of these taxa (11).

***Derris-Lonchocarpus* complex (Fabaceae).**   Gradual modification of oxidation/methylation values of flavonoid profiles suggest dispersal of original stock to have taken place from Asia to America and from forest to savanna. Similar trends are observed for other genera of Tephrosieae (*12*).

Aniba (Lauraceae). A comparison between geographical distribution and secondary metabolites of 18 species of the genus *Aniba* leads to a coherent picture of spacial and chemical evolution of these species. In chemical evolution, general mechanisms, as blocking of reaction steps leading to primary metabolites within biogenetic groups, are found to be operative. Sympatric distribution of closely related species seems to imply chemical diversity (*13*).

Conclusion

Brazilian Amazonia, populated by 18 million inhabitants, is facing with increasing frequency agonizing choices concerning what native areas can be sacrificed for social benefit, maintaining at the same time the dreaded extinctions within reasonable limits. The present chapter aimed to show that floristic inventories will continue in the realm of academic exercises until they consider morpho-chemical data within an integrated system. Methods are quoted that, for the first time, allow the attainment of this objetive. Results obtained so far seem to indicate opposed latitudinal gradients to characterize morphological and metabolic biodiversity, and seem to suggest spacial radiation of angiosperms to have followed morphological and chemical gradients. This conclusion is consistent with results obtained through the application of a quantitative measurement of angiosperm biodiversity (*14*) and dispells any doubt that plant classification must consider the different levels of manifestation of the genotype as criteria by an integrated procedure to qualify as a "natural" system (*15*). Hence much of our effort these past years centered around the development of such a system. This can be gauged by the title of the most recent contribution "Plant systematics via integration of morphology and chemistry" (Gottlieb, O.R.; Borin, M.R. de M.B.; Kaplan, M.A.C., Oswaldo Cruz Foundation and Federal University of Rio de Janeiro, unpublished paper).

Further investigation along these lines of problems concerning biodiversity and biogeography in Amazonia should comprise the following stages:
1. Selection of plant families possessing a relatively well known and reasonably well diversified micromolecular composition.
2. Organization of the chemical data referring to genera of these families (genera are here selected as basic taxonomic units since their circumscription is relatively stabilized in taxonomy).
3. Indication of habitats and habits of these genera.
4. Interpretation of eventual morphological and chemical gradients (by correlation of reaction sequences based on chemical mechanisms along the geographic routes of radiation of the genera).

Acknowledgment

The autor is grateful to Conselho Nacional de Desenvolvimento Científico e Tecnológico, Brazil, for financial support.

Literature Cited

1. Wilson, E. O . *The Diversity of Life*, Allan Lane, Ed.; The Penguin Press: London, 1992, pp. 197-199.
2. Raven, P.H.; Wilson, E.O. *Science* **1992**, *258*, 1099-1100.
3. Eigen, M.; Schuster, P. *Naturwiss.* **1977**, *64*, 541-565.
4. Gottlieb, O.R. *Phytochemistry* **1989**, *28*, 2545-2558.
5. Gottlieb, O.R. *Phytochemistry* **1990**, *29*, 1715-1724.
6. Borin, M. R. de M. B. *O valor dos Esteróides como Marcadores em Quimiossistemática*; M.Sc.-Dissertation; Universidade de São Paulo, São Paulo, SP, **1988**.
7. Borin, M.R. de M.B.; Gottlieb, O.R. *Pl. Syst. Evol.* **1993**, *184*, 41-76.
8. Kaplan, M.A.C.; Ribeiro, J.; Gottlieb, O.R. *Phytochemistry* **1991**, *30*, 2671-2676.
9. Simão, S.M.; Barreiros, E.L.; Silva, M.F. das G.F. da; Gottlieb, O.R. *Phytochemistry* **1991**, *30*, 853-856.
10. Salatino, A.; Gottlieb, O.R. *Pl. Syst.Evol.* **1983**, *143*, 167-174.
11. Gottlieb, O.R.; Kubitzki, K. *Naturwiss.* **1983**, *70*, 119-126.
12. Gomes, C.M.R.; Gottlieb, O.R.; Marini-Bettolo, G.-B.; Delle Monache, F.; Polhill, R. *Biochem. Syst. Ecol.* **1981**, *9*, 129-147.
13. Gottlieb, O.R.; Kubitzki, K. *Pl. Syst. Evol.* **1981**, *137*, 281-289.
14. Gottlieb, O.R.; Borin, M.R. de M.B. In *The Use of Biodiversity for Sustainable Development: Investigation of Bioactive Products and their Commercial Applications*; Seidl, P. R., Ed.; Associação Brasileira de Química: Rio de Janeiro, RJ, 1994, pp 23-36.
15. Gottlieb, O.R. *Micromolecular Evolution, Systematics and Ecology - An Assay into a Novel Botanical Discipline*; Springer Verlag: Berlin, 1982, pp 170.

RECEIVED September 28, 1994

AMAZON BIODIVERSITY: ENVIRONMENTAL ISSUES

Chapter 18

Advances in Amazonian Biogeochemistry

K. O. Konhauser[1], W. S. Fyfe[1], W. Zang[1], M. I. Bird[2],
and B. I. Kronberg[3]

[1]Department of Earth Sciences, University of Western Ontario, London,
Ontario N6A 5B7, Canada
[2]Research School of Earth Sciences, The Australian National University,
GPO Box 4, Canberra 2601, Australia
[3]Geology Department, Lakehead University, Thunder Bay,
Ontario P7B 5E1, Canada

Recent observations on the geochemistry, biogeochemistry and
stable isotope systematics of waters, soils and sediments of the
Amazon region indicate the profound influence of biological
processes on the nature of the surface and near surface materials.
These same materials provide evidence for major change in
vegetation types and climate during the recent past.
The chemistry of local soils and waters provides key indicators of
the agricultural and mineral resource potential of various regions.
The extremely deep bio-weathering in the Amazon system has
produced many important and new types of mineral deposits. In
regions of intense and deep leaching, the input of many chemical
species may be dominated by rain and aerosols.

Multiple Fluxes Influencing Amazonian River Chemistry

The Amazon River Basin is the largest modern river system in terms of drainage
area (6 million km^2), supplying ~20% of fresh water to the world's oceans (5.5 x
10^{12} m^3/year, Villa Nova et al.) (1) and discharging ~10^9 metric tons of sediment
to the Atlantic Ocean (Meade et al.) (2,3). The system's headwaters begin within a
few hundred kilometres of the Pacific Ocean and its tributaries extend for hundreds
of kilometres throughout central South America. The entire Basin may be divided
into the mountainous Andean regions and the much more extensive lower basins.
Chemical signatures of surface soils and sediments in lower Amazonia show these
materials have undergone intense chemical weathering, while, in the Andes
Mountains, weathered material is still being generated.
  Early studies in the Amazon Basin showed that the waters were chemically
and physically very heterogeneous. The first classification of Amazonian rivers was
based on their physical appearance (4). Whitewater rivers are rich in dissolved

0097–6156/95/0588–0208$17.00/0
© 1995 American Chemical Society

solutes and are extremely turbid owing to their high concentrations of suspended sediment. Blackwater rivers are relatively infertile rivers characterized by low sediment yields and their "tea-colored" (*5*) acidic waters that are rich in dissolved humic material derived from decaying surface vegetation (*6, 7*). Finally, clearwater rivers are relatively transparent, green-colored rivers that are neither turbid with detrital materials or colored by humic compounds (*8*).

The chemistry of Amazonian rivers have long been attributed both to the variability of the geological source and to the erosional regime through which the rivers flow. Early studies by Raimondi (*9*) and Katzer (*10*) observed that the low dissolved inorganic concentrations in some lowland rivers contrasted with those of rivers draining the Andes Mountains. It was later shown that the rivers flowing through the central lowlands were typically blackwater in composition, whitewater streams were distinctive of the waters draining the Andes, while clearwater types were characteristic of rivers originating in the Precambrian shields (*11-13*).

More recently, several papers have begun to deal specifically with the elemental composition of Amazonian rivers (*13-20*). Studies such as Stallard and Edmond (*13*) have indicated that the distribution of major cations and anions in the dissolved load were controlled by substrate lithology in the source regions. This relationship was later extended to several trace elements (*20*). The chemistry of rivers in the Amazon Basin have also been attributed to the atmospheric precipitation of cyclic salts (*14*) with the ratios of major dissolved ions in lowland rivers (which are similar to sea salt) indicative of a marine origin. Despite the compilation of published work on the chemistry of Amazonian waters, most studies have dealt specifically with 1) a limited number of major elements and trace metals, and 2) the impact of geologic, geochemical, and petrographic properties of the source regions on the chemical composition of surface waters. While weathering is undoubtedly of fundamental importance in the supply of solutes to a river system, this study suggests that other factors also have impact on Amazonian river chemistry.

Influence of Soil Geochemistry and Mineralogy. Since the earliest studies on Amazonian river chemistry, it has become apparent that the chemical and physical variability exhibited by surface waters were largely attributed to geologic heterogeneity of the Amazon Basin. On the basis of distinct differences in geology, soil types, and vegetation, Fittkau et al., (*21*) have divided the Amazon Basin into three major geochemical provinces (Figure 1): the western peripheral region, where soils (known as the varzea) and waters are fertile, the Central Amazon, an area where the soils and waters are deficient in nutrients, and the northern and southern peripheral regions (Precambrian Shields) with intermediate compositions (*20*).

Based on similarities between the major cation and trace metal levels in the soil samples (Table I), and the dissolved metal concentrations in surface waters flowing through the varzea, the Central Amazon, and the Shields (Figure 2), our research confirms that the rivers were in chemical equilibrium with their drainage

Table I. Average Chemical Composition of Soils in Study Areas with Crustal Abundance

	Crustal Abundance	Central Amazon	Shield	Varzea
SiO$_2$	58.50	70.22	51.78	63.05
TiO$_2$	1.05	0.62	1.17	...0.88
Al$_2$O$_3$	15.80	16.66	17.15	.15.47
Fe$_2$O$_3$	6.50	2.79	19.02	5.87
MnO	0.14	0.02	0.14	0.10
MgO	4.57	0.18	0.26	1.75
CaO	6.51	0.00	0.06	1.38
K$_2$O	2.22	0.04	0.22	2.21
P$_2$O$_5$	0.25	0.04	0.10	0.18
Na$_2$O	3.06	0.00	0.02	1.25
Nb	11.00	19.75	19.50	16.00
Zr	100.00	600.25	361.50	330.00
Y	20.00	8.65	25.75	39.00
Sr	260.00	9.50	22.13	214.50
Rb	32.00	5.20	15.95	99.00
Pb	8.00	<13.50	13.45	21.50
Zn	80.00	<9.75	28.75	91.75
Cu	75.00	<8.13	363.00	27.25
Ni	105.00	6.85	17.25	30.25
Co	29.00	<5.00	20.50	16.75
Cr	185.00	113.50	161.50	58.50
Ba	250.00	51.25	66.50	466.25
V	230.00	32.00	231.75	100.75
As	1.00	<5.00	16.25	<8.20
Ga	18.00	14.25	24.50	20.25
U	0.91	<3.50	<6.25	<5.00
Th	3.50	14.25	<27.75	<9.00

Major oxide compositions (in wt.%) and trace metal compositions (in Og/g). Values from Konhauser et al. (22). Crustal abundances are from Fairbridge (23) and Taylor and McLennan (24). Averages with "<" indicate concentrations below given value.

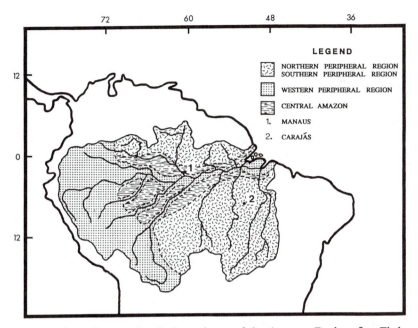

Figure 1. The major geochemical provinces of the Amazon Basin, after Fittkau et al. (*21*). Location of the study areas in Manaus and Carajás are shown (*25*).

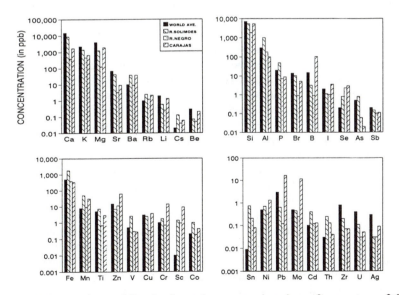

Figure 2. Comparison of dissolved metal concentrations in surface waters of the Rio Solimões, the Rio Negro, and forested streams in Carajás with world river average (*25*).

basins (25). In geologically active areas, such as the Andes Mountains, crustal elements such as Si, Al, K, Na, Ca, Be, Li, Zn, Mo, F, and B are mixed with newly extruded mantle elements such as Ca, Mg, Cu, Co, Ni, S, V, Cr, and Se (26). With high rates of erosion of the varied lithologies in the source regions, enormous quantities of unweathered minerals (feldspar and mica), metal-rich clays (smectite, chlorite, and illite), and dissolved metals are transported downstream by whitewater rivers (25), resulting in the fertile soils of the varzea that are almost exclusively derived from the annual deposition of new sediments during seasonal floods (12). The chemistry of the dissolved size fraction in the Rio Solimões, therefore, reflects the substrate lithology in the Andes, with the high concentration of calcium, for example, indicative of a limestone source (13), while continuous leaching of the varzea soils further supplies the solute-rich river with additional metals.

The high concentration of dissolved metals in the Rio Solimões is contrasted by the solute-deficient waters of the Rio Negro which drains the highly weathered lateritic and podsolitic terrains of the Central Amazon. Due to a lack of exposed rock, the intense chemical weathering of the humid tropics over millions of years, and the low rates of weathering in conjunction with the development of thick, siliceous and aluminous soils, the suspended sediment are typically cation-depleted, consisting almost entirely of quartz and kaolinite (27), while the dissolved load is dominated by silicon, with extremely low levels of major cations and trace metals.

Rivers that drain the Precambrian Shield typically carry a limited suspended and dissolved load (28), reflecting the tectonic stability which leads to low erosional rates and high leaching (26). Furthermore, in these stable cratonic regions, which are favored by sedimentary rocks such as orthoquartzites, with no nutrients, or granitic and metamorphic rocks, with a limited nutrient spectrum and low erosional rate (27), the soils are typically depleted in nutrients. Although these conditions should give rise to extremely solute-deficient rivers, the waters flowing through Carajás have an intermediate composition. This discrepancy may largely reflect the fact that Carajás is a very metallogenic area, rich in mineral deposits (29), which upon weathering could supply dissolved metals to local river systems.

These observations, collectively, suggest that we should be able to classify the chemical composition of rivers according to the geochemistry and mineralogy of the soils through which they flow. This has profound implications for using water chemisty as an indicator of the agricultural and mineral potential of a region. A comparison of three key macronutrients and the total dissolved inorganic solids (TDS) of several great rivers of the world speak eloquently about the state of the soils (Table II). The Danube, Mississippi, Yangtze, and Nile river systems all have high dissolved metal concentrations, reflecting the young, fertile soils through which they flow. Since plants require a wide range of macronutrients and micronutrients, their solute-rich waters indicate a high capacity to support bioproductivity. In contrast, the extremely low concentration of dissolved metals in the waters of the Rio Negro is indicative of the poor nutrient status of the Central Amazon. The inability of the lateritic soils to support agricultural activity (26) is probably the reason why historically this region has never been overpopulated by humans (*30*). Therefore, in regions such as the Central Amazon, chemical analyses of the river systems should reflect the incapability of the soils to sustain long-term crop production. Clearly then, using river chemistry as an agricultural indicator would be a very efficient way of assuring soil potential prior to massive development projects.

Table II. Comparison of Dissolved Solutes in Principal World Rivers (in µg/L)

	Ca	Mg	K	TDS
Danube	49.0	9.0	1.0	265.0
Mississippi	39.0	10.7	2.8	265.0
Yangtze	45.0	6.4	1.2	232.0
Nile	25.0	7.0	4.0	225.0
Lower Negro	0.4	0.1	0.5	3.4

"World river values from ref. 31".

Knowing the chemical composition of a river system may also be of use in determining the mineral potential of a region. The high concentration of a number of metals (such as Zn, Cu, Cr, Ni, Pb, Mo, Co, Mn, Rb, Ba, B, I, and Se) in the waters draining the Shield areas in Carajás (relative to both Amazonian rivers and world river average) confirms the presence of one of the greatest mining areas of the modern world. While the high aqueous concentration of some of these metals may be associated with local mining activity, many of these rivers drain completely forested areas with no anthropogenic influence. Under these conditions, river chemistry may be useful in mineral exploration.

Impact of Microbial Metal Sorption and Biomineralization.   Many of the chemical processes occurring within aqueous systems have been shown to be biologically mediated. Although all forms of life possess some characteristics that make them biogeochemically important, much of the focus today is being directed towards the role of microorganisms, such as bacteria and algae, in the cycling of elements. Their significance arises from their ability to not only live in an unparalleled variety of environments, but also their ability to change their surroundings.

In most aquatic systems, the majority of microbial activity appears to be associated with submerged solid surfaces (*32*). The complex microbial communities that develop on substrata are commonly referred to as biofilms, and consist of a consortia of bacteria, cyanobacteria, and algae held firmly together in a highly hydrated polymeric matrix of polysaccharides extruded by the cells (*33*). The formation of biofilms is generally regarded as a growth strategy that affords attached cells with a survival advantage not available to free-living cells (*33*). Benthic microorganisms benefit not only from protection against the high shear of turbulent flow (34), but also through the elevated concentrations of organic and inorganic compounds that accumulate at solid-liquid interfaces (35).

In order to utilize essential metallic ions as nutrients for growth, microorganisms must have the ability to adsorb and concentrate metal cations from solution. In many microorganisms, this may be accomplished, in part, through electrostatic interactions with reactive acidic groups (carboxyl and phosphoryl) contained within the constituent polymers of microbial cell walls (36,37). A number of experimental studies have clearly demonstrated that substantial quantities of various metals can be bound and accumulated by a variety of bacterial cells (38-40), whereas Kuyucak and Volesky (41) have shown that the cellulosic materials in filamentous algae played an important role in the biosorption of gold.

Microbial mediated physiological processes can also have profound effects on the precipitation of mineral phases. In bacteria biominerals are commonly generated as secondary events from interactions between the activity of the microorganisms and its surrounding environment (42). Based on the inherent metal-binding capacity of the anionic structural polymers that reside in the cell-wall fabric, Beveridge and Murray (39) have proposed a two-step mechanism for the development of authigenic mineral phases in association with bacterial cells. The first step involves a stoichiometric interaction between metals in solution with the cell's reactive chemical groups. Once bound to the bacteria, these metals can then serve as nucleation sites for the deposition of more metal ions from solution. The end result is a mineralized cellular matrix that contains detectable concentrations of metal ions that are not easily solubilized (43). In contrast, unicellular algae, such as diatoms, are examples of biologically-controlled mineralizers that remove silicon to fulfill physiological functions such as DNA synthesis and to form their siliceous skeletons (44). This process of biomineralization is characterized by the development of an organic framework or mold into which various ions are actively introduced. During biologically controlled mineralization an area is sealed off from the environment allowing specific ions of choice (e.g., Si) to diffuse into the space delineated for mineral precipitation. The cell therefore, creates an internal environment completely independent of the external conditions. It is under these "ideal" conditions that mineralization may occur (45).

The role of bacteria. In a recent study on the activity of benthic bacteria in Amazonian rivers (22), microbial biofilms were collected off of submerged plants and sediment, and subsequently analyzed with transmission electron microscopy (TEM), coupled with energy dispersion X-ray spectroscopy (EDS) and selected area electron diffraction (SAED). With TEM, it was observed that bacterial biomineralization involved a complex interaction between metals in solution with the reactive components of the cell. The anionically-charged cell wall and the encompassing layers provided special microenvironments for the deposition of iron and other soluble cationic species. Ferric iron, which exhibits unstable aqueous chemistries, was bound in significant amounts, and previous studies indicate that this may be sufficient to induce transformations to the insoluble hydroxide form (e.g. ferrihydrite) (*46*).

Through progressive mineralization, bound iron served as nucleation sites for the precipitation and growth of the complex (Fe, Al) - silicates (Figures 3, 4A). It is likely that the initial (Fe, Al) - silicate phases precipitated directly through the reaction of dissolved silicon and aluminum to the bound metallic cations, with continued aggregation of these hydrous precursors resulting in the formation of low-order, amorphous phases that are characterized by large surface areas with a high adsorptive affinity for additional metal cations (*47*). Because they lack a regular crystal structure, these hydrous compounds are unstable and will, over time dehydrate, converting to more stable crystalline forms. In our study area, the hydrous chamosite-like clay appeared especially reactive to silicic acid (H_4SiO_4). Continued adsorption of silicon, through hydrogen bonding of the hydroxyl groups in the bound cations with the hydroxyl groups in the soluble silica seems to have accompanied the conversion of the low-order phase to the crystalline phase (*48*). Eventually this process led to the complete encrustation of the bacterial cell within a mineralized matrix.

It is an interesting observation that all bacterial populations examined in the Rio Solimões consistently formed identical mineral phases. This suggested that the bacteria from our sampling area, regardless of their physiology, were capable of serving as passive nucleation sites for (Fe, Al) - silicate precipitation. Conversely, all bacteria from the Rio Negro showed a conspicuous absence of mineralization (Figure 4B).

It is clear then, that the differences in mineral accumulation exhibited by bacterial cells in the Rio Solimões and the Rio Negro must reflect differences in the physical and chemical conditions of their riverine environments. The Rio Solimões is a fertile river, rich in suspended sediments and dissolved inorganic solutes. As a result, the bacteria will have an abundant supply of metals in solution with which to complex and to accumulate. In contrast, the Rio Negro is an extremely infertile river characterized by low levels of major cations and trace metals. With very dilute waters, the bacteria presumably are unable to bind sufficient quantities of metals to form authigenic mineral phases, suggesting that both metal sorption and biomineralization largely reflect the availability of dissolved solutes in the water column.

In a solute-rich river system such as the Rio Solimões, the fate of the metal-loaded bacteria may have profound implications for the transfer of metals from the hydrosphere to the sediment (*49*). Given that planktonic bacterial populations in the Rio Solimões are typically on the order of 1×10^6 to 4×10^6 cells/ml (*50*), it is not difficult to imagine that these microorganisms could effectively cleanse the water of dilute metals and partition them into the sediments (*43*). Furthermore, in the microenvironment overlying a biofilm, the aqueous chemistry will be altered by the activity of the microorganisms present. Although this water layer is very thin, when one takes into consideration the large surface area of solid substratum on a river's bed that is colonized by biofilms, the volume of water that falls directly

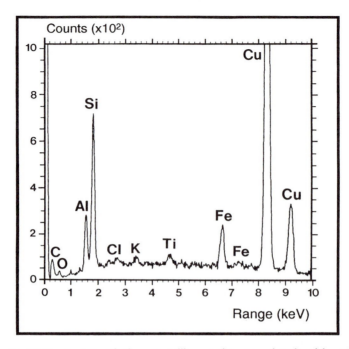

Figure 3. EDS spectra of the crystalline grains associated with epiphytic bacterial cells from the Rio Solimões. Cu peaks are from the supporting grid and the Cl peak is due to contamination from the embedding material (EPON).

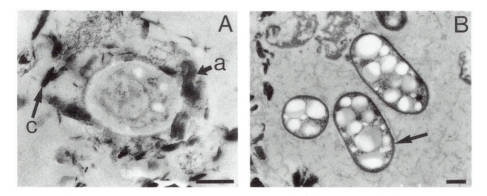

Figure 4. TEM image of (A) a completely encrusted epiphytic bacterial cell (stained with uranyl acetate and lead citrate) from the Rio Solimoes, with a complete range of morphologies including the amorphous, low-order phases (a), transition phases (t), and crystalline phases (c) and: (B) a small colony of epiphytic bacteria (stained) from the Rio Negro, loosely encapsulated by organic material. Arrow indicates the low levels of mineralization on the outer cell wall. Scale bar = 400 nm.

under microbial influence is substantial. In this regard, biofilms dominate the reactivity of the water-substrate interface and influence the transfer of metals from the hydrosphere to the sediment. Through diagenesis, the bound metals may also become immobilized as stable mineral phases and collect as sediment, suggesting the importance of microorganisms in metal deposition, low temperature clay formation, and, invariably, mudstone diagenesis.

The role of algae. Filamentous algal samples collected from different riverine environments in both the Manaus and Carajás areas indicated that metals were consistently removed from the waters by these eucaryotic microorganisms (*51*). The pattern which seems to arise is that algae effectively sequester and concentrate the metals available to them within their aqueous environment (Table III). Some metals are extracted from their surrounding environment to fulfill essential physiological functions, whereas the adsorption of other metals, without known cellular functions, largely reflects the strong complexing ability of the reactive organic components of the living cell.

If we conclude that metal concentrations within the algal biomass are a direct reflection of availability, then it is not surprising that the concentration of metals in the filamentous algae of the Rio Solimões greatly exceeded those of the Rio Negro. A similar relationship of metal concentration also arises between algal samples obtained from forested and completely deforested regions. Tropical plant species have adapted over millions of years to the highly leached and weathered soils of the tropics by developing a closed system that effectively conserves and recycles needed nutrients (*52*). Under these forested conditions the river system is continuously supplied with nutrients through the slow leaching from leaves, soils, and decaying organic matter. Removal of the vegetation through logging and slash-and-burn agricultural practices, however, disrupts the nutrient-conserving mechanisms through increased soil erosion. Over time this causes the level of soil nutrients to fall below the level of an undisturbed forest (*52*), with nutrient inputs into the river systems becoming minimal. This situation appears consistent with the data from these algal samples which show a higher metal concentration in the microorganisms from the forested streams.

The role of algae in biogeochemical cycling extends beyond trace metal extraction. Siliceous microorganisms, such as the diatoms, are commonly the most abundant freshwater eucaryotic microorganisms, and the development of large populations is invariably accompanied by a marked decline in the amount of dissolved silicon from the waters in rivers (*53*) and lakes (*54*). This situation is consistent with the surface waters of the Rio Negro, where the dissolved silicon levels were a maximum of 2100 µg/L. Despite low silicon levels, which are characteristic of Central Amazon rivers, the waters from the Rio Negro were capable of supporting prolific diatom growth, with scanning electron microscopy (SEM) indicating that submerged wood (Figure 5A), rocks, and leaves served as their solid substrates for growth (*55*).

Table III. Average Chemical Composition of Waters and Algae in Amazonian Rivers

	Rio Negro			Rio Solimões			Forested			Deforested		
	Water (ug/L)	Algae (ug/g)	Conc. Factor	Water (ug/L)	Algae (ug/g)	Conc. Factor	Water (ug/L)	Algae (ug/g)	Conc. Factor	Water (ug/L)	Algae (ug/g)	Conc. Factor
Ti	0,68	336,25	494.485	7,37	5850,00	793.758	2,86	3677,50	1.285.839	3,00	1233,33	411.111
V	0,29	12,80	44.138	2,52	210,00	83.333	0,27	108,25	400.926	0,50	70,33	141.801
Cr	0,69	9,63	13.960	1,73	141,75	81.936	14,16	86,65	6.119	12,20	48,67	3.989
Mn	9,63	87,25	9.060	50,13	8400,00	167.564	31,25	3200,00	102.400	403,67	3600,00	8.918
Fe	385,00	9475,00	24.610	1796,25	65500,00	36.465	331,00	54500,00	164.653	1596,67	52666,67	32.985
Co	0,18	1,98	10.972	1,00	57,00	57.000	0,41	92,50	225.610	2,09	57,00	27.273
Ni	0,32	4,20	13.125	0,71	67,00	94.366	1,33	66,75	50.188	2,11	32,00	15.166
Cu	0,32	7,33	22.891	2,40	94,75	39.479	3,70	309,75	83.716	8,73	80,67	9.240
Zn	11,54	48,75	4.224	7,10	207,50	29.225	61,26	150,75	2.461	170,00	138,00	812
As	0,06	1,00	16.625	0,79	26,25	33.228	0,02	7,56	377.750	0,81	6,90	8.487
Zr	0,07	8,33	118.929	0,20	108,00	540.000	0,07	154,75	2.210.714	0,06	31,67	558.495
Mo	0,25	0,72	2.890	0,45	2,53	5.611	11,61	2,20	189	0,17	2,94	17.711
Ag	0,03	0,66	21.892	0,03	6,02	200.500	0,09	0,71	7.867	0,07	0,24	3.381
Cd	0,12	0,10	863	0,40	2,11	5.281	0,13	0,41	3.173	0,15	1,30	8.475
Sn	0,21	0,55	2.607	0,75	3,85	5.133	0,08	3,09	38.594	0,08	2,10	27.632
Hg	0,17	0,09	506	0,12	0,46	3.813	0,25	0,18	711	0,24	0,18	774
Pb	0,12	7,75	64.583	0,64	39,25	61.328	16,51	8,60	521	6,27	11,90	1.898
U	0,03	0,48	16.000	0,11	4,65	42.273	0,05	5,03	100.500	0,37	3,00	8.174

Note: conc. factor = concentrations of metal in algae/concentration of metal in water. Values from Konhauser and Fyfe (51)

In accordance with low weathering rates exhibited in the Central Amazon, the source of silicon for diatom growth is largely sustained through the recycling of biogenic silica (*56*). The recycling of silicon in diatoms involves the dissolution of opaline skeletons and subsequent biological uptake. Sediment samples collected along the Rio Negro indicated the ubiquitous presence of diatom fragments which served as a major source of dissolving silica. Furthermore, within the small tributaries and flooded forests of the Rio Negro where benthic communities (e.g. *Navicula* sp.) are abundant, dissolution began in situ on the substratum. Upon death of the microorganism, the remnant siliceous frustules became fragmented, partially dissolving, and within micrometres to millimetres of the surface, reprecipitated as textureless, siliceous overgrowths (Figure 5B). These amorphous overgrowths are formed when the dissolved silicon levels become locally supersaturated before opal-A dissolution has gone to completion (*57*). Dissolution, therefore, becomes interrupted by the re-precipitation of a less soluble, nonbiogenic, siliceous ooze referred to as opal-A (*58,59*).

This dissolution-reprecipitation process may also alter the structure of the wood sample. The ubiquitous nature of the diatoms and the precipitation of the silica gel suggested that the wood samples were undergoing a void-filling process of silicification (*60*). In our samples this process began with the dissolution of diatoms on the surface and within. The silicic acid produced from the dissolving frustules may permeate throughout the wood. Hydrogen bonding between the hydroxyl groups in the silicic acid and the hydroxyl groups in the cellulose then leads to the deposition of opaline silica on the surfaces of individual wood cells (*60*). It is, therefore, no surprise to see a siliceous gel, both on the wood surface and within the wood.

The extent of this silicification process may be significant when the vast number of partially submerged trees in the study area is considered. The slow flowing movement of water within the flooded forests and the preferential attachment of diatoms to the outer wood surfaces may allow for silicon levels to build up sufficiently in microenvironments. Therefore, many of the trees within these flooded forests potentially can become reservoirs of highly reactive silica, disrupting the amount of silicon locally recycled through the freshwater system.

Metal Complexation by Organic Materials. Most studies of biogenic materials in Amazonian rivers have been confined to bulk measurements of the concentration of particulate organic carbon (POC) and dissolved organic carbon (DOC), traditionally defined by separation with a 0.5 μm filter (*61*). The POC is largely unreactive, and consists of a coarse fraction (>63 μm) comprised of tree leaf debris and some wood, with the fine fraction (<63 μm) derived primarily from soils and grasses (*5*). The DOC consists predominantly of humic substances, such as humic acids (HA) and fulvic acids (FA) (*62*). These humic substances are generally thought to be refractory, terrestrially derived organic matter leached from the surrounding soils (*63*). The remaining DOC includes potentially labile biochemical components such as proteins and carbohydrates (*64*).

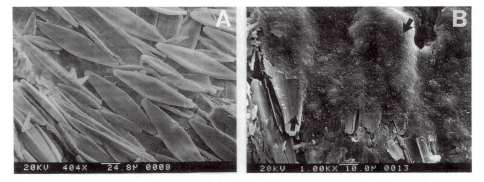

Figure 5. SEM images of (A) diatoms (*Navicula* sp.) on outer wood surface, collected upstream of Manaus on the Rio Negro and (B) the outer wood surface covered by a textureless, siliceous gel. Arrows indicate remnant frustule and siliceous gel.

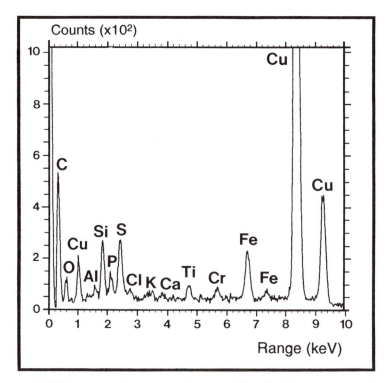

Figure 6. EDS spectra of metals complexed to "dissolved" organic material. Cu peaks are from the supporting grid and Cl peak is due to contamination from EPON.

In recent years, it has become increasingly apparent that consequential interactions exist between dissolved organic matter and the inorganic materials of freshwater systems. Studies have shown that significant quantities of trace metals are bound on the anionic surfaces of particulate humic substances (*65*). Constituent carboxyl and phenolic hydroxyl groups interact electrostatically with available cations in solution (*66*), forming organo-metallic complexes through ion exchange, surface adsorption, and chelation, with stabilities much higher than those of corresponding inorganic metal complexes (*63*).

Generally the concentration of dissolved solutes in a river system (approximately 30 mg/L, (*67*)) are more abundant than dissolved organic carbon (0.1 to 10 mg/L, with the upper value restricted to polluted systems (*68*). Under conditions of low DOC, interactions with metallic ions may cause only slight changes in the overall chemical composition of river water (*66*). Further, a majority of the reactive sites would be occupied by strongly bound metals such as Fe^{+3} or Al^{+3}, such that complexation of trace metals would be insignificant (*63*). It would, therefore, be advantageous to study a river system in which the relative abundance of dissolved organics to dissolved solutes would be much greater.

In the Amazon Basin, many of the rivers (blackwater varieties) are characterized by their high organic content and their solute-deficiency; the most notable being the Rio Negro with a TDS (total dissolved inorganic solids) of only 3.5 mg/L (*25*). The Rio Negro is the largest contributor of DOC to the Amazon River, whose average concentrations of 10.8 mg C/L greatly exceed those of the Rio Solimões with only 3.8 mg C/L (*7*). The Rio Negro is the major humic acid flux to the Amazon River system at 2.5 times the input of the Rio Solimões and also the largest fulvic acid flux amounting to 70% of the mainstream concentration. With a DOC:TDS of 3.1, the Rio Negro was chosen as an ideal study site for the role of dissolved organic matter in the complexation of dissolved metals. As the major tributary of the Amazon, which represents anywhere from 5 to 25% of the Amazon's discharge (*7*), any significant processes of metal cycling will have a marked influence on regional fluvial processes.

Samples of water were collected upstream of Manaus on the Rio Negro to identify DOC and determine its inorganic composition. Whole mounts of unfixed organic material (DOC) were prepared for electron microscopy by pipetting ~ 10 ΩL of water sample onto Formvar and carbon-coated copper grids. After allowing the water to evaporate off, the grids were analyzed by TEM and EDS.

Results from EDS analyses clearly revealed the reactive nature of the dissolved organic matter (Figure 6). Major cations in the waters of the Rio Negro, such as Fe, Si, and Al were shown to form the greater part of the organo-metallic complexes; contributing largely to the mobilization and transport of these metals in the river (*66*). The high organic-inorganic matter ratio of the river also seems to have provided sufficient reactive sites for the adsorption of Ca, K, P, and Mg, as well as trace metals, such as Ni, Cr, and Ti. Also, typical studies of dissolved metal concentration involve passing the water samples through 0.45 μm filters to remove suspended materials. Our results, however, indicated that many of the metals which presumably occurred in "solution" may exist as colloidal metal-organic complexes (*66*).

As a whole, this process has profound implications for metal cycling in rivers which drain highly leached and low relief, podsolitic terrains. The interaction between dissolved organics and dissolved solutes may, however, be limited in other Amazonian river systems. In the Rio Negro, where suspended solids are found in extremely low concentrations, the humic substances behave rather conservatively with a HA:FA of 0.64. However, in the Amazon mainstream, the hydrophobic humic acids become selectively adsorbed onto fine suspended particles from the Rio Solimões (*6*), such that the HA:FA drops to only 0.31 (*7*). Although the fulvic acids are compositionally hydrophilic, and therefore, are not readily adsorbed onto detrital material, the drastic loss of humic acids could result in a decrease in available reactive sites for organo-metallic interactions.

Carbon-Isotope Studies in the Amazon Basin: Towards a Tool for Paleovegetation

Despite the fact that the Amazon Basin represents the largest contiguous area of tropical forest in the world, there remains no consensus as to the extent of closed forest cover in the Basin in the past (*69*). This represents a major gap in our understanding of the glacial-interglacial changes in terrestrial carbon storage (e.g., Crowley, (*70*)). The existence of "centres of endemism" within the Basin has led some workers to propose that the forest contracted to isolated "refugia" during cool, dry glacial periods and expanded across much of the Basin (as today) during warm, humid interglacial periods (e.g., Bigarella and Andrade-Lima, (*71*)). This hypothesis is supported by some palynological records from the eastern (*72*) and southeastern (*73*) basin, which show that grasslands existed during the last glacial period in the now forested regions surrounding the core sites.

The refugial hypothesis has been challenged by palynological records from the western Amazon Basin which reveal an uninterrupted dominance of forest biota, and also by new hypotheses which favor the maintenance of a high degree of endemism by hydraulic disturbance of the floodplain (*74,75*). Nelson et al. (*76*) have also demonstrated that some centres of endemism are an artifact of specimen collection densities.

The major barrier to producing a consistent record of vegetation change has been that many proxy vegetation and climate records yield only fragmentary information which is poorly time constrained, and which pertains only to a localized region. The carbon-isotope composition of organic matter preserved in ancient sediments offers the possibility of investigating past vegetation changes on a basin-wide scale. This is because closed forests are almost exclusively dominated by plants which photosynthesize via the C_3 photosynthetic pathway, with $\delta^{13}C$ values of ~-28‰. In contrast, tropical grasses which are common in open grassland/woodland (cerrado) areas, photosynthesize via the C_4 pathway, with

$\Omega^{13}C$ values of ~-12‰ (*77*). Therefore, the $\Omega^{13}C$ values of organic matter in river sediments will depend (among other things) upon the proportion of forest (C_3-) and grassland (C_4-) - derived carbon contributing to the carbon pool. A full discussion of the factors controlling the carbon-isotope composition of organic matter in fluvial sediments is provided by Bird et al. (*78*).

Terrestrial Sediments. In order to determine the range of values likely to be encountered in rivers with forested and non-forested catchments, suspended sediment and bottom sediment samples were collected and analyzed from the Tocantins/Araguaia River system, on the eastern margin of the Basin (*78*). This is the only major tributary of the Amazon River with a catchment that is not predominantly forested, and one of the few tributaries for which there is not already a substantial $\Omega^{13}C$ database (*5,79*). In addition, rivers in the States of Rondonia and Acre, with catchments partly deforested (in the last twenty years) were also sampled to determine how rapidly changes in the catchment vegetation are registered in the $\Omega^{13}C$ value of riverine POC.

Figure 7 shows that there is a difference of 2-4‰ between the $\Omega^{13}C$ values of organic matter from rivers draining forest versus cerrado catchments during the wet season, but no difference during the dry season. The lack of difference in the dry season implies that most rivers in the cerrado region are fringed by a C_3-dominated strip of forest, which provides the bulk of the carbon exported during the dry season (-27 to -30‰). During the wet season, high runoff rates transport C_4 carbon to the river from more remote areas leading to $\Omega^{13}C$ values as high as -24.6‰.

The $\Omega^{13}C$ value of the POC is rivers draining forested catchments is the same as that of the forest vegetation during both the wet and dry seasons. However, in rivers draining cerrado catchments, the $\Omega^{13}C$ values of POC are consistently lower than in the bulk vegetation (cerrado soils have $\Omega^{13}C$ values between -15 and -25‰). This is due to the afore-mentioned bias toward C_3 carbon derived from close to the river, possible preferential metabolism of C_4 carbon by micro-organisms in the river, and some contribution from low-$\Omega^{13}C$ algal biomass (*78*).

The $\Omega^{13}C$ values of the POC in rivers draining deforested catchments span almost the entire range of values found in the forested and non-forested catchments (-23.5 to -29.7‰), with the coarse (>63 Ωm) fractions of the samples tending to be 1-2‰ enriched in ^{13}C compared with the fine (<63 Ωm) fraction.

The results from the study suggest that, while the $\Omega^{13}C$ value of POC from rivers draining non-forested catchments does not accurately reflect the bulk $\Omega^{13}C$ value of biomass in the catchment, the non-forest $\Omega^{13}C$ signature can nonetheless be readily distinguished from the $\Omega^{13}C$ signature of POC in rivers draining forested catchments. These findings are in accord with data from the Zaire (*80*) and Sanaga (*81*) River systems in Africa, where $\Omega^{13}C$ values from -14.6‰ have been recorded from rivers draining non-forested catchments. The isotopic composition of POC from rivers with deforested catchments in the Amazon Basin suggests that the $\Omega^{13}C$ value of coarse POC responds to changes in catchment

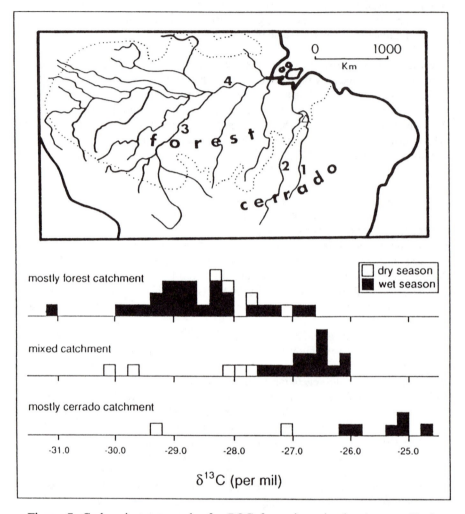

Figure 7. Carbon isotope results for POC from rivers in the Amazon Basin [additional data from Hedges et al. (*5*) and Cai et al. (*79*)]. 1. Tocantins R.; 2. Araguaia R.; 3. Madeira R., 4. Amazon R. Dotted line indicates present extent of forest. General regions in which samples were collected for this study indicated by letters: (A) Marabá-Tocantinopolis, Pará; (B) Carajás, Pará; (C) Ariquemes, Rondonia; (D) Rio Branco, Acre. δ^{13}C results reported in per mil, relative to the PDB standard (from Bird et al. (*78*)).

vegetation on timescales of years to decades, while that of fine POC may take longer to respond.

Biomarker Carbon-Isotope Analyses. While a forest/non-forest carbon-isotope signature is readily identifiable in terrestrial environments, it is very difficult to obtain complete, well dated fluvial sequences. Complete sedimentary sequences are readily available from the submarine fan deposits which lie off the mouth of the Amazon River (*82*), but the analysis of bulk sediments from the marine environment is not appropriate because terrestrially-derived riverborne carbon in the sample rapidly becomes mixed with marine-derived carbon offshore from the river mouth.

One way of circumventing these problems is to utilize biomarker compounds for isotopic analysis which are specific to terrestrial vegetation. There are many such compounds, and two classes which offer considerable potential are the long-chain n-alkanes and the lignin-derived phenols.

N-alkanes.   Straight chain hydrocarbons (n-alkanes) are common minor components of algae, bacteria and vascular plants (*83*), and are comparatively resistant to diagenetic alteration (*84*). Algae and bacteria are generally characterized by n-alkane predominances at low carbon numbers (C_{16}-C_{24}), usually with little difference in the relative abundance of n-alkanes with odd or even carbon numbers (low carbon preference index - CPI). In contrast, vascular plants are characterized by peak n-alkane abundances at high carbon numbers (C_{27}-C_{33}), with a strong preference for odd over even carbon numbers (high CPI's) (*83*).

Gas-chromatography-combustion-mass-spectrometry (GC-C-MS) enables the carbon-isotope analysis of n-alkanes of each carbon number, and therefore, the separate analysis of alkanes derived from algal/bacterial and vascular plant sources. This ability has previously been utilized to determine the origin of waxy n-alkanes in a variety of sediments (*85,86*). The carbon-isotope composition of n-alkanes derived from vascular plants can, therefore, be used to determine the relative contributions of C_3- and C_4-derived carbon to organic matter in marine, terrestrial or composite sediment samples. This approach has been applied in a preliminary manner to representative samples from the Amazon Basin (Bird and Summons, unpublished data).

The river sediment samples (RIV-NF, catchment not forested, i.e. woodland/grassland dominated; RIV-F, catchment forested) both show strong modes at C_{29}- C_{31} with high CPI values (Figure 8A) consistent with organic matter in the samples being derived predominantly from vascular plant sources. In addition, both samples contain shorter chain n-alkanes (down to C_{15}), suggestive of an algal/bacterial contribution to the bulk carbon in the samples.

The marine samples (MAR-P, proximal marine, close to the river mouth; MAR-D distal marine) have bimodal n-alkane distributions (Figure 8B), with peaks at C_{29}-C_{31}, from terrestrial carbon discharged by the Amazon River, and broader

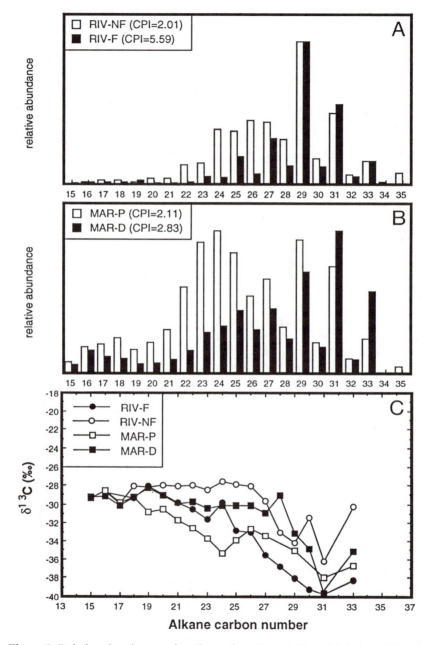

Figure 8. Relative abundances of n-alkanes in sediments from (a) forested-F and non-forested-NF rivers in the Amazon Basin and (b) distal-D and proximal-P marine sediments off the mouth of the Amazon River. $\delta^{13}C$ values for individual n-alkanes in both riverine and marine samples is given in (c), relative to the PDB standard (from Bird et al., in prep.).

peaks around C_{24} due to algal/bacterial input from the marine environment. The shape of the mode at around C_{24}, in both samples, suggests that n-alkanes derived from algae/bacteria probably contributes significantly to the predominantly terrestrial C_{27} peak and possibly a small amount to the C_{29} peak. In both the river and marine sediments, there is a strong preference for odd over even carbon numbers at high carbon numbers, which become progressively weaker below $\sim C_{25}$.

The $\delta^{13}C$ values for n-alkanes from river and marine sediment samples are shown in Figure 8C. They exhibit approximately constant values between -28 and -30‰ at low carbon number, consistent with a mixed algal/bacterial source, decreasing to variably lower values with increasing carbon number. The $\delta^{13}C$ values for n-alkanes from RIV-NF (non-forested catchment) are consistently higher than for the same n-alkane in RIV-F (forested catchment), as expected from the bulk $\delta^{13}C$ values of the samples, and consistent with the higher proportion of C_4 biomass in the non-forested catchment.

The marine sediment samples have n-alkane $\delta^{13}C$ values which are slightly higher than the sample from the forested catchment (RIV-F) at high carbon numbers, tending to progressively higher values at lower carbon numbers. The amount-weighted mean $\delta^{13}C$ value of the odd-numbered n-alkanes at carbon numbers of 29 or higher (terrestrially-derived) in both the marine samples are intermediate between the corresponding values for the river samples. This is to be expected, as the carbon currently exported from the Amazon Basin is derived from a mix of forested and non-forested sources. Sample MAR-P exhibits comparatively low $\delta^{13}C$ values at carbon numbers between C_{19} and C_{25}, with a pronounced minimum at C_{24}, corresponding to the lower carbon number peak in n-alkane abundance in the sample. It is possible that these low values result from a comparatively large bacterial hydrocarbon component in this sample.

The results discussed above should be considered preliminary, but they do suggest that the carbon-isotope composition of terrestrially derived n-alkanes accurately reflect the isotopic composition of the bulk vegetation from which they were derived. They also suggest that this terrestrial isotopic signature can be resolved in marine sediments, where the bulk terrestrial carbon-isotope signature is obscured by admixture with marine-derived carbon.

Lignin-derived phenols. Lignin compounds are phenolic polymers that form a major component of the cell walls of vascular plants. They are comparatively resistant to degradation, and are common constituents of soil and sediments, including those of the Amazon Basin.

Upon oxidative degradation, the complex lignin polymers break down into a suite of simple lignin-derived phenols which are indicative of their source. Of these, p-coumaric acid occurs only in non-woody angiosperm and gymnosperm tissues (*87*), and is therefore, of considerable interest as a potential carrier of a carbon isotope signal for C_3/C_4 ratios in the catchment vegetation. Both compounds have been identified as constituents of terrestrial and marine sediments (5,88).

Preliminary results have been obtained for both p-coumaric acid extracted from plant samples using a 1N NaOH extraction under nitrogen, followed by separation of the phenolic component by either extraction and isolation of the individual lignin-derived phenols using reversed-phase preparative-scale HPLC (e.g., Hartley and Buchan (89)). The results, detailed in Table IV, are consistent with previous findings that the bulk "lignin" component of plants is depleted in ^{13}C relative to the bulk plant tissue by 2-7‰ (90). Comparison of the results for p-coumaric acid with the bulk lignin-derived phenol fraction suggests that p-coumaric acid may be less depleted in ^{13}C than bulk lignin.

Table IV. Carbon-Isotope Composition of Bulk Vegetation, Bulk Lignin-Derived Phenol (LDP) Fraction and Isolated p-Coumaric Acid (PCA)

Sample	$\Omega^{13}C$ bulk	$\Omega^{13}C_{LDP}$ bulk	$\Omega^{13}C_{PCA}$ 1st extinction	$\Omega^{13}C_{PCA}$ 2nd extraction
Corn Cob (C4)	-12.1	n.d.	-15.4	-15.8
Sugar Cane Stalk (C4)	-12.8	-15.4	-15.2	n.d.
Bamboo Stalk (C3)	-27.8	-31.2	-30.3	-30.7
Wheat Stalk (C3)	-25.5	n.d.	-28.3	-29.3

Chemical Evidence in the Acre Subbasin for Arid Climate Conditions During the Last Glacial Cycle. The Acre subbasin of western Amazonia contains mainly fine-grained sediments (representing a fluvial-lacustrine system) with veins of calcium sulfate (gypsum) and calcium carbonate (aragonite) concretions. The presence of these precipitates is considered strong supporting evidence for arid conditions. Radiocarbon ages of carbonate samples, ranging up to 53,000 yr BP, indicate arid climate conditions during the Last Glacial Cycle. The preservation of these older sediments on the surface is attributed to the lack of contemporary input of Andean sediments. This is due to the separation of rivers in the Acre region from adjacent upper Andean tributaries by a small mountain range (Serra do Divisor) along the contemporary Peruvian/Brazilian border. Thus, the modern rivers flowing through the Acre region comprise of reworked sediments, deposited prior to the formation of the Serra do Divisor. Satellite imagery revealed the original system as a mega alluvial fan extending from the Andean highlands to the Acre region (~1,000,000 km^2), with the Acre subbasin representing the distal portion of the original fan system. (91).

Deciphering Chemical Signatures on Amazonian Surface Materials

Chemical weathering refers to the dissolution of rocks by surface waters. Rocks are physical mixtures of silicate minerals, characterized by their symmetrical arrangements of silicon and oxygen. Aluminum, Ca, Mg, Na and K are also present in surface rocks at per cent levels. Chemical weathering of continental rocks may be modelled in terms of feldspar dissolution, the most abundant minerals in the earth's crust (*92*):

$$3H_2O + 2KAlSi_3O_8 \rightarrow Al_2Si_2O_5(OH)_4 + 4SiO_2 + 2K^+ + 2OH^- \quad (1)$$
(aq)   K-feldspar(s)   kaolinite (s)   (aq)   (aq)   (aq)

$$H_2O + Al_2Si_2O_5(OH)_4 \rightarrow 2Al(OH)_3 + 2SiO_2 \quad\quad\quad (2)$$
(aq)   kaolinite (s)   gibbsite(s)   (aq)

These reactions are chemically driven by the high degree of disequilibrium of the "feldspar-rainwater" system. Feldspars crystallize from molten magmas at ~1000°C and thus are not in equilibrium with earth surface conditions. On a global scale the influence of chemical weathering is reflected in the differences in concentration of rock-derived solutes between average rain water and average river water (Table V a). The incongruent dissolution of primary rock minerals, such as feldspars, is accompanied by the formation of secondary minerals, typically clay mineral groups. The fundamental difference between primary aluminosilicate-silicate minerals and clay minerals is the degree of long-range order: the former having strong chemical bonds in three directions, while in clay minerals, strong bonds between the major components (O, Al, Si) exist in only two directions. The variations in spacing between weakly bonded layers is typically used to characterize clay mineralogy.

Table V(a). Compositions of surface waters (in $\bigcirc$g/ml)

	World Average		Central Amazonia	
	[1]Rain	[2]Rain	[3]Rain	[3]Rivers
Na	2.0	6.0	0.007	0.2-2
SiO$_2$	<0.01	13.0	<0.01	3.0-12.0
K	0.3	2.0	0.008	0.2-2.0
Ca	0.09	15.0	0.004	0.2-2.0

Note: values from [1]Garrels and Mackenzie (*93*), [2]Livingston (*94*), and [3]Stallard and Edmond (*13*).

In our approach, three types of minerals accumulate as weathering proceeds and serve as indicators of regional chemical weathering intensity:
1. Complex clay minerals incorporate cations (Na^+, K^+, Mg^{2+}, Ca^{2+}) into their lattices and have high capacities to exchange these ions (mainly for H^+ ions). Their lattices also have significant hydration capacities by incorporating hydroxyl ions.
2. In kaolin group minerals, aluminosilicate sheets are more closely spaced than in complex clays. Due to fewer interstitial and surface sites for incorporating and exchanging major element cations (Na^+, K^+, Mg^{2+}, Ca^{2+}) their compositions are expressed as ratios of Al, Si, O and OH.

3. Aluminum oxide/hydroxide phases, such as gibbsite [$Al(OH)_3$], accumulate from the dissolution of kaolin phases [equation (2)] during the final stages of weathering.

It is important to note that any of these mineral phases may form transiently (as weathering proceeds) depending on local pore and ground water salinities.

During the initial and intermediate stages of weathering, the formation of complex clay minerals is sustained by high salinity surface waters that are in contact with the primary rock minerals (Figure 9, dashed line). This situation is prevalent in much of the northern hemisphere where continental-scale glacial advances (over the past two million years) renewed land surfaces with abundant fresh rock debris. Depending on regional rainfall and the extent of mineral surfaces available for reaction, these conditions may prevail for thousands of years or until weathering profiles are deep enough to remove surface waters from contact with primary rock minerals. At this stage the diminished salinities in pore waters in the weather cover favour the formation of kaolin phases. The dominance of kaolinite throughout much of lower Amazonia suggests that this region is at an advanced stage of weathering (Figure 9), with weathering profiles often exceeding 100m.

In general, the gradual mineralogical and chemical changes that arise in surface soil and sediment as weathering proceeds, is reflected by the chemical composition of regional river systems. For example, the rivers draining the highly weathered Amazonian lowlands have significantly lower rock-derived constituents than the mountainous Andean tributaries (Table V b). Data from central Amazonian rivers show higher silica/cation ratios than in average river water, indicating these waters are in contact with materials in transition from advanced to extreme weathering stages (Table V a). Kaolin phases are buffered by high levels of silica (e.g., 10 Og/ml), derived mainly from feldspars in the initial and intermediate stages of weathering, and from quartz during advanced to extreme weathering. That feldspar dissolution is virtually complete in central Amazonia, is indicated by the low levels of the major cations in feldspars (Na, K and Ca). In these rainforested areas there is evidence that biological particulates are the main sources of Na, K and Ca ions (13,96).

Table V(b). Salinity data for Amazonian River waters (in Og/ml)

Drainage Basin	Dry Season	Wet Season
Mountainous Andean Streams	36-248	84-144
Tributaries draining both highlands and lowlands	17-84	8-50
Central Amazonia	5-51	4-22
Amazon River mouth	48	28

Values taken from Kronberg et al. (97) and Gibbs (28).

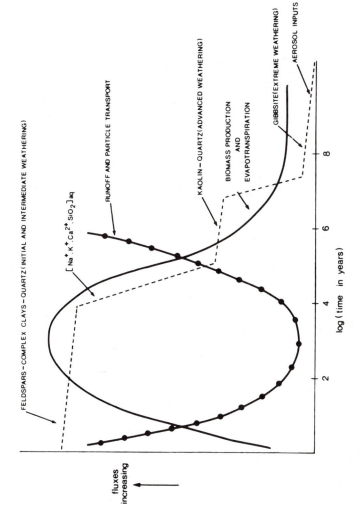

Figure 9. Interplay of variables during weathering. (1) In intensely weathered rainforested regions evapotranspiration is maintained although there is no net biomass productivity. (2) As weathering intensifies, soil nutrient reserves become depleted and in the normal course of events, both net biomass productivity and evapotranspiration fall (from Kronberg et al. (95)).

In eastern Amazonia massive Al accumulations (the Paragominas and Trombetas bauxite deposits) provide evidence that the erosional regime in these areas are extremely weathered. Such intense weathering requires extensive water/mineral interaction over long periods of time. Aluminum accumulation may be facilitated by increasing porosities (if low overburden pressures are maintained as weathering proceeds) due to decreases in molar volumes of up to 70 % in aluminum hydroxide phases relative to complex clay minerals. Isotope studies of gibbsite phases from the Paragominas area (98) indicate extremely low ^{18}O values (-9‰) relative to the modern ^{18}O values of the Carajás region samples. This implies that the formation of aluminum oxide phases occurred in the former area during a strongly monsoonal climate regime. $^{40}Ar/^{39}Ar$ dating of K-bearing manganese from the Carajás region further suggests a complex weathering history dating from the mid-Tertiary (~30 million years ago). Therefore, the massive bauxite deposits to the north and east of Carajás likely formed earlier. It is then possible that weathering over tens of millions of years were required to produce the quasi continuous layers of bauxite extending over ~100,000 ha at the Paragominas deposit (99).

Weathering Time Constants and Geological Constraints. A minimum rate of chemical weathering can be ascertained by estimating the time required for the dissolution of 1 m of continental crust (density, $2g/cm^3$) that receives 1 m rainfall per year, with rock mineral constituents removed in solution at levels found in average river water (~100 ug/ml). Under these conditions, continental surfaces would be lowered by 1m every 20,000 yr (50 m/million years). This figure falls within the range (10-100 m/million years) of other estimates of continental erosion (100).

Consideration of these time constants and the average elevation above sea level (500 m) of continental crust underscores the importance of both geologically constructive processes, such as volcanic activity, which adds ~ 1 billion tons of new rock material to continental surfaces every year, or major episodic glacial events. Therefore, the degree of advanced weathering found in lower Amazonia indicates that this region was both isolated (for tens of millions of years) from constructive geological processes (i.e. by being remote from tectonic plate margins) and was located within humid, tropical latitudes.

Chemical Trends During Weathering. As chemical weathering progresses there is an overall transfer of rock mineral constituents to the hydrosphere. In the initial and intermediate stages of weathering elements may be transiently stored in complex clay phases, until weathering depths prevent surface waters from maintaining contact with primary minerals. Hence, chemical compositions of young soils are typically similar to those of surface rocks (Table VI), however, with different mineral assemblages. As weathering advances and kaolin phases accumulate, levels of most major elements diminish substantially (see Amazonian soils, Table VI) relative to crustal abundances. Data for 30-40 minor and trace elements obtained using spark source mass spectrometry showed that as weathering intensifies some elements (e.g., Cl, As, Sb, Sn, I, Cs, Pb) became enriched relative to their crustal abundances, whereby most others became depleted. Elements with the most strongly depleted concentrations included the alkali metals (except Cs), the alkaline earths and most first row transition metals. Elements displaying enrichment may participate in sorption processes. For example, laboratory studies showed that iodine could be strongly sorbed onto iron hydroxide surfaces. Some refractory elements, such as Ti and Zr, considered to have relatively immobile oxide phases, exhibited depletion with depth in some bauxite profiles. Concentration trends among minor and trace elements during weathering cannot be readily explained in terms of their physico-chemical properties in bulk phases. Their chemical pathways during weathering appear to involve an array of processes including, formation of complex ions, surface reactions, dissolution and reprecipitation, as well as processes that may be mediated by the microbiosphere (*99,100*).

Table VI. Compositions of Rocks and Weathered Materials (in wt%)

	[1]*Surface Rocks*	[2]*Young Soils*	[3]*Amazonian Soils (Lower Basin)*	[4]*Amazonian Bauxite (Paragominas)*	*Andean-Derived Sediments*
Na	3.0	0.5	0.03	0.002-0.006	0.3
Al	9.0	5.3	11.0	20.0-32.0	10.0
Si	31.0	28.0	13.0	1.0-5.0	27.0
K	3.0	2.0	0.2	0.002-0.02	2.2
Ca	3.0	2.0	0.005	0.0002-0.0005	0.3
	f,qz	qz,f,cl,ca	ko,qz,go	gi,hte,ko	qz,ko,il

Values from [1]Taylor and MacLennan (*24*), [2]Kronberg (unpub. data), [3]Kronberg et al. (*99*), [4]Kronberg et al. (*100*), and [5]Kronberg et al. (*97*).
f=feldspar,   qz=quartz,   cl=chlorite,   ca=calcite,   ko=kaolinite,   go=goethite, hte=hematite, il=ilmenite

Iron also accumulates during extreme weathering processes. The original source of iron is from silicate minerals, such as pyroxene and amphibole minerals, while in highly weathered terrains Fe appears mainly as oxides, with highly variable degrees of crystallinity, colours, morphologies and levels of aggregation.

Weathering and Soil Fertility.   The general infertility of soils in the lower Amazonian Basin is shown by the depletion (often by 1 or 2 orders of magnitude relative to young soils) of major rock-derived nutrients such as K, Ca, Mg and most first row transition elements. In the past, the vast rainforests covering much of the lower Basin have raised hopes of agricultural potential in Amazonia. However, considering the Ca value (0.005%, Table VI) for a typical Amazonian soil (density 2.0 g/cm^3), the growing layer (upper 10 cm) of 1 ha of such land would merely contain 100 kg total of Ca. This compares to 40,000 kg in young soils (Table VI). Given that a typical grain crop will remove 20-40 kg of Ca in the harvested biomass, only 1-2 harvests could be theoretically possible. In addition, at these concentration levels (50 uOg/g) laboratory experiments have shown that Ca is not exchangeable. The infertility of these soils becomes abundantly clear when similar depletion levels are extended to the 20-30 other bioessential elements (*99*).

The explanation for extensive Amazonian rainforests overlying infertile soils is their capacity to store and recycle nutrients within the biomass. Stark and Jordan (*101*) have shown that nutrient losses to ground and stream waters are balanced by rain and aerosol additions. Some atmospheric inputs are derived from as far away as the Sahara, estimated to supply Amazonia with ~1 kg P/ha/yr. The Amazonian forests also participate in regional and global water and energy cycles by virtue of the large quantities of latent heat transferred to the atmosphere as water is vaporized from leaf surfaces (*102*). Tropical regions, such as Amazonia, also provide much of the sensible heat (i.e. heat exchanged as air passes over the Earth's surface) that is transported poleward and contributes to balancing negative radiation losses at higher latitudes (*103*).

Weathering of Mineral Deposits

Over the past decade there has been great interest in the weathering of ores and areas of metal enrichment, particularly in the Carajás region of the Amazon. The Carajás region is located in an Archean continental rift basin, where the late Archean and early Proterozoic rift-filling sequences are represented by metavolcanic and metasedimentary rocks. These rocks contain the world's largest high-grade iron deposits, together with significant resources of Cu, Mn, Au, Ag and Cr. Intense chemical weathering of these rocks (or deposits) resulted in further enrichment of metals near the surface in the form of lateritic deposits (*104*), with accumulation of Fe-oxides (mainly goethite and hematite), Mn-oxides (mainly cryptomelane, hollandite and lithiophorite) and kaolinite.

The Igarapé Bahia lateritic gold deposit is located in northwest part of the Carajás plateau at an elevation of 650 m. This plateau is incised to elevations of 400 m by the steep-sided valley of Igarapé (stream) Bahia and several small tributary streams (Figure 10). Due to the strong weathering over millions of years, and the resultant formation of advanced laterite profiles, this region provided a unique opportunity to study the geochemical behavior of various elements during laterisation.

Laterite Profiles. The laterite profiles can be divided into six horizons (*105,106*). From unweathered bedrock to top, the profiles comprise of a saprolite zone, pallid zone, mottled zone, ferruginous zone and topsoil (Figure 11). The unweathered bedrock is mainly composed of quartz and chlorite, with sulfide-quartz veinlets and impregnated sulfide minerals (mainly chalcopyrite and bornite). Other minerals, as identified by electron microprobe, include calcite, albite, micas, and accessory minerals. The saprolite is as much as 40 m thick, consisting of clay minerals, Fe-oxides, Mn-oxides and some relict primary minerals. The pallid zone is as much as 20 m thick and is characterized by pure white to multicolored kaolinitic clays with quartz, ilmenite and goethite. The mottled zone is as much as 100 m thick, and is composed of yellow to brown kaolinitic clays with mottled and blocky ferruginous lumps. Lastly, the ferruginous zone is as much as 25 m thick, and is dominated by Fe-oxide nodules and Mn-oxides in the lower level, with small amounts of kaolinite, quartz and gibbsite.

Laterite formation is attributed to continuous leaching of alkalis, alkaline earths, and some SiO_2 (*105-109*). A mottled zone is developed by local redistribution of Fe, which overlies the saprolite above the water table level. As weathering proceeds Al and Si are progressively removed from the upper part of the profile whereas Fe-oxides are retained to form a ferruginous zone. At Igarapé Bahia, however, the profile is characterized by a major pallid zone, suggesting that this profile presumably developed through two main stages: (1) development of the ferruginous zone and mottled zone with saprolite over the unweathered rocks, and (2) incision of landscape and modification of the profile to form the pallid zone and more saprolite.

Major-Oxide Geochemistry. The major-oxide data of bedrock and laterite samples were analyzed by X-ray fluorescence spectrometry (XRF), and are presented in Table VII. In the profiles, from base to top, Fe_2O_3 increases, whereas in the pallid zone it becomes depleted relative to the other weathered zones. The SiO_2 decreases from base to top. The Al_2O_3 increases from base to the mottled zone, where kaolinite is predominant, then decreases to the ferruginous zone. The MnO is enriched in the lower levels of the ferruginous zone. The TiO_2 is mainly enriched in the mottled zone as relict ilmenite. The P_2O_5 shows a progressive enrichment trend, due both to the sorption by goethite, the stability of phosphate minerals (monazite, churchite, apatite), and its biological affinity (*110*). Alkaline and alkaline earth oxides are depleted in most laterite samples with trace amounts of K_2O (in some samples) due to presence of cryptomelane. The MgO in upper

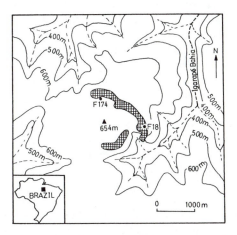

Figure 10. Topographic map of the Igarapé Bahia lateritic gold deposit, showing locations of boreholes F18 and F174. Test pit 12 is located 25m west of borehole F18. Shaded areas represent gold ore zones (from Zang and Fyfe (*105*)).

levels are mainly attributed to relict tourmaline. Na_2O values are commonly below the detection limit for XRF.

Trace-Element Geochemistry. Trace element concentration in the profile are given in Table VIII. The Euclidean distances between elements are rescaled and plotted in a dendrogram (Figure 12) that revealed four distinct element associations, namely Fe-, Mn-, Al- and Si-correlated elements.

Fe-Correlated Elements. This association includes Fe, P, Nb, Au, Ag, Sr, Y, La, Ce, Pb, Pt. Pd and Mo. These elements are characterized by process enrichment, with an increasing trend from base to top of profile. They are also enriched in the Fe-rich breccia zone, reflecting their associations with Fe-oxides. Palladium and Pt, which are not included in the cluster analysis, also show a similar enrichment pattern as Fe, suggesting that they may also belong to this association. Two Ag-Pd alloy grains were identified in the ferruginous zone (*111*).

Immobility of Nb has also been documented in other laterite profiles (*100,112,113*). The correlation of Nb with Ti and Fe suggests that Nb is hosted by relict rutile and Fe-oxides. Molybdenum and Pb are probably adsorbed by goethite (*114*), whereas the accumulation of phosphate minerals in the ferruginous zone leads to the enrichment of Y and some light rare earth elements (La, Ce). However, the positive Ce anomaly is mainly due to formation of cerianite (CeO_2) in laterite (*106*).

Native gold and electrum (Ag-Au alloy) are found within Fe-oxide nodules in the ferruginous zone. Most of the gold grains are less than 10 microns in size. The fineness [$1000 \times Au/(Au+Ag)$] of gold increases downwards in the profile, suggesting a three-stage genetic model for the lateritic deposit (*105*). In the first stage thiosulfate ions from weathering of sulfides are believed to transport both Au and Ag into the profiles to form the electrum. In the second stage, both Au and Ag are then remobilized and incorporated into the ferruginous zone as the erosion surface was lowered by weathering. In the third stage, after uplift and incision of the landscape, chloride leaching separated Ag from the gold grains to purify the electrum. The survival of electrum in the top of the profile can be attributed to lower chloride concentrations and enrichment of organic matter in the near-surface environment.

Mn-Correlated Elements. This association includes most of the first-row transition metals (Sc, Mn, Co, Ni, Cu, Zn), as well as Ba and Cd. These elements are characterized by an enrichment in the lower level of the ferruginous zone and the upper level of the mottled zone. They are enriched in either Mn-oxides or Fe-oxides. Ba (in the form of hollandite) comprises up to 4.93 wt%. Concentrations of trace elements in goethite decrease from the base to the top in the profile, due to increase of crystallinity of goethite or reworking of the mineral (*106*). Thus some trace elements follow the pattern of Mn enrichment although they may be enriched in goethite.

Table VII. Major-oxide compositions of the laterite profile (wt%)

Sample	depth(m)	Fe_2O_3	Al_2O_3	SiO_2	MnO	TiO_2	MgO	CaO	K_2O	P_2O_5	LOI	total
ferruginous zone												
B95	0.0-2.0	74.83	5.87	3.27	2.21	0.03	-	-	0.02	0.55	12.10	98.88
F18-7	3.5-3.7	70.94	11.74	4.62	0.22	0.28	0.70	-	-	0.90	10.49	99.94
P-1	6.9-10.5	75.34	5.98	5.60	1.90	0.38	0.76	0.02	0.08	1.20	8.40	99.73
P-3	12.0-14.0	76.32	4.19	3.67	4.90	0.42	0.66	-	0.06	0.93	8.83	100.04
P-6	18.0-20.0	86.56	0.96	3.59	0.22	0.16	0.74	-	0.02	0.98	7.01	100.29
P-7	20.0-21.0	58.94	12.64	12.32	1.14	1.44	0.68	-	0.04	0.78	12.41	100.47
mottled zone												
F18-16	24.7-25.0	44.52	20.28	21.81	0.21	1.68	0.36	-	-	0.53	10.44	99.87
F18-20	39.1-39.3	39.83	20.98	22.60	0.58	3.77	0.30	-	0.01	0.44	11.52	100.08
F18-27	70.2-70.4	39.22	22.53	25.04	0.66	1.30	0.33	-	-	0.28	10.55	99.94
F18-32	82.0-82.2	58.29	13.83	13.84	2.03	0.73	0.70	0.02	0.05	0.89	10.00	100.41
pallid zone												
F18-41	118-119	20.35	22.85	42.35	0.44	4.10	0.22	-	0.02	0.13	9.65	100.15
saprolite zone												
F18-50	148-162	4.77	13.88	47.54	0.14	0.92	8.60	0.23	0.01	0.19	5.52	100.01
unweathered rock												
F18-55	217-226	4.74	5.03	86.12	0.03	0.11	1.36	0.06	0.84	0.06	1.32	99.73
F10	241.9	13.64	14.30	47.64	0.22	1.32	4.59	6.86	0.55	0.10	2.83	100.38
F14	261.6	14.09	13.76	52.06	0.15	1.47	6.52	3.09	0.51	0.20	4.18	100.08

Total Fe content as Fe_2O_3; LOI=loss on ignition. Na_2O is 3.49 and 2.58 in F10 and F14, respectively, and is lower than 0.01 in other samples.   Values from Zang and Fyfe (*105*).

Table VIII. Trace-element concentrations in the laterite profile (ppm)

Sample	Sr	Ba	Sc	V	Cr	Co	Ni	La	Ce	Y	Zr	Nb	Mo
B95	11	461	23	218	<1	96	51	150	1321	30	15	18	139
F18-7	52	43	19	149	59	44	77	792	399	22	37	18	149
P-1	327	561	50	135	34	105	69	>2000	>2000	110	54	27	288
P-3	75	475	40	132	21	76	58	1262	>2000	78	55	32	245
P-6	10	27	14	70	70	34	77	417	379	24	26	26	110
P-7	23	148	47	153	63	32	65	987	>2000	33	181	37	173
F18-16	3	82	19	257	162	16	82	49	154	10	113	13	17
F18-20	8	458	67	358	24	27	32	51	67	23	277	21	9
F18-27	3	44	50	298	42	16	44	499	58	12	110	13	7
F18-32	59	734	68	245	45	106	158	307	324	44	81	16	76
F18-41	11	45	62	320	32	30	30	63	123	31	288	6	5
F18-50	2	7	20	133	157	45	117	45	60	13	117	14	4
F18-55	5	96	3	24	119	8	49	20	27	2	56	4	6
F10	53	71	26	236	107	35	42	16	21	8	59	12	3
F14	62	112	26	225	36	37	26	41	51	23	105	15	3

Sample	Ga	Cu	Pb	Zn	Cd	Ag	Au	Pt	Pd	S	Cl
B95	<2	6891	154	85	2	4.9	10.730	0.010	0.083	440	141
F18-7	<2	2006	62	17	3	3.9	1.930	<0.005	<0.001	210	126
P-1	<2	2867	160	17	3	25.0	2.860	<0.005	0.004	540	63
P-3	3	7080	136	16	2	22.1	0.314	<0.005	<0.001	640	90
P-6	<2	1913	112	29	2	40.3	11.160	<0.005	0.006	560	90
P-7	8	4997	116	26	2	11.5	6.980	<0.005	0.027	560	180
F18-16	15	946	11	14	1	<0.5	0.300	<0.005	<0.001	220	99
F18-20	33	2114	89	24	2	<0.5	0.070	<0.005	<0.001	220	216
F18-27	20	738	<2	9	2	<0.5	0.060	0.009	0.021	100	90
F18-32	9	5816	36	60	3	3.2	1.891	<0.005	<0.001	110	126
F18-41	39	368	6	53	2	<0.5	0.007	<0.005	<0.001	100	171
F18-50	17	145	2	45	<1	0.5	0.003	0.005	0.002	220	99
F18-55	14	1126	<2	13	<1	0.5	0.014	<0.005	<0.001	1340	135
F10	28	31	9	65	<1	0.7	<0.001	<0.005	<0.001	110	171
F14	30	170	40	40	<1	1.1	<0.001	<0.005	<0.001	1560	126

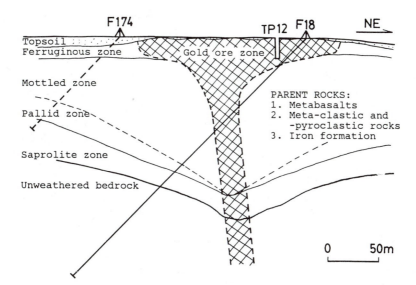

Figure 11. Schematic cross-section of the Igarapé Bahia lateritic gold deposit. Shaded areas represent gold ore zone which is developed on a near-vertical hydrothermally mineralized breccia zone (from Zang and Fyfe (*105*)).

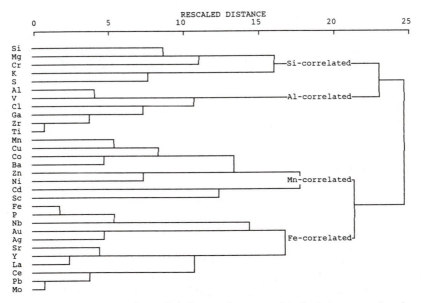

Figure 12: Dendogram of rescaled distance between chemical elements, showing Fe-, Mn-, Al-, and Si-correlated elements.

During incipient weathering Cu, Zn and Cd released from sulfides may be precipitated as carbonates (Cu also as native copper) or adsorbed by Fe-oxides (CuO 4.09 wt%) and Mn-oxides (CuO 2.09 wt%). As weathering proceeds the native copper and carbonates are dissolved and leached while some adsorbed elements are retained. Copper may replace Fe in the goethite lattice (*115*). Both Ni and Co can be enriched in Mn-oxides (CoO 1.47 wt%). It is well known that Ni occurs as an hydrated silicate (garnierite) in lateritic deposits. In this profile, however, only small amounts of Ni-carbonates were observed in fractures of the saprolite zone. Nickel is also found to be concentrated in goethite (NiO 0.09 wt%). The association of Ni with goethite was also reported in other laterite profiles (*116,117*). Cobalt has also been suggested to replace Fe^{3+} in Fe-oxides (CoO 0.18 wt%) (*118*). This correlates well with McKenzie (*119*) who noted that Mn-oxides contain relatively large amounts of Co, Ni and Cu, while Burns (*120*) suggested that Co coexists in Mn-nodules as Co^{3+} substituting for Mn^{4+}.

Al-Correlated Elements. Several elements, including Al, Ga, Cl, V, Ti, Zr are shown to be enriched in the mottled zone. Among these elements Ga probably replaces Al^{3+} in gibbsite and kaolinite. Titanium and Zr are frequently used as index elements to evaluate mobilities of other elements and calculate quantity of rocks consumed in weathering processes. In our profile, Ti and V are mainly present in relict ilmenite, whereas Zr occurs in zircon. Mobilities of Ti and Zr during intense weathering are also documented by Kronberg et al. (*100*), Zeissink (*116*) and Tampoe (*113*). The calculation, however, can not be applied to laterisation processes. Chlorine is readily leached out when the parent rock is weathered. The enrichment of Cl in the mottled zone is probably due to atmospheric addition rather than from weathering of parent rocks, as suggested by Kronberg et al. (*95*). Therefore, the association of Cl with Al could reflect porosity of the laterite profile, especially in the mottled zone.

Si-Correlated Elements. These elements include Si, Mg, K, Cr and S and are characterized by a gradual leaching from the profile as weathering proceeds, such that their concentrations decrease from the base to the top of the profile. Both Mg and K can be strongly enriched in smectite in the saprolite zone during incipient weathering, but are eventually leached out when smectite is replaced by kaolinite. Chromium is mainly enriched in silicates of parent rocks, such as chlorite (Cr_2O_3 0.12 wt%) and biotite (Cr_2O_3 0.23 wt%). It is, however, dissolved as the silicates are weathered. Chromium can be dissolved as $CrOH^{2+}$ under acid conditions at 25°C and 1 atm. (*121*). Sulfur is leached during sulfide dissolution.

Copper Dynamics in Two Profiles from the Carajás Mineral Province. In the Cu-mineralized area of the Carajás region two weathering profiles that developed in contrasting rock types were investigated. One profile, developed in silicate rocks, contained complex clay minerals at the base, an intermediate kaolinitic zone and an upper iron-rich zone. The second profile developed in iron-rich silicate and oxide phases. In the latter, weathering proceeded slowly and was confined mainly to fissures extending tens of meters into the weathered zone. Because physical and mineralogical interactions, such as clogging of fissure systems by expanding clays, had reduced leaching and mineral alteration, there was extensive preservation of original mineralogy and retention of elements (e.g., Mg and Ca) that are normally rapidly leached.

The chemical dynamics of Cu in these profiles was studied using a suite of extractive techniques as well as bulk and surface analyses. The differences in styles of Cu partitioning gives some insight into the complexity of element partitioning as weathering proceeds. In the iron-rich profile, the style and extent of weathering appeared to be controlled by the kinetics of hematite (Fe_2O_3) alteration, with Cu hosted primarily in the fissure system (mainly in oxide and sulfide phases), while some Cu was associated with iron silicate phases (nontronite). In the silicate-rich profile only a small fraction of Cu was exchangeable. It appeared that Cu was immobilized in amorphous and crystalline iron oxide phases (ferrihydrite and goethite respectively) through surface complexation, followed by incorporation into bulk precipitates. Microscale observations in samples from both profiles also indicated the possible influence of microbial activity.

Conclusions

Rivers transport most of the materials which enter the oceans, yet our knowledge of the processes influencing their chemical composition remains inadequate. Much of what is known has been based on a compilation of data from extremely diverse sources. However, the inconsistencies suffered by fragmented data can be overcome through the study of large river systems which encompass a range of geological, climatic, and ecological variability to permit study of the concomitant effects determining river chemistry on a continental scale. Unfortunately, the difficulty encountered with most river basins is that much of their catchment has been influenced by man through agricultural and pastoral activities, industry, logging, hydroelectric projects, and settlement. The Amazon Basin, however, provided (by today's standards) an unusually ideal study area primarily because much of it remains largely undisturbed.

This study considered some of the influences which effect not only the source of solutes to Amazonian river systems, but also some of the factors which control the distribution of metals within their water columns and sediments. Without question, the largest source of metals to the river systems has proven to be weathering of the substrate lithology in source regions and within the erosional regime. Weathering not only governs the distribution of major elements and trace metals, but it also determines the concentration of organic matter in the river and the activity of microbial populations. These factors have proven to be of profound implication since organic material, both cellular and non-cellular, were shown to partially control the concentration of dissolved metals in both solute-rich rivers and solute-deficient rivers, respectively.

The preliminary results detailed above provide support for the proposition that it should be possible to obtain a record of vegetation change in the Amazon Basin from the carbon-isotope composition of terrestrial biomarker compounds extracted from marine sediments off the mouth of the river. The advantage of this approach is that it will yield a complete record from at least the last glacial maximum, and one which will be a record of large-scale changes in vegetation integrated across the entire basin.

It is clear that the detailed soil and water chemistry of the region and the stable isotope systematics provide indicators of the agricultural and mineral resource potential of this vast region and the climate history that has influenced the Amazon Basin. Recent data also shows that the chemistry and physical state and complexity of the weathered materials can only be explained by consideration of specific processes involving microorganisms. From surface to depths of 100's of meters, the weathered rocks are porous and permeable and obviously host living organisms. Modern observations show that the biosphere can extend to depths of several kilometers.

In Amazonia, humid climates and tectonic quiescence over millions of years have resulted in a high degree of partitioning of chemical elements in surface soils and sediments. The significant variation in the fate of elements at profile and regional scales underscores the influence of biological, physical and geological processes on chemical weathering pathways.

Literature Cited

1. Villa Nova, N.A.; Salati, E.; Matsui, E. *Acta Amazonica*1976, *6*, 215-228.
2. Meade, R.H.; Nordin, C.F.; Curtis Rodrigues, F.M.C.; Do Vale, C.M. *Nature,* 1979, *278*, 161-163.
3. Meade, R.H.; Dunne, T.; Richey, J.E.; Santos, de U.M.; Salati, E. *Science* 1985, *228*, 488-490.
4. Sioli, H. *Hydrobiol.* 1950, *43*, 267-283.
5. Hedges, J.I.; Clark, W.A.; Quay, P.D.; Richey, J.E.; Devol, A.H.; Santos, U. de M.. *Limnol. Oceanogr.* 1986, *3*, 717-738.
6. Leenheer, J.A. *Acta Amazonica* 1980, *10*, 513-526.
7. Ertel, J.R.; Hedges, J.I.; Devol, A.H.; Richey, J.E.; Ribeiro, M.N.G. *Limnol. Oceanogr.* 1986, *31*, 739-754.
8. Sioli, H. *The Amazon and its Main Affluents: Hydrography, Morphology of the River Courses, and River Types*. In *The Amazon*; Sioli, H., Ed.; W. Junk, Dordrecht, 1984; pp 127-166.
9. Raimondi, A. *Aguas Potables del Perù*. F. Masias, Lima, **1884**, pp. 127-134.
10. Katzer, F. *Wiss. Math. naturw.* 1897, *17*, 1-38.
11. Sioli, H. *Amazoniana* 1968, *1*, 267-277.
12. Sioli, H. *Tropical Rivers as Expressions of Their Terrestrial Environments*. In *Tropical Ecological Systems. Trends in Terrestrial and Aquatic Research;* Golley, F.B;. Medina, E., Eds.; Springer: New York, 1975; pp 275-288.

13. Stallard, R.F.; Edmond, J.M.. *J. Geophys. Res.* **1983**, *88*, 9671-9688.
14. Gibbs, R.J. *Science* **1970**, *170*, 1088-1090.
15. Gibbs, R.J. *Geochim. Cosmochim. Acta* **1972**, *36*, 1061-1066.
16. Gibbs, R.J. *Science* **1973**, *180*, 71-73.
17. Gibbs, R.J. *Geol. Soc. Amer. Bull.* **1977**, *88*, 829-843.
18. Stallard, R.F. *Eos* **1978**, *59*, 276.
19. Furch, K.; Junk, W.J.; Klinge, H. *Acta Cient. Venez.* **1982**, *33*, 269-273.
20. Furch, K. *Water Chemistry of the Amazon Basin: the Distribution of Chemical Elements Among Freshwaters.* In *The Amazon*; Sioli, H., Ed; Junk, W., Dordrecht, 1984; pp 167-200.
21. Fittkau, E.J.; Irmler, U.; Junk, W.J.; Reiss, F.; Schmidt, G.W.. *Productivity, Biomass, and Population Dynamics in Amazonian Water Bodies.* In *Tropical Ecological Systems: Trends in Terrestrial and Aquatic Research.*; Golley, F.B.; Medina, E., Eds.; Springer: New York, 1975; pp 289-311.
22. Konhauser, K.O.; Fyfe, W.S.; Ferris, F.G.; Beveridge, T.J. *Geology* **1993**, *21*, 1103-1106.
23. Fairbridge, R.W. *Encyclopedia of Earth Sciences Series*; Van Nostrand and Reinhold, New York, 1972, vol. 1VA; 1321 pp.
24. Taylor, S.R.; McLennan, S.M.. *The Continental Crust: its Composition and Evolution;* Blackwell Scientific Publications: Oxford, 1985; 312 pp.
25. Konhauser, K.O.; Fyfe, W.S.; Kronberg, B.I. *Chem.Geol.* **1994**, *111*, 155-175.
26. Leonardos, O.H.; Fyfe, W.S.; Kronberg, B.I. *Chem. Geol.* **1987**, 60, 361-370.
27. Stallard, R.F. *Weathering and Erosion in the Humid Tropics.* In *Physical and Chemical Weathering in Geochemical Cycles;* Lerman, A.; Meybeck, M., Eds; Kluwer Academic: Dordrecht, 1988; pp 225-246.
28. Gibbs, R.J. *Geol. Soc. Amer. Bull.* **1967**, *78*, 1203-1232.
29. Shaw, A. *Mining Magazine, August* **1990**, 90-97.
30. Fyfe, W.S. *Episodes* **1989**, *12*, 249-254.
31. Berner, E.K.; Berner, R.A. *The Global Water Cycle;* Prentice-Hall: New Jersey, 1987.
32. Fletcher, M. *Effect of Solid Surfaces on the Activity of Attached Bacteria.* In *Bacterial adhesion*; Savage, D.C; Fletcher, M., Eds;. Plenum; New York, 1985; pp 339-362.
33. Costerton, J.W.; Cheng, K.J.; Geesey, G.G.; Ladd, T.I.; Nickel, J.C.; Dasgupta, M.; Marrie, T.J. *Ann.Rev. Microbiol.* **1987**, *41*, 435-464.
34. Shimp, R.J.; Pfaender, F.K. *Appl. Environ. Microbiol.* **1982**, *44*, 471-477.
35. Neihof, R.A.; Loeb, G.I. *Limnol Oceanogr.* **1972**, *17*, 7-16.
36. Beveridge, T.J.; Murray, R.G.E. *J. Bacteriol.* **1980**, *141*, 876-887.
37. Ferris, F.G.; Beveridge, T.J. *FEMS Microbiol. Lett.* **1984**, *24*, 43-46.
38. Beveridge, T.J. *Can. J. Microbiol.* **1978**, *24*, 89-104.
39. Beveridge, T.J.; Murray, R.G.E. *J. Bacteriol.* **1976**, *127*, 1502-1518.
40. Mullen, M.D.; Wolf, D.C.; Ferris, F.G.; Beveridge, T.J.; Flemming, C.A.; Baily, G.W. *Appl. Environ. Microbiol.* **1989**, *55*, 3143-3149.

41. Kuyucak, N.; Volesky, B. *Biorecovery* **1989**, *1*, 219-235.
42. Mann, S. *Biomineralization in Lower Plants and Animals-Chemical Perspectives.* In *Biomineralization in Lower Plants and Animals;* Leadbeater, B.S.C.; Riding, R., Eds.; The Systematics Association. Special Volume No. *30*, Clarendon: Oxford, 1986, pp 39-54.
43. Beveridge, T.J.; Fyfe, W.S. *Can. J. Earth Sci.* **1985**, *22*, 1893-1898.
44. Sullivan, C.W.; Volcani, B.E. *Silicon in the Cellular Metabolism of Diatoms,* In *Silicon and Siliceous Structures in Biological Systems;* Simpson, T.L.; Volcani, B.E., Eds.; Springer: New York, 1981; pp. 15-42.
45. Simkiss, K. *The processes of Biomineralization in Lower Plants and Animals.* In: *Biomineralization in Lower Plants and Animals*; Leadbeater, B.S.C.; Riding, R.; The Systematics Association. Special Volume No. 30, Clarendon Press: Oxford, 1986; pp 19-38.
46. Ferris, F.G.; Beveridge, T.J.; Fyfe, W.S. *Nature* **1986** *320*, 609-611.
47. Wada, K. *Amorphous Clay Minerals-Chemical Composition, Crystalline State, Synthesis, and Surface Properties.* In Developments in Sedimentologoy; van Olphen, H.; F. Veniale, F., Eds.; *Proceedings of the VII International Clay Conference.* Elsevier Scientific.; Amsterdam, 1981, Vol. 35 pp 385-398.
48. Nadeau, P.H. *Clay Miner.* **1985**, *20*, 499-514.
49. Beveridge, T.J.; Meloche, J.D.; Fyfe, W.S.; Murray, R.G.E. *Appl. Environ. Microbiol.* **1983**, *45*, 1094-1108.
50. Wissmar, R.C.; Richey, J.E.; Stallard, R.F.; Edmond, J.M. *Ecology* **1981**, *62*, 1622-1633.
51. Konhauser, K.O.; Fyfe, W.S. *Energy Sources* **1993**, *15*, 3-16.
52. Jordan, C.F. *Soils of the Amazon Rainforest.* In *Amazonia*;. Prance, G.T.; Lovely, T.E., Eds.; Pergamon Press: Oxford, 1985; pp 83-94.
53. Lack, T.J. *Freshwat. Biol.* **1971**, *1*, 213-224.
54. Schelske, C.L.; Stoermer, E.F. *Science* **1971**, *173*, 423-424.
55. Konhauser, K.O.; Mann, H.; Fyfe, W.S. *Geology.* **1992**, *20*, 227-230.
56. Reynolds, C.S. *Diatoms and the Geochemical Cycling of Silicon.* In *Biomineralization in Lower Plants and Animals;* Leadbeater, B.S.C.; Riding, R., Eds.; Systematics Association Special *Volume 30*. Claredon: Oxford, 1986, pp 269-289.
57. Hesse, R.. *Origin of Chert: Diagenesis of Biogenic Siliceous Sediments.* In *Diagenesis*;. McIlreath, I.A.; Morrow, D.W., Eds.; Geological Association of Canada, Geoscience Canada Reprint Series 4, 1990; pp 227-251.
58. Hein, J.R.; Scholl, D.W.; Barron, J.A.; Jones, M.G.; Miller, J. *Sedimentology* **1978**, *25*, 155-181.
59. Williams, L.A.; Parks, G.A.; Crerar, D.A. *J. Sediment Petrol.* **1985**, *55*, 301-311.
60. Sigleo, A.C. *Geochim. Cosmochim. Acta* **1978**, *42*, 1397-1405.
61. Schlesinger, W.H.; Melack, J.M. *Tellus* **1981**, *33*, 172-187.
62. Lamar, W.L. *U.S. Geol. Surv. Prof. Paper 600-D*, D24-D29 **1986**.

63. Reuter, J.H.; Perdue, E.M. *Geochim. Cosmochim. Acta* **1977**, *41*, 325-334.
64. Degens, E.T. *Mitt. Geol. Palaont. Inst. Univ. Hamburg SCOPE/UNEP Sonderb* **1982**, *52*, 1-12.
65. Rashid, M.A.,. Absorption of metals on sedimentary and peat humic acids. *Chem. Geol.* **1974**, *13*, 115-123.
66. Beck, K.C.; Reuter, J.H.; Perdue, E.M.. *Geochim. Cosmochim. Acta* **1974**, *38*, 341-364.
67. Bowen, H.J.M. *Environmental Chemistry of the Elements.*; Academic Press: London, 1979; pp 237-272.
68. Stumm, W.; Morgan, J.J. *Aquatic Chemistry*; Wiley-Interscience: New York, 1970.
69. Colinvauz, P.A. *Nature* **1989**, *340*, 188-189.
70. Crowley, T.J. *Nature* **1991**, *352*, 575-576.
71. Bigarella, J.J.; Andrade-Lima, D. de *Paleoenvironmental Changes in Brazil.* In *Biological Diversification in the Tropics.*; Prance, G.T., Ed.; Columbia University Press: New York, 1982; pp 27-40.
72. Absy, M.L.; Clef, A.; Fournier, M.; Martin, L.; Servant, M.; Siffedine, A.; Ferreira da Silva, M.; Soubies, S.; Suguio, K.; Turcq, B.; Van der Hammen, T. *Mise en Evidence de Quatre Phases d'ouverture de la Forêt Dense dans le sud-est de l'Amazonie au cours des 60000 Derniéres Années. Premiere Comparaison avec d'autres Régions Tropicales.*; C.R. Acad.Sci. Paris, 1991; 312, pp 673-678.
73. Van der Hammen, T. *J. Biogeogr.* **1974**, *1*, 3-26.
74. Räsänen, M.E.; Salo, J.S.; Kalliola, R.J. *Science* **1987**, *238*, 1398-1401.
75. Bush, M.B.; Colinvaux, P.A.; Weimann, M.C.; Piperno, D.R.; Liu, K-B. *Quat. Res.* **1990**, *34*, 330-345.
76. Nelson, B.W.; Ferreira, C.A.C.; da Silva, M.F.; Kawasaki, M.L. *Nature* **1990**, *345*, 714-716.
77. Smith, B.N.; Epstein, S. *Plant Physiolo.* **1971**, *47*, 380-384.
78. Bird, M.I.; Fyfe, W.S.; Pinheiro-Dick, D.; Chivas, A.R. *Global Biogeochem. Cycles* **1992**, *6*, 293-306.
79. Cai, D.-L.; Tan, F.C.; Edmond, J.M. *Estuarine Coastal Shelf Sci.* **1988**, *26*, 1-14.
80. Mariotti, A.; Gadel, F.; Giresse, P.; Kinga-Mouzeo *Chem. Geol.* **1991**, *86*, 345-357.
81. Bird, M.I.; Giresse, P.; Chivas, A.R. *Chem. Geol.* **1993**, *107*, 211.
82. Showers, W.J.; Angle, D.G. *Continent. Shelf. Res.* **1986**, *6*, 227-244.
83. Kolattukudy, P.E. *Chemistry and Biochemistry of Natural Waxes;* Elsevier: New York, 1976.
84. Van Fleet, E.S.; Quinn, J.G. *Deep Sea Res.* **1979**, *26A*, 1225-1236.
85. Reiley, G.; Collier, R.J.; Jones, D.M.; Eglington, G.; Eakin, P.A.; Fallick, A.E. *Nature* **1991**, *352*, 425-427.
86. Hayes, J.M.; Freeman, K.H.; Popp, B.N.; Hoham, C.H. *Org.Geochem.* **1990**, *16*, 1115-1128.

87. Ertel, J.R.; Hedges, J.I. *Geochim Cosmochim. Acta.* **1985**, *49*, 2097-2107.
88. Hedges, J.I.; van Geen, A. *Mar. Chem.* **1982**, *11*, 43-54.
89. Hartley, R.D.; Buchan, H. *J. Chromatogr.* **1979**, *180*, 139-143.
90. Benner, R.; Fogel, M.L.; Sprague, E.K.; Hodson, R.E. *Nature* **1987**, *329*, 708-710.
91. Kronberg, B.I.; Benchimol, R.E.; Bird, M.I. *Interciencia* **1991**, *16*, 138-141.
92. Kronberg B.I.; Nesbitt, H.W. *J. Soil Sci.* **1981**, *32*, 453-459.
93. Garrels R.M.; Mackenzie, F.T. *Evolution of Sedimentary Rocks;* W.W. Norton: New York, 1971.
94. Livingstone, D.A.,. Chemical Composition of Rivers and Lakes. *U.S. Geol. Surv. Prof. Pap. 440-G,* **1963**, pp 1-63.
95. Kronberg, B.I.; Nesbitt, H.W.; Fyfe, W.S. *Chem. Geol.* **1987**, *60*, 41-49.
96. Stallard, R.F.; Edmond, J.M. *J. Geophys. Res.* **1981**, *86;* 9844-9858.
97. Kronberg B.I.; Nesbitt, H.W.; Lam, W.W. *Chem.Geol.* **1986**, *54*, 283-294.
98. Bird, M.I.; Longstaffe, F.J.; Fyfe, W.S.; Kronberg, B.I.; Kishida, A. *Geophysical Monograph 78,* American Geophysical Union. **1993**.
99. Kronberg, B.I.; Couston, J.; Stilianidi, B.; Fyfe, W.S.; Nash, R.A.; Sugden, D. *Econ. Geol.* **1979**, *74*, 1869-1875.
100.Kronberg, B.I.; Fyfe, W.S.; McKinnon, B.J.; Couston, J.F.; Stilianidi Filho, B.; Nash, R.A. *Chem. Geol.* **1982**, *35*, 311-320.
101.Stark, N.M.; Jordan, C.F. *Ecology* **1978**, *59*, 434-437.
102.Newell, R.C. *The Amazonas Forest and the Atmospheric General Circulation.* In *Man's Impact on Climate;* Kellog, W.H.; Robinson, G.D., Eds.; MIT Press: Boston, **1971;** pp 457-459.
103.Molion, L.C.B.; Betancurt, J.J.U. *Land Use and Agrosystem Management in Humid Tropics.* In *Woodpower: New Perspectives in Forest Usage;*.Talbot, J.J.; Swanson, W., Eds.;. Pergamon: Oxford, 1981; pp 119-128.
104.Melfi, A.J.; Trescases, J-J.; Carvalho, A.; Barros de Oliveira, S.M.; Ribeiro Filho, E.; Laquintine Formoso, M.L. *Sci. Geol. Bull.,* **1988**, *41*, 5-36.
105.Zang, W.; Fyfe, W.S. *Econ. Geol.* **1993**, *88*, 1768-1779.
106.Zang, W. Ph.D. Thesis, University of Western Ontario, Canada, **1993**
107.McFarlane, M.J. *Laterite and Landscape;* Academic Press: London; 1976.
108.Tardy, Y.; Nahon, D. *Amer. J. Sci.* **1985**, *285;* 865-903.
109.Nahon, D.B. *Evolution of Iron Crusts in Tropical Landscapes.* In *Rates of Chemical Eeathering of Rocks and Minerals;*. Colman, S.M.; Dethier, D.P. Eds.; Academic: Orlando, 1986; pp 169-191.
110.Nickel, E.H. *Geochem. Explor* **1984**, *22*, 239-264.
111.Zang, W.; Fyfe, W.S.; Barnett, R.L. *Mineral. Mag,* **1992**, *56*, 47-51.
112.Melfi, A.J.; Cerri, C.C.; Kronberg, B.I.; Fyfe, W.S.; McKinnon, B. *J. Soil Sci.* **1983**, *34*, 841-851.
113.Tampoe, T.J. *Ph.D. Thesis,* University of Western Ontario, Canada. 1989.
114.Jones, L.H.P. *J. Soil. Sci.* **1957**, *8*, 313-327.
115.Lesarge, K.G. M.Sc. Thesis, University of Western Ontario, Canada, 1991.
116.Zeissink, H.E. *Chem.Geol.* **1971**, *7*, 25-36.

117 Schellmann, W. *Geochemical Principles of Lateritic Nickel Ore Formation.* Proc. 2nd Int. Semin. on Laterization Processes. Univ. São Paulo, São Paulo, Brazil, 1983; pp 119-135.

118. Topp, S.E.; Salbu, B.; Roaldset, E.; Jorgensen, P. *Chem. Geol.* **1984**, *47*, 159-174.

119. McKenzie, R.M. *Australian J. Soil Res.* **1967**, *5*, 235-246.

120. Burns, R.G. *Geochim. Cosmochim. Acta* **1976**, *40*, 95-102.

121. Brookins, D.G. *Eh-pH Diagrams for Geochemistry.*; Springer: Berlin, 1988.

RECEIVED December 28, 1994

Chapter 19

The Chemistry of Headwater Streams in the Rio das Mortes System and Its Effect on the Structure of the Biotic Community

Charles W. Heckman

Max-Planck-Institut für Limnologie, Postfach 165, D-24302 Plön, Germany

Investigations of the headwater streams located north and west of Cuiabá in the regions of Chapada dos Guimarães and Jaciara have been conducted for a period of about two and a half years. Most of these streams flow toward the Pantanal, but a few flow to the Amazon through the Rio das Mortes system. The chemistry of these streams is remarkable due to the extremely low concentrations of ions, reflected by the very low electrical conductivity. Of all the substances tested, only ferrous ions seem to be present in appreciable amounts. Silica, probably in colloidal form, was found to be very abundant in these waters, while calcium and other alkali earth elements are nearly completely absent. The species diversity in these streams is obviously influenced by the low mineral content, and a very interesting aquatic flora and fauna are present. Physiological and ecological adaptations and survival strategies of these species are discussed.

The Rio das Mortes is a river in the Amazon Basin that is part of the Araguaia and Tocantin River System, flowing over 2000 km to the sea (Figure 1). It arises from three springs on the plateau to the east of Cuiabá, the capital of the Brazilian State of Mato Grosso. The region was included in the southern peripheral geochemical province of Amazonia by Fittkau (1). Many streams and rivers arise on this plateau, but nearly all of them flow southward into the Pantanal, a wetland in the Rio Paraguai System. The water courses arising from this region of Brazil, which is situated on the Precambrian Central Brazilian Shield (2), are referred to as clearwater rivers, as defined by Sioli (3). The water is characterized by its great clarity and very low content of dissolved substances.

This study was undertaken to determine the water chemistry at the springs feeding the river and in the uppermost section of the river itself. A survey of the

0097–6156/95/0588–0248$12.00/0
© 1995 American Chemical Society

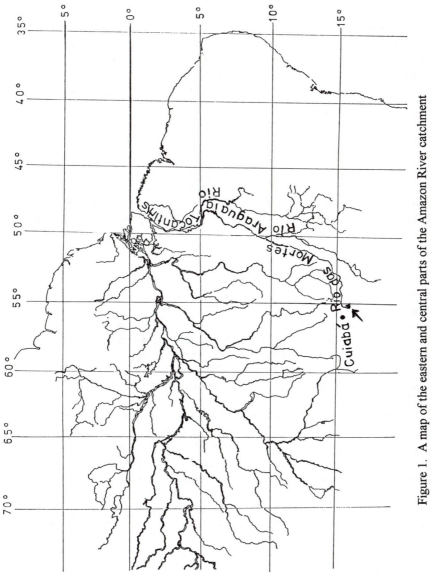

Figure 1. A map of the eastern and central parts of the Amazon River catchment area with the location of the sampling site marked by an asterisk.

biotic community in the river was also undertaken, and the strategies by which the individual aquatic species utilize the available nutrients were investigated.

Location

The three springs from which the Rio das Mortes arises are located on the plateau just south of the highway between Cuiabá and Jaciara, Mato Grosso, at 15°50'S, 55°17'W. These springs (Figure 2) feed streams bordered by gallery forests (Figure 3). They converge and flow northward. Samples of the water and specimens of the flora and fauna were collected at four points: one at each of the springs and the fourth in the river where it crosses beneath the highway (Figure 4).

Materials and methods

To determine water temperature, conductivity, pH, and the concentration and saturation of oxygen, electronic sensors were used in the field. The WTW LF 196, pH 196, and OXI 196 devices were employed for these analyses. The measurement of pH in water with an electric conductivity as low as that in the Rio das Mortes is highly problematic, and the values recorded contribute little to an understanding of the water chemistry. The electronic sensor for pH had to be left in the water for 30 to 45 minutes to permit it to stabilize.

Most analyses were carried out in the field immediately after the water samples were collected to eliminate the possibility of changes during transport. Total hardness was determined by complexometric titration against Merck Titriplex III® using a mixed indicator with a maximum error of -0 and +0.1 mmol/ml. Carbonate hardness is a parameter formulated according to the conditions usually encountered in waters of the North Temperate Zone. It is theoretically never higher than total hardness, but in tropical waters, the values recorded are frequently higher than those of total hardness. The values for this parameter must therefore be interpreted with great caution. It was determined by acidometric titration using a mixed indicator with a maximum error of -0 and +0.2 mmol/ml.

The calcium ion concentration was determined by chromatic comparison of the red-violet complex produced by the reaction with Merck Calcospectral® in an aequeous medium with a maximum error of ±1.5 mg/l and titrimetrically utilizing the disodium salt of ethylene-dinitrilotetraacetic acid to chelate the calcium and calconcarboxylic acid as the calcium specific indicator. The titrimetric method was preferred.

Chlorosity was determined by chromatic comparison of ferric thiocyanate produced by the reaction of chloride and mercuric thiocyanate in the persence of ferric ions; the maximum error is ±1.5 mg/l.

To determine dissolved iron, a chromatic comparison was made of the violet complex produced by the reaction of ferrous ions with a buffer of thioglycolate. The maximum error of ±30%.

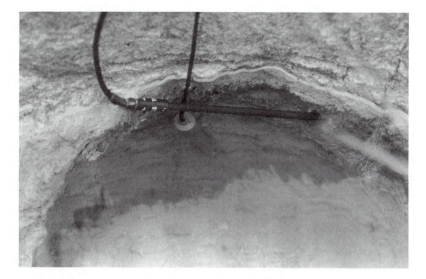

Figure 2. A pool just below one of the springs at the westernmost source of the Rio das Mortes; the electrodes to record water temperature and pH are immersed in the water demonstrating its great clarity.

Figure 3. The course of a stream flowing from the westernmost source of the Rio das Mortes is lined by a narrow, dense gallery forest, seen in the background.

The concentration of aluminum ions was determined by colorimetric comparison of the color formed after the reaction of Chromazurol S$^®$ and aluminum in an acetate buffer, with a maximum error of ±0.04 mg/l

Dissolved chromium was determined by colorimetric comparison of the red-violet chromous diphenylcarbazone complex formed by the simultaneous oxidation of diphenylcarbazide and reduction of the chromic ions. The maximum error is ±0.005 mg/l.

The ammonium concentration was determined by a chromatic comparison after the Berthelot Reaction with a maximum error of ±0.015, while the nitrite analysis was based on a chromatic comparison after the Griess Reaction with a maximum error of ±0.01.

The values for free phosphate-P were recorded after chromatic comparison of the phosphomolybdenum blue produced by the reaction of the phosphate with molybdate ions in a sulfuric acid solution and the reduction of the molybdosulfuric acid produced by ascorbic acid; the maximum error is ±0.01.

The sulfate concentrations determined in the field proved to be too inaccurate to accept. Chromatic comparison of the brownish red complex produced by the reaction of iodate with tannin in a weak acid medium provided values much higher than those determined for the samples brought to the laboratory and analyzed by a spectrophotometric comparison of barium sulfate turbidity produced in a gelatine sol.

The test for hydrogen sulfide employed a chromatic comparison after the reaction of hydrogen sulfide with N,N'-dimethyl-1,4-phenylene-diamine dihydrochloride and oxidation with ferric ions to methylene blue with a maximum error of ±0.02.

Several tests that could not be carried out in the field were completed in the laboratory within five hours after the samples were collected. These included chemical oxygen demand (COD), which was determined photometrically after oxidation of the oombustible organic material in chromate sulfuric acid in the presence of silver sulfate. The nitrate concentration was calculated from the nitrite concentration determined photometrically employing the Griess Reaction after the reduction of the nitrate to nitrite by metallic cadmium. The nitrite-N concentration determined in the field was subtracted from the result to calculate the nitrate-N content of the water, which was then converted to the nitrate concentration by multiplication. Fluoride was determined photometrically based on the decoloration of Spadns reagent.

The time of day and the weather conditions at the time of sampling were noted, because in many water bodies, the illumination and weather conditions have a great effect on various parameters, especially the oxygen concentration and pH.

Results

Physical and Chemical Data. The physical and chemical data recorded at the four sampling points on October 22, 1993 are shown in Table I. To determine the

Figure 4. The Rio das Mortes near the bridge on the highway from Cuiabá to Jaciara is characterized by a fast-flowing, clear water.

Table I. Some Chemical and Physical Characteristics of the Water at the Three Sources of the Rio das Mortes and at the Point Where the River Crosses the Highway Between Cuiabá and Rondonópolis, Downstream from the Confluence of the Three Streams, on October 22, 1993.

	East Spring	Center Spring	West Spring	At Highway
Time (GMT + 3)[*]	1445	1330	1630	0930
Water temperature (°C)	24.2	23.4	24.1	23.1
Conductivity (µS/cm)	13.18	6.88-9.66	4.72	3.32
pH	5.40	4.85	5.17	4.47
O_2 concentration (mg/l)	4.4	3.2	4.4	3.9
O_2 saturation(%)	58	39	59	63
Total hardness (mmol/l)	N.D.	N.D.	N.D.	N.D.
CO_3 hardness (mmol/l)	0.5	<0.1	0.1	0.2
Concentrations (mg/l) of:				
Cl	<3 (t.)	<3 (t.)	<3 (t.)	<3 (t.)
Ca	<1	<1	<1	<1
Fe	<0.1	1.2 ± 0.2	0.15 ± 0.05	0.1 ± 0.05
NH_4	<0.012 (t.)	0.025 ± 0.01	<0.035 ± 0.01	<0.12
NO_2	<0.005 (t.)	0.005 ± 0.003	<0.005	<0.005 (t.)
PO_4 -P	<0.015	<0.015	<0.015	<0.015

[*] Local daylight savings time is three hours earlier than Greenwich Mean Time.
N.D. = Not detectable (the color of the indicator suggested that the end point had been reached before any of the titration reagent has been added).
t = Trace (denotes the presence of a substance at a concentration below the determinable limit).

variability of the physical and chemical parameters of the emerging ground water, samples were analyzed at the westernmost spring on five dates under different weather conditions (Table II). Five sets of data were also collected from the Rio das Mortes where it crosses under the highway (Table III).

The data require careful evaluation. Little useful information is provided by the pH and oxygen concentration values. As already mentioned, the conductivity of the water is extremely low, making the determination of a pH value problematic. The fact that the water seeps out of the ground at many points and trickles down into small pools makes rapid changes in the oxygen concentration possible. Thus, the values recorded are usually dependent on the distance between the electrode and the point at which the ground water emerges and does not reflect the amount of oxygen dissolved in the water while it is still underground.

The rapid changes that occur in the chemical parameters are reflected by the changes in conductivity at the southernmost spring feeding the river. At that location, the conductivity recorded on October 22, 1993 decreased from 9.66 to 6.88 within a distance of 3 m. This is obviously related to the oxidation of the ferrous ions present in the emerging spring water at a concentration of 1.2 mg/l near the source. The uppermost pools of water fed by the spring contain much yellowish precipitate, obviously ferric hydroxide with which numerous iron bacteria are associated. At this spring. the lowest oxygen concentration of 3.2 mg/l was recorded, suggesting that oxygen was being consumed for the oxidation of the ferrous ions.

No trace of hydrogen sulfide was found in any samples, and due to the very low COD and moderately high oxygen concentrations, none would be expected. No trace of aluminum ions was recorded, either. No fluoride was detected at the westernmost spring, but a concentration of 0.1 mg/l was recorded in the river near the highway.

Only the amount of silicon compounds in the water is considerable. It seems likely that most of this occurs in the form of colloidal polysilicic acids, which should not increase the conductivity of the water.

Determinants of the Community Structure. Although the water from the springs flows across bare sediment that appears to harbor few organisms, closer examination reveals that many species belonging to a variety of higher taxa inhabit the water bodies and adjacent waterlogged soils. Thus, in spite of the great shortage of both inorganic and organic nutrients, a considerable number of species inhabit these and other headwater streams on the plateau near Jaciara. An analysis of the trophic relationships among these and the neighboring terrestrrial communities is necessary to reveal their survival strategies. The findings revealed that the nutrient sources for both the plants and animals in the streams are terrestrial organisms which fall into the water or otherwise contribute allochthonous detritus. Because the headwaters have eroded deep depressions on the plateau and are partially filled with a very soft sediment, they can function as traps for both large and small animals. These are fed upon by the numerous aquatic

Table II. Some Chemical and Physical Characteristics of the Water in the Westernmost Spring Feeding the Rio das Mortes on Three Dates During Various Seasons in 1993 and 1994.

	1993			1994	
	Oct. 12	Oct. 22	Nov. 19	Feb. 2	July 9
Time (GMT + 3)	1545	1630	1330	1640	1105
Weather	sunny	overcast	light rain	rain	sunny
Water temperature ($^\circ$C)	24.3	24.1	24.3	25.3	19.6
Conductivity (μS/cm)	5.39	4.52	6.97	6.62	3.16
pH	4.96	5.17	5.25	5.52	5.31
O_2 concentration (mg/l)	5.5	4.4	6.3	5.1	8.4
O_2 saturation (%)	70	59	83	70	100
COD (mg O_2/l)	--	--	--	<15 (t.)	15
Total hardness (mmol/l)	N.D.	N.D.	<0.1 (t.)	N.D.	N.D.
CO_3 hardnesse (mmol/l)	<0.1	0.1	<0.1 (t.)	<0.1	<0.1
Concentrations (mg/1) of:					
Cl	<3 (t.)	<3 (t.)	<3 (t.)	<3 (t.)	<3 (t.)
Ca	¾ 1	<1	<1	<1	<1
Fe	<0.1	0.15 ± 0.05	<0.1	0.1 ± 0.05	<0.1 (t.)
Al	--	--	--	<0.07	<0.07
Cr	--	--	--	<0.005	<0.005 (t.)
NH_4	0.012 (t.)	<0.035 ± 0.01	<0.02 ± 0.01	<0.012	<0.012
NO_2	<0.005 (t.)	<0.005 (t.)	<0.005	<0.005 (t.)	0.01 ± 0.003
NO_3	--	--	--	<0.08	0.25
PO_4 -P	<0.015 (t.)	<0.015	<0.015	<0.015	<0.015
Si	--	--	--	1.3	1.6

The notation used is explained in Table I.

Table III. Some Chemical and Physical Characteristics of the Water in the Rio das Mortes at the Highway Between Cuiabá and Rondonópolis on Five Dates During Various Seasons in 1993 and 1994.

	1993			1994	
	Sept. 12	Oct. 22	Nov. 19	Feb. 2	July 12
Time (GMT +3)	1530	0930	1630	1755	1445
Weather	sunny	overcast	heavy rain	rain	sunny
Water temperature ($^\circ$C)	23.1	23.1	22.9	23./9	20.6
Conductivity (μS/cm)	6.04	3.32	5.07	2.27	2.21
pH	5.28	4.47	5.29	6.50	4.33
O_2 concentration (mg/l)	5.5	4.9	7.8	7.2	8.0
O_2 saturation (%)	71	63	100	94	96
COD (mg O_2/1)	<15 (t.)	--	--	<15 (t.)	<15 (t.)
Total hardness (mmol/l)	N.D.	N.D.	N.D.	N.D.	N.D.
CO_3 hardness (mmol/l)	0.1	0.2	<0.1	N.D.	<0.1
Concentrations (mg/l) of:					
Cl	<3 (t.)	<3 (t.)	<3 (t.)	<3 (t.)	<3(t.)
Ca	¾ 1	<1	<1	<1	<1
Fe	<0.2	0.1 ± 0.05	0.7 ± 0.2	<0.1	<0.1
Al	--	--	--	<0.07	<0.07
Cr	--	--	--	<0.005	<0.005
NH_4	0.025 ± 0.01	<0.012	<0.012	<0.012	<0.012
NO_2	<0.005	<0.005 (t.)	<0.005 (t.)	<0.005	<0.005 (t.)
NO_3	--	--	--	<0.04	0.2
PO_4-P	<0.015 (t.)	<0.015	<0.015	<0.015	<0.015
Si	--	--	--	1.3	3.5

The notation used is explained in Table I.

predators or decompose through the action of the saprobic community to release inorganic nutrients for the aquatic primary producers.

In the forefront of the nutrient collecters in the headwater streams are the predators and scavengers, which consume the terrestrial organisms that fall into the water. The most obvious ot these are the surface-dwelling heteropterans. Species of various sizes are specialized for the capture of prey in different size classes. The largest predator on the surface of the water is the gerrid, *Cylindrostethus palmaris*, which reaches a length of 17 mm. In the medium size class, there are several other gerrid species, including *Limnogonus lotus* White, 1879, and possibly undescribed species of *Limnogonus* and *Ovatametra*. Among the rather small surface predators is *Rhagovelia* cf. *whitei*. Sharing the small terrestrial prey that approaches the streams is the toad bug, *Gelastocoris willineri*, which conceals itself in the sand along the shores. Insects too large to be captured by the heteropterans can be consumed by various littoral spiders, some of which are quite large. The commonest of these are species of *Tetragnatha*. Predators living below the surface also share the prey falling into the water. These include species of *Buenoa* and *Belostoma*, as well as various dytiscids of the genera *Laccophilus* and *Laccodytes*.

The organisms that the surface dwelling insects fail to capture or find unpalatable, such as millipedes that migrate from the gallery forests bordering the streams, are consumed by various aquatic arthropods, such as the very large gyrinid, *Enhydrus tibialis*, smaller species of *Gyretes*, and hydrophilids of the genus *Berosus*. The nutrients released through the action of predators and scavengers serve to support the growth of algae in the streams. These algae and other microorganisms support a variety of chironomid larvae, which, in turn, form a basis for nutrition of the dragonfly larvae inhabiting the stream. Microorganisms and small insects are also the most probable nutritional sources for a species of fish and several species of larval anurans inhabiting the stream, in spite of its apparent unsuitability for vertebrates due to the shortage of minerals. After the stream enters the gallery forest, detritus from the forest plants provides an additional source of nutrients for mayfly larvae and other detritivores.

Significant for the ecology of the aquatic community in the headwaters of the Rio das Mortes and neighboring rivers is the observation of cow skeletons in two of the three springs (Figure 5). The eroded pits filled with very soft mud formed by the outflowing water form natural death traps for large animals (Figure 6). While their cadavers are decomposing, enormous amounts of essential inorganic nutrients are released into the streams. Terrestrial insect scavengers that come to feed on the carrion also provide a rich food supply for the aquatic predators. In the easternmost spring, there were large masses of filamentous green algae, mainly *Spirogyra* spp. These algae appeared yellowish and rather moribund. It is therefore very likely that the mineral nutrients released into the water while the cow was decomposing supported the luxurient growth of the *Spirogyra* a few meters downstream. This growth ceased after the nutrient supply was exhausted, but the masses of the algae persisted. This indicates that the nutrition of the aquatic community is sporadic, and the species present have the tendency to absorb and

Figure 5. The bones of a cow in the stream a few meters from the westernmost source of the Rio das Mortes. The minerals released during the decomposition of the carcass apparently supported the development of a dense mat of *Spirogyra* sp. and other algae, which were moribund at the time of the analyses.

Figure 6. The deep eroded ravine at the westernmost source is barely visible where the stream enters the gallery forest. It is therefore likely that occasionally, passing animals will fall into this ravine and become trapped in the soft sediment at the bottom. Such animals are a major source of nutrients for the community of aquatic biota.

conserve available nutrients to assure survival of the species during long periods of famine.

The efficient gathering of sporadically available nutrients and an adaptation to long periods of fasting seem to be basic characteristics of communities inhabiting headwater streams on this plateau. How its conservative function can prevent or retard the downstream transport of nutrient elements in discussed in the next section.

Discussion

The data on the water chemistry indicate that the subterranean aquifers have been leached of nearly all soluble substances. This should not be unexpected considering that the geological formations of the Central Brazilian Shield are very ancient (2), and that springs arising from this kraton have almost certainly been feeding rivers for several hundred million years. The importance of the geology and geochemistry as determinants of water chemistry in the Amazon drainage system was discussed by Gibbs (4) and Fittkau (5).

A survey of the streams and rivers crossing the highway between Cuiabá and Porto Velho revealed that the conductivity and calcium concentration of waters from the sandstones of the Parecis Formation between Diamentino and Pimenta Bueno are similarly low (6). This region of Mato Grosso is also located in the region of the Central Brazilian Shield. There are few water bodies in the world with such low mineral nutrient content, but values nearly as low have been recorded in New Guinea, where the conductivity of the rainwater was sometimes equivalent to that in the Rio das Mortes (7).

In contrast, the conductivity of the waters in the Pantanal to the south varies greatly with the season but remains roughly one to two orders of magnitude higher than those recorded in the headwater streams (8). In this respect, the waters of the Pantanal are low but similar to those in many other tropical surface water bodies fed by rainfall in other parts of the world (9,10). However, the conductivity of this water is far below that in various tropical African lakes (11-13).

The nearest sites from which stream sediments were collected for analysis by atomic absorption spectrometry are two streams near Chapada dos Guimarães, for which data on the content of various metals, are reported in Table IV. The water chemistry of these streams, especially the conductivity and water hardness, is similar in that in the headwaters of the Rio das Mortes. Due to differences in the methods of collection and fractionation of the sediment samples by different researchers, much of the data in the literature are not strictly comparable. Nevertheless, from various data compiled by Petr and Irion (9), it is evident that the amounts of most metals in these samples are much less than those in the clay fraction of river sediments in many parts of the world. Unfortunately, these authors provided no comparative data for iron, which was present in rather large amounts in the Chapada sediment samples. Especially notable is the very small amount of calcuim present, which accounts for the undetectable water hardness. The Ca content of the clay fraction of the sediment in various rivers of the world is 90 to

Table IV. Content (mg/kg dry weight) of Various Elements in the Sediment of two Streams at the Edge of the High Plateau Near Chapada dos Guimarães.

Ag	<0.05	<0.05
Au	<0.25	<0.25
Ca	12.98	71.99
Cd	<0.01	0.01
Co	0.69	3.66
Cr	2.50	6.90
Cu	2.13	3.12
Fe	4.516 17	11.169.86
Hg	<0.010	1.019
Mg	24.83	127.49
Mn	10.69	45.21
Ni	1.18	5.27
Pb	2.38	4.18
Zn	4.59	11.55

The sediment samples, taken on July 5, 1992, were analyzed by atomic absorption spectrometry through the courtesy of Dipl. Biol. J. Kaiser in Pinneberg, Germany more than 3000 times greater than that in one of the streams near Chapada. Only in Rio Negro sediments is calcium at a level nearly as low as that recorded for the Chapada samples.

The tendency of water courses to carry the essential nutrient elements downstream makes it necessary for the aquatic biotic communities to develop very efficient ways to retard this process. In spite of these conservative processes, some losses are unavoidable, and the longer or more repidly the leaching occurs, the more oligotrophic the waters become. The headwaters of the Rio das Mortes and neighboring rivers provide extreme examples of mineral depletion.   Table V outlines the conditions in rivers flowing southward into the Rio Paraguay System. As this table shows, the relatively great variability in the pH (14) and concentrations of various inorganic ions (15) reported for the class of water courses designated "clearwater rivers" is not observed in the streams arising from the plateau region near Jaciara, except in the outflow of a hot spring at Aguas Quentes, which flows into the Corrego Cupim. All of the water samples collected are extremely poor in all ions analyzed, except occasionally Fe^{2+}. The variability elsewhere led Sioli (3) to conclude that the clearwater river class is not a well defined grouping, as the black and white water classes seem to be. It is rather a heterogeneous grouping varying considerably according to the geological features of the region of origin. In contrast, the streams arising on the plateau near Jaciara form a rather homogeneous class characterized by extremely low concentrations of ions.

Under these circumstances, it is necessary for the organisms inhabiting these waters, especially elements of the microbiota, to be physiologically adapted to absorb and efficiently store every available trace of nutrient entering the water. That this is happening is indicated by the decrease in the conductivity from values of 13.18, 9.66, and 4.72 μS/cm in the three headwater springs to 3.32 μS/cm in the river downstream from the confluence of the three streams they feed.

Small amounts of the nutrients enter the water courses with rainwater runoff, as shown by sharp increases in the conductivity during heavy rain showers (Wantzen, M., personal communication). The conductivity in the surface waters of Amazonian rivers (16) as well as in the Pantanal to the south during the rainy season (8) is considered low, but it is generally at least 25 μS/cm. In the groundwater feeding the headwater streams, the conductivity is only about a tenth to a fifth of that in the rain-fed surface water bodies of the region. Thunderstorms can be expected to provide considerable amounts of nitrogen compounds (17) as well as particulate matter suspended in the air as a result of the frequent fires in the region. The nutrients falling with the rainfall may not enter the water bodies, or if they do, they may not remain long in the water. This is indicated by the analytical results, which show that the conductivity decreased between the headwater springs and the point along the river near the highway, even though a heavy rain had been falling just before the sample was taken, and a light rain was still falling when the samples were being analyzed.

The main source of nutrients for the aquatic community in the uppermost section of the Rio das Mortes is the terrestrial community. Streams and rivers are

Table V. Some Chemical and Physical Characteristics of the Water in Headwater Streams Flowing into the Rio Paraguai Watershed from the Plateau to the Northwest and West of Cuiabá, Mato Grosso.

Name of the stream	Filipa	Filipa	Ajudicaba	Correia	Formoso			Aguas Quentes
					1	2	3	
Date	July,19	July,28	July,19	July,19	July,28	July,28	Sept.,14	July,28
Time (EDT)	1055	1100	1600	1730	1435	1500	1140	1815
Weather	sunny	sunny	sunset	sunny	sunny	sunny	cloudy	night
Water temperature (°C)	23.9	24.3	20.0	22.6	23.9	--	24.9	40.2
Conductivity (µS/cm)	8.54	6.43	5.16	2.49	3.62	5.15	2.93	48.3
pH	4.91	5.13	4.45	5.07	4.55	3.95	--	5.52
O_2 concentration (mg/l)	6.1	6.0	3.7	7.6	5.8	5.0	7.6	6.8
O_2 saturation (%)	77	78	43	95	85	63	99	106
Total hardness (mmol/l)	N.D.	N.D.	N.D.	N.D.	N.D.	N.D.	N.D.	0.15
CO_3 hardness (mmol/l)	0.15	0.15	<0.1	<0.1	0.2	0.15	<0.2	0.85
Concentrations (mg/l) of:								
Cl	<3	<3	<3	<3	<3	<3	<3	<3
Ca	<1	6 ± 1.5	<1	<1	5 ± 1.5	5 ± 1.5	--	2 ± 1.5
Fe	1.0	1.3	0.3	0.5	0.3	0.3	0.2	<0.1
NH_4	0.04	0.025	0.020	<0.012	<0.012	<0.012	0.03	0.08
NO_2	0.012	0.005	<0.005	<0.005	<0.005	<0.005	<0.005	<0.005
PO_4-P	<0.015	<0.015	<0.015	<0.015	<0.015	<0.015	<0.015	0.025

The three streams that converge to form the Rio Formoso are shown separately.

generally important one-way avenues of transport for nutrient substances from highlands to lowlands and finally to the sea. The terrestrial soils of ancient highland areas can therefore also be assumed to be depleted of minerals, just as the streams are. The physiological adaptations of the species present in the water courses to conserve the smallest amounts of available nutrient elements therefore deserve closer study.

Nutrient depletion is seldom considered by limnologists to be an ecological problem because most waters in the temperate zones are presently being subjected to increasing eutrophication, and this has captured the attention of scientists to such an extent that few people even consider the negative consequences of processes causing an opposite effect. Normally, the further loss of vital nutrients is minimized by the conservative action of the biotic community, which obviously has the capability of absorbing and storing the smallest amounts of nutrients entering the water.

Coming to terms with long periods of famine has obviously been possible for many species because their diversity in the headwater streams is surprisingly great. More precise information about the species concerned is unavailable, however, and a review of the literature confirms that there has been practically nothing reported about the species identified in the headwater streams. The knowledge of these species should be improved so that the impact of agricultural activities in regions depleted of minerals can be better understood. Because the biotic communities, both terrestrial and aquatic, seem to be the only entities preventing the complete loss of all soluble materials from the region, it can be concluded that any activities stripping the land in such regions of its vegetation would lead to a complete impoverishment in mineral nutrients and the elimination of nearly all productivity within a very short period of time. This is, in fact, being observed in many parts of Brazil, where agricultural methods suitable only for the temperate zones are being introduced.

Summary

Basic data on the physical and chemical conditions in the springs and headwater streams of the Rio das Mortes System in Mato Grosso, Brazil, reveal that this clearwater river offers its biotic community practically no essential nutrient elements whatsoever. The extremely low conductivity and the undetectable degree of water hardness provide the most obvious evidence of the poverty of this water in dissolved minerals. Iron was the only common metal that appeared to be present in moderate quantities in any of the springs.

In spite of the poverty in nutrients, the biotic communities of the headwater streams are moderately rich in species, some of which consume terrestrial organisms or detritus entering the water while others utilize nutrients that are introduced with rainfall or from decomposing animals that die near the streams. The activity of the biota serves to prevent or retard the further loss of nutrients from these ancient water courses, This explains why even temporary destruction of the biota in or bordering the streams on the Central Brazilian Shield can result in irreplaceable losses of nutrient elements from an already impoverished system and permanent reductions in fertility and productivity.

Acknowledgements

This paper resulted from the cooperation between the Max-Planck-Institut für Limnologie in Plön and the Universidade Federal de Mato Grosso in Cuiabá under the Governmental Agreement on Technological Development between Germany and Brazil as part of the SHIFT program financed by the Bundesministerium für Forschung und Technologie (BMFT), the Conselho Nacional de Desenvolvimento Científico e Tecnólogico (CNPq), and the Instituto Brasileiro de Meio Ambiente e Recursos Naturais Renováveis (IBAMA). Thanks are also due to Dipl. Biol. Joachim Kaiser for analyzing the sediment samples and to Dipl. Biol. Matthias Wantzen for providing supplemental information on the effects of rainfall in streams of the region.

Literature Cited

1.   Fittkau, E. J. *Münster. Forsch. Geol. Paläontol.* **1971**, *20/21*, 35-50.
2.   Putzer, H. *The Geological Evolution of the Amazon Basin and its Mineral Resources.* In *The Amazon*; Sioli, H, Ed.; W. Junk, Dordrecht, 1984; pp 15-46.
3.   Sioli, H. *The Amazon and its Main Affluents: Hydrography, Morphology of the River Courses, and River Types.* In *The Amazon*; Sioli, H., Ed.; W. Junk, Dordrecht, 1984; pp 127-165.
4.   Gibbs, R. J. *Geol. Soc. Amer. Bull.* **1967**, *78*, 1203-1232.
5.   Fittkau, E. J. *Amazoniana* **1974**, *5*, 77-134.
6.   Furch, K.; Junk, W. J. *Water Chemistry and Macrophytes of Creeks and Rivers of Southern Amazonia and the Central Brazilian Shield.* In *Tropical Ecology and Development*; Furtado, J. I., Ed.; Int. Soc. Trop. Ecol., Kuala Lumpur, 1980; pp 771-796.
7.   Petr, T. *Limnology of the Purari Basin. Part 1. The Catchment Above the Delta.* In *The Purari - Tropical Environment of a High Rainfall River Basin*; Petr, T., Ed.; W. Junk, The Hague, 1983; pp 141-177.
8.   Heckman, C. W. *Int. Revue Ges. Hydrobiol.* **1994**, *79*, 397-421.
9.   Petr, T.; Irion, G. *Geochemistry of Soil and Sediments of the Purari River Basin.* In *The Purari - Tropical Environment of a High Rainfall River Basin*; Petr, T., Ed.; W. Junk, The Hague, 1983; pp 109-122.
10. Saeni, M. S.; Sutamihardja, R. T. M.; Sukra, J. *Water Quality of the Musi River in the City Area of Palembang.* In *Tropical Ecology and Development*; Furtado, J. I., Ed.; Int. Soc. Trop. Ecol., Kuala Lumpur, 1980; pp 717-724.
11. McLachlan, A. J. *The Aquatic Environment: I. Chemical and Physical Characteristics of Lake Chilwa.* In *Lake Chilwa*;. Kalk,M.; McLachlan, A. J.; Howard-Williams, C.,Eds.; W. Junk, The Hague, 1979; pp 59-78.

12.   Talling, J. F.; Talling, I. B. *Int. Rev. Ges. Hydrobiol.* **1965**, *50*, 421-463.
13.   Allanson, B. R. *The Physico-Chemical Limnology of Lake Sibaya.* In *Lake Sibaya*; Allanson, B.R., Ed.; W. Junk, The Hague, 1979; pp 42-74.
14.   Sioli, H. *Tropical Rivers as Expressions of their Terrestrial Environments.* In *Tropical Ecosystems; Trends in Terrestrial and Aquatic Research*; Golley, F. B.; Medina, E., Eds.;.Springer, New York, 1957; pp 275-287.
15.   Junk, W.; Furch, K. *Acta Amazonica* **1980**, *10*, 611-633.
16.   Furch, K. *Water Chemistry of the Amazon Basin: The Distribution of Chemical Elements among Freshwaters.* In: *The Amazon*; Sioli, H., Ed.; W. Junk, Dordrecht, 1984; pp 167-199.
17.   Jones, M. J.; Bromfield, A. R. *Nature (London)* **1970**

RECEIVED November 29, 1994

Chapter 20

The Chemistry of Atmospheric Aerosol Particles in the Amazon Basin

Paulo Artaxo, Fábio Gerab, Márcia A. Yamasoe, and José V. Martins

Instituto de Fisica, Universidade de São Paulo, Caixa Postal 20516, CEP 01452–990, São Paulo, SP, Brazil

Amazon Basin tropical rainforest is a key region to study the processes that are responsible for global atmospheric changes. There are large emissions of primary biogenic aerosol particles released naturally by the vegetation, and also large amounts of fine mode aerosol particles emitted during biomass burning. Fine and coarse mode aerosol particles were collected at three sites using Stacked Filter Units. Particle induced X-ray emission (PIXE) measured concentrations of up to 20 elements in the fine mode: Al, Si, P, S, Cl, K, Ca, Ti, V, Cr, Mn, Fe, Ni, Cu, Zn, Br, Rb, Sr, Zr, and Pb.

Biogenic and biomass burning aerosol particles dominate the fine mode mass concentration, with the presence of K, P, S, Cl, Zn, Br, and FPM (Fine mode mass concentration). During the dry season, at two sites, a strong component of biomass burning is observed. Inhalable particulate matter ($d_p < 10\mu m$) mass concentration up to 700 $\mu g/m^3$ was measured. Absolute Principal Factor Analysis (APFA) showed four components: soil dust (Al, Ca, Ti, Mn, Fe), biomass burning (Black carbon, FPM, K, Cl), natural biogenic particles (K, S, Ca, Mn, Zn), and marine aerosol (Cl). Samples collected directly over biomass burning fires in cerrado and primary forest show large amounts of organic acids, in addition to K, Cl, P, black carbon and other species.

Biogenic and Biomass Burning Aerosol Particles in Tropical Rainforest Regions

Tropical rainforest vegetation formations are characterized by intense sources of biogenic gases and aerosols. The Amazon Basin has the world's largest rainforest, and is a region with intense convective activity (1), resulting in rapid vertical

0097–6156/95/0588–0265$12.00/0

mixing of biogenic gases and aerosols to high altitudes where they can be transported over long distances and have an impact on the global tropospheric chemistry. It is becoming clear that it is necessary to increase our knowledge of the chemical processes that determine the composition of the atmosphere in background areas, and to understand biosphere-atmosphere interactions. It also is necessary to obtain a better understanding of the alterations in the atmospheric composition due to changes in land use in tropical rainforests. The tropical rainforests of the world are in a delicate nutrient-limited environment (2). Due to nutrient poor oxisols and ultisols in these regions the deposition of airborne aerosol particles is needed to achieve a nutrient balance (2, 3). Several elements, like phosphorus, are a critical airborne element that could limit annual net primary production of the Amazon basin tropical rainforest (4).

In a tropical forest atmosphere, vegetation plays a major role in controlling the airborne particle concentration. The forest vegetation is the principal global source of atmospheric organic particles (5, 6). Only few studies involving natural released biogenic aerosols have been conducted in tropical rainforests (7-16). The natural biogenic aerosol particles consist of many different types of particles, including pollen, spores, bacteria, algae, protozoa, fungi, fragments of leaves, excrement and fragments of insects. A significant fraction also comprises secondary aerosol particles formed by gas-to-particle conversion of organic and sulfur-related biogenic gases. These biogenic particles can be sub-micrometric in size. Bacteria in forested areas were found in the size range of 0.5 to 2.5 μm in size. Biological particles exhibit cloud influencing and can act as cloud condensation nuclei, potentially affecting the cloud formation mechanisms.

The rapid deforestation now occurring in tropical regions has the potential of changing the atmospheric composition and has a regional climatic impact that can affect a large portion of the equatorial region. Biomass burning is a major source of particulate matter and gaseous emissions into the atmosphere (17). More than 80% of the emissions from biomass burning originate in the Tropics. The high rate of tropical biomass burning in the last decade is mainly a result of burning of cerrado in Africa and deforestation in the Amazon Basin (18). Estimates of total biomass consumed on a global basis range from 2 to 10Pg (1 Petagram=10^{15}g) per year (17). In terms of total particulate matter (TPM), emissions are around 104 Tg (1 Teragram = 10^{12}g) per year (19). For particulate matter in the fine mode (FPM, d_p<2.0 μm), emissions are estimated as 49 Tg of fine particles per year, accounting for about 7 % of the global fine mode aerosol particle emission rate. For elemental carbon, the emission of 19 Tg/year could account for a very high (86 % of the total) anthropogenic emissions (19).

Experimental Methods

Fine and coarse aerosol particles were sampled using stacked filter units (SFU) (20). The SFU was fitted with a specially designed inlet, which provided a 50 % cutoff diameter of 10 μm (21). The SFU collects coarse mode particles

$(2.0 < d_p < 10 \ \mu m)$ on a 47-mm-diameter, 8 μm pore-size Nuclepore filter while a 0.4 μm pore-size Nuclepore filter collects the fine mode particles $(d_p < 2.0 \ \mu m)$ (*22*). The flow rate was typically 16 liters per minute, and sampling time varied from 24 to 72 hours. Particle bounce is not a problem in the Amazon basin due to the high relative humidity (70 to 95%).

Three monitoring stations are being operated continuously in different vegetation formations. Figure 1 shows a map of South America with the location of the three background sampling stations. The first aerosol sampling station is situated in Cuiabá, at the Brazilian cerrado, south of the Amazon basin rainforest. The Cuiabá sampling station allows the analyses of regional effects of biomass burning emissions due to the location of the station. The site is heavily affected by regional cerrado biomass burning. A second aerosol sampling station was installed at the "Serra do Navio," in the Northern part of the Amazon basin. This station is located 190 Km North of the equator, in a primary tropical rainforest. The Serra do Navio sampling site is relatively free from regional biomass burning emissions, and there are no industrial activities for at least a thousand kilometers around the sampling site. The third sampling site is in the region of "Alta Floresta", near the border of Mato Grosso and Amazonas states. The Amazon basin dry season starts in March, extending until late September. Biomass burning occurs mainly in the end of the dry season, in August and September.

Direct biomass burning emissions were measured directly over the fires in two different vegetation formations. Fires in primary forest were measured in the state of Rondônia, near the village of Ariquemes. Fires in the cerrado vegetation formation were measured at the Brasília Ecological Station, administrated by IBGE (Instituto Brasileiro de Geografia e Estatística). Fine and coarse mode aerosols were collected using Stacked Filter Units with battery operated vacuum pumps. Sampling time varied from 2 to 10 minutes, since samples were collected a few meters over the fires. A 10 μm inlet was used to collect only inhalable particles.

The elemental concentrations were measured with the Particle-Induced X-ray emission (PIXE) (*23*) method. It was possible to determine the concentrations of up to 20 elements (Al, Si, P, S, Cl, K, Ca, Ti, V, Cr, Mn, Fe, Ni, Cu, Zn, Br, Rb, Sr, Zr, and Pb). A dedicated 5SDH tandem Pelletron accelerator facility, the LAMFI (Laboratório de Análise de Materiais por Feixes Iônicos) from the University of São Paulo was used for the PIXE analyses. Detection limits are typically 5 ng/m^3 for elements in the range $13 < Z < 22$ and 0.4 ng/m^3 for elements with $Z > 23$. The precision of the elemental concentration measurements is typically less than 10%, with 20% for elements with concentration near the detection limit. The fine and coarse fraction aerosol mass concentrations are obtained through gravimetric analyses of the Nuclepore filters. Detection limit for the aerosol mass concentration is 0.3$\mu g \ m^{-3}$. Precision is estimated at about 15%. Black carbon concentration was measured using a reflectance technique using a photometer.

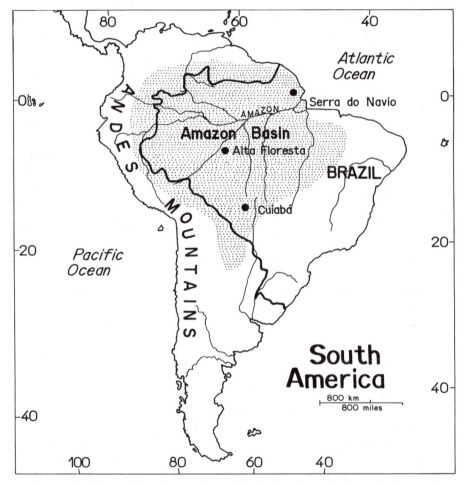

Figure 1. Map of South America with the location of three background monitoring stations in the Amazon Basin: Cuiabá, Alta Floresta and Serra do Navio.

Absolute Principal Factor Analyses

To separate the different components of natural biogenic aerosols using the elemental composition, absolute principal factor analyses (APFA) was used (*24, 25*). APFA offers the possibility to obtain a quantitative elemental source profile instead of only a qualitative factor loading matrix as in traditional applications of factor analyses. In principal factor analyses a model of the variability of the trace element concentrations is constructed so that the set of intercorrelated variables is transformed into a set of independent, uncorrelated variables. The APFA procedure obtains the elemental mass contribution of each identified component by calculating the absolute principal factor scores (APFS) for each sample (*12, 13*). The elemental concentrations are subsequently regressed on the APFS to obtain the contribution of each element for each component. The measured aerosol mass concentration can also be regressed on the APFS to obtain the aerosol total mass source apportionment.

Background Aerosol Monitoring - Results and Discussion

A large number of samples were collected (150 samples in Alta Floresta, 183 samples in Cuiabá and 99 samples in Serra do Navio), allowing a detailed analysis of dry and wet season atmospheric conditions. Biomass burning season occurs mainly in August and September. Due to logistical reasons, a fraction of the fine mode filters was analyzed by PIXE. From the total number of aerosol samples, 116 fine mode filters were analyzed for elemental composition in Alta Floresta, 136 in Cuiabá, and 48 fine mode samples in Serra do Navio. The aerosol mass concentration for the fine, coarse and inhalable particulate matter, and the fine fraction (dp<2mm) elemental data obtained from the fine mode SFU filters collected at the three sampling stations will be discussed in this paper.

Figure 2 presents the time series of the inhalable particulate matter concentration for the 150 SFU collected at Alta Floresta, 183 SFU collected at Cuiabá and the 99 SFU collected in the Serra do Navio sampling site. There is very clearly a large increase in aerosol loading in the atmosphere during the biomass burning season (August and September) for the Alta Floresta and Cuiabá sampling sites. In Cuiabá, from an inhalable particulate matter (IPM) concentration of about 10-20 $\mu g/m^3$ during the wet season, the concentration goes as high as 100 to 150 $\mu g/m^3$ during the biomass burning season. Aircraft measurements over large areas of the Amazon Basin show very high IPM concentrations up to 300$\mu g/m^3$ (*16*). These high concentrations are observed in areas as large as 2 million square kilometers, using aircraft and remote sensing measurements, and they generally last for about two months, August and September. For the Alta Floresta site, concentrations higher than 600 $\mu g/m^3$ were observed. During the dry season in Alta Floresta and Cuiabá, soil dust concentrations can be relatively high, being suppressed when the heavy rains arrive, generally in October. Only in December and January, with heavy rains, aerosol concentrations dropped below 20$\mu g/m^3$. In

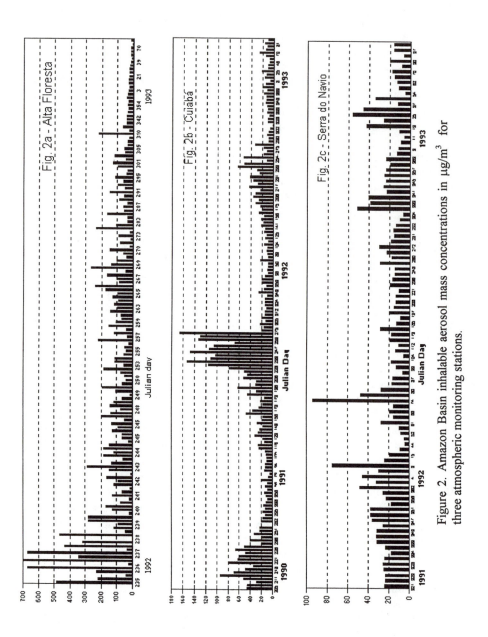

Figure 2. Amazon Basin inhalable aerosol mass concentrations in $\mu g/m^3$ for three atmospheric monitoring stations.

Serra do Navio, much lower inhalable particulate mass concentration was observed. A background value of 10-20 $\mu g/m^3$ is observed. Higher concentrations (40 to 80 $\mu g/m^3$) were observed in January and February, possibly an effect of the long range transport of Saharan dust. The geographical location in the North of the Amazon basin of the Serra do Navio sampling site makes it less influenced by regional biomass burning. Some increase during the biomass burning season can be seen in Serra do Navio time series of fine mass concentration, but not so pronounced as in Cuiabá and Alta Floresta. The coarse particle mass concentration shows some clear episodes of high concentration during January and February. During this time of the year, the position of the ITCZ (Intertropical Convergence Zone) allows the intrusion of Sahara dust into the Amazon basin, as discussed in detail in Prospero et al. (*4,26*).

The fine mode aerosol samples were analyzed by the PIXE method to obtain the elemental composition of the airborne particles. Table I presents the average elemental concentration in ng/m^3 for the fine mode aerosol at the three sampling sites. FPM represents the fine mode aerosol mass concentration in $\mu g/m^3$. It is possible to observe the higher effect of biomass burning at the Alta Floresta site, with the higher concentrations for K, S, Zn, FPM and other elements. The higher soil dust concentrations in Alta Floresta can also be deduced from the high concentration of Al, Si, Ti, and Fe. The concentration of some elements like P, Cl, Ca, Mn, and Br are very similar for the three sites. The average concentration of heavy metals, indicative of industrial contributions as Cr, Ni, Cu, Pb, is very low, generally bellow $5ng/m^3$. This fact indicates the absence of industrial or other anthropogenic emissions at the sites, with the exception of biomass burning.

The variability of the elemental concentrations at the three sampling sites was analyzed using the Absolute Principal Factor Analysis (APFA) technique. Table II presents the VARIMAX rotated factor loading matrix for the three sampling sites. For Alta Floresta samples, only two factors explain most of the data variability. The first factor clearly represents soil dust aerosol, with high loadings for Al, Si, Ca, Ti, and Fe. The second factor, representing natural biogenic aerosol mixed with biomass burning aerosol particles, has high loadings for K, FPM, S, Cl, Ca, and Zn. Unfortunately it was not possible to measure black carbon concentrations for the Alta Floresta fine fraction samples, not allowing the discrimination between the biomass burning component and the natural released biogenic particles. For the Cuiabá sampling site, three factors explain most of the data variability. The first factor has high loadings for Si, Ti, Fe, Al, Ca and Mn, representing soil dust particles. The second factor has high loadings for FPM, black carbon, Cl and K, representing biomass burning particles. The third factor with S, Zn, K, Ca and Mn represents the naturally released biogenic particles. There are several arguments that support this interpretation for the Cuiabá factor analyses results. Black carbon is essentially emitted by biomass burning, as well as chlorine and organic material responsible for the FPM concentrations (*16*). Zinc, sulfur and potassium were observed as naturally emitted by the vegetation in background areas not affected by biomass burning in the Amazon basin (*11, 13*).

Table I. Average Elemental Concentration in ng/m³ for Fine Mode Aerosol Particle (d_p<2μm) in three Remote Atmospheric Monitoring Stations in the Amazon Basin (*).

	Alta Floresta		Cuiabá		Serra do Navio	
	Average	Std. Dev. (n)	Average	Std. Dev. (n)	Average	Std. Dev. (n)
Al	398	468 (116)	91.5	98.1 (136)	170	201 (48)
Si	485	508 (116)	134	129 (109)	414	496 (48)
P	16.0	18.6 (12)	10.0	7.6 (72)	15.5	10.8 (20)
S	948	741 (116)	389	339 (136)	526	397 (48)
Cl	12.9	10.0 (70)	10.4	11.9 (136)	13.2	10.4 (48)
K	701	728 (116)	326	353 (136)	251	205 (48)
Ca	43.6	37.3 (116)	29.1	24.2 (136)	54.1	40.9 (48)
Ti	23.6	26.0 (116)	6.92	7.36 (136)	14.2	15.9 (48)
V	4.28	4.53 (86)	0.86	0.61 (25)	1.68	1.79 (17)
Cr	5.09	3.98 (67)	3.42	2.89 (66)	2.23	0.98 (17)
Mn	3.53	3.20 (46)	3.60	3.26 (136)	3.78	1.98 (48)
Fe	249	269 (116)	175	170 (136)	120	119 (48)
Ni	1.98	1.52 (12)	1.09	0.67 (8)	0.21	0.06 (2)
Cu	3.22	3.12 (53)	1.55	1.19 (69)	1.65	1.78 (25)
Zn	8.21	6.91 (116)	5.81	4.84 (136)	3.08	1.85 (48)
Br	5.86	3.27 (4)	5.61	5.86 (87)	5.26	3.14 (28)
Rb	-	-	1.32	1.02 (41)	1.07	0.54 (19)
Sr	-	-	0.70	0.59 (49)	1.06	0.56 (20)
Zr	-	-	1.22	0.92 (95)	1.40	0.59 (23)
Pb	-	-	1.68	1.35 (87)	0.82	0.21 (16)
Black	-	-	2051	1922 (136)	-	-
FPM(*)	49.9	54.8 (116)	10.5	10.7 (136)	9.87	7.47 (48)

(*) Average and standard deviation (Std.Dev.) are shown. Numbers in parentheses are the number of samples in which the concentrations of the element was above the detection limit. Only samples with values above the detection limit were used in calculating the average. Black is the black carbon concentration. FPM is the fine particle gravimetric mass concentration expressed in μg/m³.

Table II. Factor Analysis Results for the Amazon Basin Fine Aerosol Fraction: VARIMAX Rotated Factor Loading Matrices for each of the Three Sampling Locations

	Alta Floresta		Cuiabá			Serra do Navio		
	Soil	Biog./B.	Soil	Burning	Biog.	Soil	Biog.	Marine
Al	**0.94**	0.32	**0.86**	0.34	0.32	**0.99**	0.01	-0.06
Si	**0.94**	0.32	**0.93**	0.14	0.26	**0.98**	-0.05	0.02
S	0.32	**0.87**	0.40	0.37	**0.77**	-0.13	**0.97**	0.08
Cl	0.25	**0.85**	0.17	**0.87**	0.23	0.04	**0.51**	**0.79**
K	0.35	**0.92**	0.42	**0.59**	**0.64**	0.11	**0.98**	0.13
Ca	**0.74**	**0.62**	**0.68**	0.20	**0.64**	**0.91**	0.08	0.30
Ti	**0.92**	0.37	**0.91**	0.25	0.29	**0.99**	-0.08	0.02
Mn	-	-	**0.63**	0.41	**0.63**	**0.79**	0.06	-0.08
Fe	**0.94**	0.33	**0.89**	0.23	0.36	**0.99**	-0.01	0.09
Zn	0.53	**0.67**	0.54	0.42	**0.66**	0.25	**0.68**	**0.64**
Br	-	-	-	-	-	-0.23	**0.79**	**0.47**
FPM	0.35	**0.90**	0.18	**0.93**	0.22	0.07	**0.95**	0.22
Black	-	-	0.29	**0.90**	0.26	-	-	-
λ	4.7	4.4	4.9	3.6	2.8	5.5	4.2	1.4
Var.	47	44	41	30	23	46	35	12

Values in bold represents the statistically significant factor loadings. λ is the eigenvalue for each retained factor; Var. represents the percentage of the variance explained by each factor.

At the Serra do Navio sampling site, three factors were obtained. The first one with high loading for Al, Fe, Ti, Si, Ca and Mn, represents soil dust particles. Studies of long range transport of aerosol particles in the Amazon Basin have indicated the possibility that these particles are Sahara desert dust particles being injected into the Amazon Basin (*4,27*). The second factor at Serra do Navio with high loadings for K, FPM, S, Br, and Zn represents natural biogenic aerosol. The third factor with Cl, Zn and Br represents the marine aerosol particles. The Serra do Navio sampling site is about 300 km away from the ocean, and the prevailing wind direction is from the Atlantic Ocean to the sampling site. Generally, more than 90% of the element variability is explained by the factor analyses model. This indicates the adequacy of the factor model, and the number of factors retained in each analyses.

The APFA procedure allows obtaining absolute source profiles in units of ng/m^3. Figure 3 shows the absolute elemental source profiles for the biogenic plus the biomass burning aerosol component for the three sites. The elemental composition is shown normalized to the fraction of the FPM concentration apportioned to the components for each sampling site. Potassium, for example, appears at 1.4 %, 3.2 %, 2.7 % of the FPM respectively for the Alta Floresta, Cuiabá and Serra do Navio sampling sites. These values agree quite well with the presence of K in plants that is about 1.8 % (*28*). Sulfur appears enriched at the three sites (2 to 6% of FPM), due to gas-to-particle conversion of sulfur gaseous compounds. Zinc appears at 0.01, 0.05, 0.02% respectively for the Alta Floresta, Cuiabá and Serra do Navio sampling sites, whereas Zn in plants appears at an average of 0.02% (*28*), a value very close to the ones obtained in this work. The elemental profiles for the three sites agree relatively well for several elements such as P, Cl, K, Ca, Cr, Zn and Br. Black carbon at Cuiabá appears at 19% of the FPM, whereas direct measurements show a highly variable value, depending on the physical parameters of the burns, varying from 4% for low intensity smoldering fires, to 25% for flaming combustion (*29, 30*). The remaining 70 to 80% of the aerosol mass is organic carbon. Most of the sulfur concentration is of biogenic origin as there are no large power plants or industries in the Amazon basin. The absolute elemental source profile for the soil dust component is presented in Figure 4. The absolute concentration in ng/m^3 for each element apportioned to the soil dust component shows similar concentrations for Fe, Mn, Ca and Zn for the three sampling sites. Alta Floresta and Serra do Navio show similar concentrations for Al, Si, Ti and Cu.

These results shows that fine mode aerosol particles in the Amazon basin can be classified mainly in four groups: naturally released biogenic particles, biomass burning emitted particles, soil dust particles and marine aerosol particles. From a background concentration of about 10 μg/m³ for inhalable particulate matter, the mass concentration goes as high as 150 μg/m³ during the biomass burning season at the Cuiabá sampling station, and more than 600 μg/m³ in Alta Floresta. Large amounts of fine particles are injected in the atmosphere, where they can travel for long distances. The composition of these biomass burning particles is

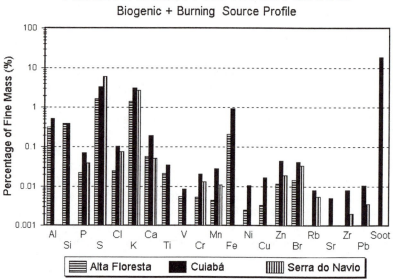

Figure 3. Biogenic and biomass burning elemental profile for the three sites.

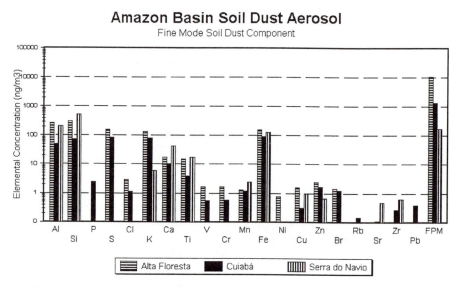

Figure 4. Source profile of the fine mode soil dust component for the three sites.

dominated by organic and black carbon, with the presence of K, Cl, S, Ca, Mn, Zn, and other elements. The natural biogenic component emitted by the Amazon tropical rainforest is rich in K, P, Zn, Mn, S, Cl, Ca and organic components. There are close similarities between the biogenic and biomass burning elemental profiles in the three sampling sites in the Amazon Basin, despite the large differences between the sampling areas and pluviometric regimes. The high concentrations measured indicate the possibility of regional and global importance of fine mode biomass burning emissions.

Aerosol Measurements Directly over Fires in the Amazon Basin

In two different vegetation formations - cerrado and primary forest, aerosol samples were collected directly over the fires. For the cerrado fires, sampling occurred in the Brasília ecological reserve of IBGE (Instituto Brasileiro de Geografia e Estatística). They were planned prescribed fires. For the primary tropical forest emissions, sampling was done in the state of Rondônia, near the village of Ariquemes. Stacked Filter Units were used for sampling, and normal PIXE and Ion Chromatography (IC) was used to measure the elemental and ionic component. Table III shows the fine mode average concentration for the measured species separated for ecosystem and fire type. In order to be able to compare concentrations with large spatial variability, the elemental and ionic concentration were normalized to the total fine mode mass concentration. Figure 4 shows the average elemental composition for each fire phase and each type of ecosystem. For some elements like phosphorus, the concentrations are very similar for all four situations, whereas for Zn the emissions factors differ significantly. Emissions of nitrates and chlorine are much higher in the cerrado ecosystem than in the tropical forest. It is difficult to know if these differences are due to changes in the biomass composition or distinct processes regulating the emissions.

The comparison of our measurements with similar data collected by other authors is shown in Table IV. The emission factor for Cl, SO_4 and Ca is normalized to potassium emissions. It is possible to observe that emissions for sulfates, reported by several authors, varies by a factor of 8. Crutzen et al. *(31)* and Bingemer et al. *(32)* have studied the sulfur cycle over the tropical forest ecosystem and found large emissions of organic sulfur compounds in both natural release and biomass burning emissions. The results from Ward and Hardy *(33)* are similar to ours. Gaudichet et al. (Gaudichet, A.; Echalar, F.; Chatenet, B.; Quisefit, J.P.; Malingre, G.; Cachier, H.; Buat Menard, P.; Artaxo, P.; Maenhaut, W. *J. Atmos. Chem.*, in press.) have also compared emissions from Africa and Amazonia, observing similar differences in the emissions as those observed in this work. Table V shows the factor analyses calculations for the elements and ionic species measured over the fires in the tropical cerrado and forest. Potassium and Chlorine are always closely associated with Zn. Magnesium and calcium are closely related for both cerrado and forest fires emissions. The organic acids (formic and acetic) are not directly associated with any other element or ion. Some factors are similar

Table III. Amazon Basin Biomass Burning Aerosol Direct Emissions: Ionic and Elemental Composition. Samples from Savanna and Tropical Forest for Different Fire Combustion. Values Expressed as Ratios to Fine Particle Mass in Percentage

	Savanna (% of fine mass)		Tropical Forest (% of fine mass)	
	Phase flaming	Phase smoldering	Phase flaming	Phase smoldering
Ac^-	0.28 ± 0.11 (56)	0.25 ± 0.11 (32)	0.23 ± 0.10 (21)	0.27 ± 0.08 (39)
Fo^-	0.024 ± 0.012 (55)	0.023 ± 0.013 (32)	0.019 ± 0.008 (13)	0.025 ± 0.014 (29)
NO_3^-	0.61 ± 0.36 (54)	0.36 ± 0.32 (31)	0.11 ± 0.08 (15)	0.12 ± 0.07 (25)
SO_4^{2-}	0.74 ± 0.36 (56)	0.35 ± 0.33 (32)	0.87 ± 0.64 (21)	0.36 ± 0.31 (39)
$C_2O_4^{2-}$	0.08 ± 0.05 (56)	0.06 ± 0.05 (30)	0.06 ± 0.06 (19)	0.04 ± 0.04 (34)
Na^+	0.022 ± 0.016 (39)	0.024 ± 0.020 (20)	0.016 ± 0.11 (17)	0.020 ± 0.037 (25)
NH_4^+	0.10 ± 0.08 (28)	0.047 ± 0.037 (11)	0.09 ± 0.06 (14)	0.046 ± 0.039 (23)
K^+	2.9 ± 2.0 (56)	1.3 ± 1.7 (32)	0.7 ± 0.6 (21)	0.43 ± 0.25 (39)
Mg^{2+}	0.035 ± 0.024 (36)	0.032 ± 0.019 (20)	0.024 ± 0.021 (18)	0.026 ± 0.028 (20)
Ca^{2+}	0.10 ± 0.09 (42)	0.09 ± 0.09 (24)	0.08 ± 0.03 (20)	0.06 ± 0.04 (37)
Al	2.3 ± 6.0 (14)	0.7 ± 1.2 (11)	0.4 ± 0.4 (8)	0.5 ± 0.5 (18)
Si	4.2 ± 1.6 (17)	2.3 ± 1.3 (10)	---	---
P	0.049 ± 0.019 (9)	0.043 ± 0.012 (16)	0.048 ± 0.011 (11)	0.033 ± 0.010 (20)
Cl	1.5 ± 1.4 (53)	0.7 ± 1.0 (27)	0.12 ± 0.14 (8)	0.08 ± 0.05 (16)
Fe	0.15 ± 0.26 (13)	0.09 ± 0.15 (12)	0.031 (1)	0.048 ± 0.027 (8)
Cu	0.006 ± 0.004 (54)	0.004 ± 0.002 (26)	0.004 ± 0.001 (20)	0.003 ± 0.002 (36)
Zn	0.019 ± 0.015 (54)	0.012 ± 0.015 (12)	0.006 ± 0.004 (20)	0.004 ± 0.002 (25)
Br	0.06 ± 0.04 (35)	0.04 ± 0.02 (8)	0.05 ± 0.02 (8)	0.05 ± 0.06 (9)
Black	13 ± 7 (56)	8 ± 7 (32)	8 ± 7 (21)	4.4 ± 2.6 (39)

For each element or ion, averages, standard deviation and number of samples in which the element or ion was measured above analytical detection limits are presented. Averages were calculated using only values above the detection limit.

Table IV. Ratios for Potassium of the Average Values for Emissions of Cl, SO_4 and Ca Observed in Direct Emissions of Biomass Burning Plumes. Samples from Savanna and Tropical Forest (Values Observed by Other Authors Are Also Shown)

Type of Vegetation	Cl	SO_4^-	Ca
Savanna (Brasília) ("flaming")[a]	0.52	0.26	0.03
Savanna (Brasília) ("smolderinng")[a]	0.54	0.27	0.07
Tropical Forest (Rondônia) ("flaming")[a]	0.17	1.24	0.11
Tropical Forest (Rondônia) ("smolderinng")[a]	0.19	0.84	0.14
Savanna (Brasília) ("flaming")[b]	0.33	0.1	0.1
Tropical Forest (Rondônia) ("flaming")[b]	0.48	0.54	0.16
Tropical Forest (Rondônia) ("smoldering")[b]	0.25	0.38	0.21
Savanna[c]	0.17	0.95	0.05
Savanna[d]	0.59	0.18	0.18

Data from: a) This work; b) Ref. 33 (BASE-B-Brasília, Marabá); c) Ref. 14 (Brushfire/80-Brasília); d) Ref. 30 (DECAFE/FOS - Ivory Coast, Africa).

Table Va. Principal Factor Analysis Results for Biomass Burning Direct Emissions in the Amazon Basin Savanna - Brasília

Var.	Fac. 1	Fac. 2	Fac. 3	Fac. 4	Fac. 5	Fac. 6	Fac. 7	Comm.
K	0.97	-	-	-	-	-	-	0.96
Cl	0.93	-	-	-	-	-	-	0.90
Zn	0.55	-0.30	0.26	-0.25	-	0.23	0.42	0.75
Cu	-	-	-	0.23	-	-	0.87	0.87
Mg	-	-	-	-	0.82	-	-	0.73
Br	-	-	0.24	-	-	0.71	-	0.65
Ca	-	-	-0.20	-	0.79	-	-	0.71
Black C	-	-	-	0.85	-	-	0.29	0.84
FPM	-0.50	0.74	-	-0.25	-	-	0.25	0.93
Ac$^-$	-	0.94	-	-	-	-	-	0.92
Fo$^-$	-	-	0.86	-	-	-	-	0.78
NO$_3^-$	0.53	-0.21	0.44	-	-0.21	-	0.24	0.65
SO$_4^-$	0.78	-	-	-	-	-	0.24	0.71
C$_2$O$_4^{2-}$	0.62	0.55	-	0.35	-	-0.25	-	0.90
Na$^+$	-0.25	-	-0.55	-0.58	-	-	-	0.78
NH$_4^+$	-	-	-	-	-	0.83	-	0.75

Table Vb. Factor Analysis Results for Biomass Burning Direct Emissions in Tropical Rainforest - Rondônia

Var.	Fac. 1	Fac. 2	Fac. 3	Fac. 4	Fac. 5	Fac. 6	Fac. 7	Comm.
K	0.83	-0.31	-	-	-	-	-	0.87
Cl	0.75	-	-	-	-0.20	0.21	-	0.72
Zn	0.79	-0.80	-	-0.25	0.20	-	0.25	0.82
Cu	-	-	-	-	-	-	0.87	0.80
Mg	-	0.27	0.80	-0.26	-	-	-0.24	0.91
Br	0.85	-	0.35	0.21	-	-	-	0.92
Ca	-	-	0.91	0.21	-	-	-	0.92
Black C	0.36	-0.56	-0.35	-	0.30	0.30	-	0.76
FPM	-	0.86	-	-	-	-	-	0.86
Ac$^-$	-	0.86	-	0.26	-	-	-	0.87
Fo$^-$	-	0.27	-	0.92	-	-	-	0.94
NO$_3^-$	-	-	-	-	-	0.70	0.49	0.76
SO$_4^-$	-	-0.69	-	-	0.58	-	-	0.89
C$_2$O$_4^{2-}$	0.21	0.21	0.24	0.30	0.70	-0.20	-	0.79
Na$^+$	-	-	-	-	-	-0.82	-	0.71
NH$_4^+$	-0.44	-	0.21	-0.47	0.58	-	-	0.81

Only eigenvalues larger than 0.20 are shown. The last column shows the communalities.

for cerrado and forest, like the FPM (fine particle mass) and acetate, indicating the high organic influence in the FPM.

Acknowledgments

We would like to acknowledge Paulo Roberto Neme do Amorim, José Roberto Chagas, Saturnino José da Silva Filho, INPE and ICOMI for support during the sampling program in the Amazon basin. We thank Alcides Camargo Ribeiro, Ana Lúcia Loureiro, Francisco Echalar, Társis Germano and Márcio Sacconi for assistance during sampling and PIXE analyses. This work was financed through grant 90/2950-2 from "FAPESP-Fundação de Amparo à Pesquisa do Estado de São Paulo", and contract number CI1*-CT92-0082 from the Commission of the European Communities.

Literature Cited

1.   Garstang, M.; Scala, J.; Greco, S.; Harriss, R.; Beck, S.; Browell, E.; Sachse, G.; Gregory, G.; Hill, G.; Simpson, J.; Tao, W.K.; Torres, A. *J. Geophys. Res.* **1988**, *93*, 1528-1550.
2.   Vitousek, P.M.; Sanford, R.L. *Ann. Rev. Ecol. Syst.* **1986**, *17*, 137-167..
3.   Salati, E.; Vose, P.B. *Science* **1984**, *225*, 129-138.
4.   Swap, R.; Garstang, M.;. Greco, S.;. Talbot, R.; Kallberg, P. *Tellus* **1992**, *44B*, 133-149.
5.   Cachier, H.; Buat-Menard, P.; Fontugne, M.; Rancher, J. *J. Atmos. Chem.* **1985**, *3*, 469-489.
6.   Crozat, G.; Domergue, J.L.; Baudet, J.; Bogui, V. *Atmos. Environ.* **1978**, *12*, 1917-1920.
7.   Lawson, D.R.; Winchester, J.W. *Science* **1979**, *205*, 1267-1269.
8.   Orsini, C.; Artaxo, P.; Tabacnicks, M. *Atmos. Environ.* **1982**, *16*, 2177-2181.
9.   Artaxo, P.; Orsini, C. In *Aerosols: Formation and Reactivity*; Israel, G., Ed.; Pergamon Journals, Great Britain, 1986; pp 148-151.
10.   Artaxo, P.; Orsini, C. *Nucl. Instrum. Meth. in Physics Res., B22*, 1987; pp 259-263.
11.   Artaxo, P.; Maenhaut, W. *Nucl. Instrum. and Meth. in Physics Res., B49;* 1990, 366-371.
12.   Artaxo, P.; Storms, H.; Bruynseels, F.; Van Grieken, R.; Maenhaut, W. *J. Geophys. Res.* **1988**, *93*, 1605-1615.
13.   Artaxo, P.; Maenhaut, W.; Storms, H.; Van Grieken, R. *J. Geophys. Res.* **1990**, *95*, *D10*, 16971-16985.
14.   Artaxo, P.; Yamasoe, M.; Martins, J.V.; Kocinas, S.; Carvalho, S.; Maenhaut, W. In: *Fire in the Environment: The Ecological, Atmospheric and Climatic Importance of Vegetation Fires;* Crutzen, P.J.; Goldammer, J-G., Eds.; Dahlem Konferenzen ES13, Chichester: John Wiley & Sons, 1993; pp. 139-158.

15.   Artaxo, P.; Rabello, F. Watt, M.L.C.; Grime, G.; Swietlicki, E. *Nucl. Instrum. and Meth. in Physics Res., B75* **1993**, pp. 521-525.

16.   Artaxo, P.; Gerab, F.; Rabello, M.L.C. *Nucl. Instrum. and Meth. in Physics Res.* , *B75* **1993**, pp. 277-281.

17.   Crutzen, P.; Andreae, M. O. *Science* **1990**, 250, 1669-1678.

18.   Setzer, A.W.; Pereira, M.C. *Ambio* **1991**, *Vol. 20*, 19-22.

19.   Levine, J., *EOS* **1990**, *vol. 71(37)*, 1075-1077.

20.   Parker, R.D.; Buzzard, G.H.; Dzubay, T.G.; Bell, J.P. *Atmos. Environ.* **1977**, *11*, 617-621.

21.   Cahill, T.A.; Eldred, R.A.; Barone, J.; Ashbaugh, L. *Report Federal Highway Administration FHW-RD-78-178*; ., Air Quality Group, University of California: Davis, USA, 1979; 78 pp.

22.   John, W.; Hering, S.; Reischl, G.; Sasaki, G. *Atmos. Environ.* **1983**, *17*, 373-382.

23.   Johansson, S.A.E.; Campbell, J.L. *PIXE - A Novel Technique for Elemental Analysis;* John Wiley & Sons: New York, 1988.

24.   Thurston, G.C.; Spengler, J.D. *Atmos. Environ.* **1985**, *19*, 9-25.

25.   Hopke, P.K. *Receptor Modeling in Environment Chemistry*; John Wiley: New York, 1985.

26.   Prospero, J.M.; Glaccum, R.A.; Nees, R.T. *Nature* **1981**, *289*, 570-572.

27.   Talbot, R.W.; Andreae, M.O.; Berresheim, H.; Artaxo, P.; Garstang, M.; Harriss, R.C.; Beecher, K.M.; Li, S.M. *J. Geophys. Res.* **1990**, *95*, 16955-16970.

28.   Bowen, H.J.M. *Environmental Chemistry of the Elements*; Academic Press, N.Y., 1979.

29.   Lobert, J.M.; Warnatz, J In *Fire in the Environment: The Ecological, Atmospheric, and Climatic Importance of Vegetation Fires;* Crutzen, P.J.; Goldammer, J.-G., Eds.; Dahlem Konferenzen, *ES13.*, Chichester: John Wiley & Sons. 1993; pp 15-37.

30.   Lacaux, J.-P; Cachier, H.; Delmas, R. In *Fire in the Environment: The Ecological, Atmospheric, and Climatic Importance of Vegetation Fires;* Crutzen, P.J.; Goldammer, J.-G., Eds.; Dahlem Konferenzen, *ES13.*, Chichester: John Wiley & Sons. 1993; pp 159-191.

31.   Crutzen, P.J.; Heidt, L.E.; Krasnec, J.P.; Pollock, W.H.; Seiler, W. *Nature* **1979,** *282*, 253-256,.

32.   Bingemer, H.G.; Andreae, M.O.; Andreae, T.W.; Artaxo, P.; Helas, G.; Jacob, D.J.; Mihalopoulos, N.; Nguyen, B.C. *J. Geophys. Res.* **1992**, *97*, *D6*, 6207-6217.

33.   Ward, D.E.; Hardy, C.C. *International* **1991**, *17*, 117-134.

RECEIVED November 28, 1994

Chapter 21

The "Garimpo" Problem in the Amazon Region

Alexandre Pessoa, Gildo Sá Albuquerque, and Maria Laura Barreto

Centro de Tecnologia Mineral, Cidade Universitária, Rua 4, Quadra D, Ilha do Fundão, CEP 21941–590 Rio de Janeiro, RJ, Brazil

This paper deals with questions relating to a controversial subject: "garimpo" in the Amazon region. Besides the general considerations on the subject, specific approaches are made on the state-of-the-art of the "garimpo", its economic and social aspects, its past and present legal concepts as well as its impact on the environment.

The number of active small-scale miners, known as "garimpeiros" in the Amazon region is not precisely known. Figures show estimates which vary from 400,000 to 1,000,000 people who work directly with "garimpo" and it has also been estimated that each "garimpeiro" is indirectly responsible for another four to five people.

Another interesting fact is the estimate of machinery and equipment involved in "garimpo" work (numbers date back to mid-1980's) 25,000 items of production equipment: dredges, ferries and concentration mills; 20 helicopters, 750 aircraft of different types and sizes, 10,000 small boats or engine-driven canoes and around 1,100 items of digging and excavation equipment.

"Garimpo" practice for concentrating gold bearing material uses mercury as one of its main inputs. Data from Brazilian studies indicates an average usage of 1.3 kg of mercury for each kg of gold produced. The estimated gold production in the "garimpos" from 1983 to 1990 is 80 tons average, so that 100 tons of mercury are annually discharged into the environment due to "garimpo" activities.

Studies confirm that "garimpo" activities, mainly in the northern and midwestern regions of Brazil, have been responsible for more than 100 tons mercury emitted per year into the environment, 55% reaching the atmosphere as vapor and the remainder into the rivers as metallic mercury (1). These numbers match the estimates of mercury consumption by the "garimpos" based on importation figures and its usage in other activities. According to this study, in 1991, 168 tons were used by the "garimpos" for a total of 337 tons of imported mercury (2).

Depending on how mercury is disposed of, it can spread throughout vast regions or remain practically unchanged close to the areas where it was discharged (3). Mercury finds its way into the environment mainly during amalgamation procedures, disposal of its tailings and amalgam pyrolysis. Another source of emission occurs in the urban centers at the gold buying agents, where the bullion containing approximately 5% of its weight in mercury, is purified.

Besides the chemical pollution, the "garimpos" represent a serious threat to the rivers, because of the great quantity of fine particulate material (clay size) produced, which prevents the plancton photosynthesis processes, as well as reduction of the oxygen content in the river waters.The silting up conditions are visible and pervasive not only in gold "garimpos", but also in the cassiterite "garimpos" (tin "garimpos") as well as those of other economically valuable heavy minerals.

Mining and Concentration Methods Used in the Various "Garimpos" in the Amazon Region

There is no doubt that gold predominates in "garimpo" activities in the Amazon region and, therefore, in this paper general aspects of gold explotation are emphasized in any discussion of the impacts of "garimpos" in the Amazon region. The types of gold "garimpo" are directly related to the types of gold deposits to be worked. The gold deposits are divided into two main types: primary and secondary.

"Garimpos" which work with primary gold deposits are typical of the State of Mato Grosso, mainly in the border regions of the Mato Grosso Wetlands (Pantanal), such as Poconé, Barão de Melgaço, Cangas and Nossa Senhora do Livramento. "Garimpeiros" of the Tapajós region have revealed the discovery of primary gold deposits in several places. The aim of this type of "garimpo" is to recover gold contained in rocks and soils that show a gold content which varies between 0.6 to 20g/ton. In this case, large amounts of gold rich soils and soft rocks - usually containing quartz veins - are excavated.

In the secondary gold deposits the work of extracting the gold-bearing material is done directly on the active river beds or on the steep banks of the "igarapés" (small waterways) and may be subdivided into: raft "garimpos" and shallow river "garimpos". The alluvium of the active river beds is the feed used by the rafts. In the shallow rivers, the steep banks are broken down with jets of water, and are excavated from soil level to an approximate depth of five to eight meters. According to the "garimpeiros", the greatest concentration of gold occurs in the gravel layer. After the ore is broken up and pulped, it is manually separated using a pitchfork, so that the pulp is drained into a pit from which it is pumped into a gravimetric concentration chute (4).

The procedures used by the "garimpeiros" in mining and concentration activities in gold "garimpos" as well as cassiterite "garimpos" (second in importance economically) and also those of other heavy minerals that occur in the Amazon region can be basically classified as follows:

Manual. Uses primitive equipment such as hoes, shovels, pick-axes, kerosene cans (to carry the material), etc. In deeper locations, diesel oil pumps are used do dry the excavation. The work is normally done by two people who share costs and profits. It is estimated that 15% of the "garimpeiros" work this way. Due to this primitive mining process, very few tailings are discharged into the rivers. The gold is recovered in small concentration boxes with crossed riffles.

Dredges. This process uses floating dredges with 10 in. or 12 in. hydraulically controlled suction pumps. There are several types of such equipment and the losses due to excess water can reach 50 to 60% in the case of fine gold, dropping considerably in cassiterite mining. Amalgamation is carried out directly on the dredge in the concentration equipment, which receives the pulp. Besides the limited space on the dredge, and the lack of control over the pulp flow, the amalgamation is done in a very rudimentary way and very often poorly. Consequently, the gold losses are considerable, although, even worse, the mercury losses are pulverized by intense shaking of the pulp and the ore.

Rafts. This system is a similar to the dredge, although the hydraulic control is substituted by an underwater operation, where a diver manually directs the suction process. This is extremely dangerous, and many people have lost their lives, hence its gradual disuse. The engine-driven pumps have a 5 in. or 6 in. diameter and the gold concentration also uses mercury, causing well-known environmental problems.

Hydraulic Disintegration. Breaking down steep banks is done using a high pressure water jet pump. In fact two jet nozzles are used - one cutting and washing material down the steep bank and the other adjusting the pulp dilution and directing it to a suction pump which, in turn, directs it to the concentration unit. This unit can be assembled on a raft or on the ground. The water and pulp pumps are powered by diesel oil engines.

In the case of cassiterite, the concentration units are three stage jigs of different types and sizes. In most cases, this equipment belongs to legally constituted mining companies which have given up work and rent it out to the "garimpeiros".

This is certainly the most widely used method in the Amazon region.

Concentration mills. This method is employed by small mining companies which recover gold from veins from underground mining. The mining uses caterpillar tractors and shovels, etc. when in an open pit, while the ore processing uses hammer mills, jigs, centrifuges, etc . In some cases, cyanide heap leaching is used.

The intense mechanization in the alluvium mining systems used in the Amazon region, does not belong to the "garimpo" and/or to small companies, but is restricted to larger operations carried out by medium-size or big companies.

The rudimentary "garimpo" uses simple equipment such as the "lontona" or the "cobra fumando" (cradle or rocker) for gold recovery. This equipment uses mercury and its loss to the environment has a great impact on nature.

Environmental Impacts

"Garimpo" activity causes strong physical and chemical environmental impacts. The physical impacts are more readily apparent (the most common are deforestation, silting-up, degradation of river banks, etc.), while the chemical impacts are often not perceived immediately, endangering the ecosystem and human beings as seriously or even more so.

The mining and concentration procedures employed in the cassiterite "garimpos" and in those of other heavy metal minerals, cause stronger physical impacts. In the case of gold, along with these physical impacts there is the pollution by mercury, which is widely used for recovering fine gold. Amalgamation is generally employed on alluvium gravimetric concentrates or on primary ores with free gold.

Gold in contact with metallic mercury forms amalgams, the main forms being: $AuHg_2$, Au_2Hg and Au_3Hg. The amalgam resulting from a gold-mercury contact is not uniform. In practice, the amalgam resulting from mining-metallurgical processes contains from 0 to 20% gold. As seen above, regardless of the type of "garimpo", the mercury emited into the environment occurs mainly when amalgamation tailings are discarded and when the amalgam is burned.

On the other hand, during amalgamation the excess mercury is filtered and reused, although it acquires characteristics that are not suitable for efficient recovery of precious metals. Therefore it is usually discarded when it starts to "harden", and becomes another source of pollution.

The main reason for mercury emission into the environment is that retorts are not used when the amalgam is burned. The "garimpeiros" burn the amalgam in open frying pans using butane/propane gas torches, at a temperature of 400 °C to 450 °C, resulting in more than 70% loss of the mercury. Hence, approximately 20-25% of the mercury emitted into the environment comes from the final deposits of amalgam tailings, and 75-80% from the open air burning process (*4*). The emission factor (ratio of mercury used per quantity of gold produced) is approximately 1:1.35 (*1*).

Besides the losses due to the processes used by the "garimpeiros", whether in the final deposit of amalgamation tailings ("resumo"), or in the volatilization of mercury when burning the amalgam, there is also a third source of loss: gold buying agents in the towns near the "garimpo" areas. During the burning processes at the site, the "bullion" produced by the "garimpeiros" still contains some impurities which are removed by the gold buying agents. The gold produced by the first burn can still have up to 5% mercury. The impurities are removed at a high temperature using an acetylene burner and purifiers (scorifying agents). As hoods are not available to retain vapor, these gold buying agents are an important source of mercury loss (*5,6*).

Some areas are already affected by the mercury discharged into the rivers. Table I shows mercury concentration in fish of selected rivers in the Brazilian Amazon.

Table I. Mercury Concentration in Fish from Rivers of the Brazilian Amazon Region

Area	Hg (μg/g, wet weight)	Ref.
Madeira river basin (Rondônia) uncontaminated	0.07 - 0.17	7
Madeira river basin (Rondônia) contaminated	0.21 - 2.70	7
Madeira river basin (Rondônia) contaminated	0.32 - 2.89	8
Teles Pires river	0.03 - 0.28	4
Tapajós river basin	0.01 - 0.68	9
Contaminated rivers (world)	1.3 - 24.8	10

Table II, gives an idea of the mercury concentraton in different gold "garimpo" areas, also outside the Amazon region, although the greatest concentration of gold "garimpo" is effectively in that region.

Table II. Mercury Concentration in Water and Sediments in Different Areas Contaminated by Gold "Garimpo" in Uncontaminated Tropical Areas and Global Averages

Area	Hg (sediments) (ppm/dry weight)	Hg (water) (ppb)	Ref.
Madeira river (Rondônia)	0.05 - 0.28	0.04 - 0.46	11
Mutum-Paraná river (Rondônia)	0.21-19.8	0.02 - 8.6	7
Paraíba do Sul river (Rio de Janeiro)	0.3 - 0.9	< 0.04 - 0.48	12
Jamari river (Rondônia) uncontaminated	<0.04	<0.03	7
Peixoto de Azevedo river (Mato Grosso) contaminated	0.27-5.24	---	4
Tanque dos Padres (Poconé, MT)	<0.05 - 0.51	<0.04 - 0.2	13
Uncontaminated rivers (global interval)	<0.3	---	14

Mercury is not the only source of pollution in the mining and processing of gold and cassiterite, when suitable environmental care is not taken. The silting up of the rivers caused by the accumulation of tailings, in the case of gold as well as in the case of cassiterite, is as serious a problem as that of mercury, and must be solved in a short period of time. It could even be said that the silting up processes cause more immediate and stronger effects on the biota. In fact, it was found that rivers with the same characteristics as the Amazonian rivers, such as the Teles Pires in Mato Grosso, for example, may become dead rivers due exclusively to the silting up of their river beds, caused by mining and concentration processes carried out without the necessary technical support.

Legal Aspects

The Brazilian Mining Law, approved in 1967, defines the profile of the "garimpeiro" as a professional who works the outcropping deposits (typically alluvium, eluvium and colluvial deposits) manually (with the help of tools). Ideally, he should be an individual professional without economic and technical resources, who would make "garimpo" mining his means of subsistence. Because of this technical and economic limitation, the damage to the mineral reserves, even in the case of ambitious mining practice (predatory mining), would be negligible (15).

The most recent Brazilian Constitution (1988), favors the "garimpeiro"- even in detriment to the constituted mining activity - according to many - and gives the Federal Government the power to establish areas and conditions for the "garimpo" activity (Art. 21, XXV and Art. 174, paragraphs 3 and 4). The aim is to encourage the "garimpeiros" to associate in cooperatives, and to do so, gives them priority for prospecting and mining the deposits that could be exploited by the "garimpo"- in areas where they are already working at in other areas that may be legally determined (16).

Up to 1988, there was no reference made whatsoever in any legal document, to the "garimpo", as a mining activity with rights and responsibilities, rather than a mining activity always subordinate to the prospecting and mining systems. The Constitution raised the "garimpo" activity to mining system "status", recognizing it as an economically profitable and socially desirable activity.

The "Regime de Permissão de Lavra Garimpeira" ("Garimpo" Mining Permit System) was instituted as a result of these constitutional provisions and can basically be defined as the system to be applied to the eluvium, alluvium, colluvial or other deposits as defined by the DNPM ("Bureau of Mines") that may be mined without the need for previous prospecting work. This law can only be applied inside well-defined areas (17,18).

The "Garimpo" Mining Permit introduced a new mining system, with rights and responsibilities, defining the difference between the Concession and the Permit systems as: the type of deposits that may be worked by the "garimpo", the individual work, and the absence of mineral prospecting studies.The cooperative was chosen as the type of organization, because in the constitutional legislator's evaluation, this would hasten the social and economic development of the "garimpeiros" and make environmental preservation feasible.

These distinctions between the systems are, in fact, strictly partial, which means, for example, eluvium, alluvium and colluvial mineral deposits can be the subject of concessions under the mining concession system. On the other hand, the "Garimpo" mining permit, although a simplified mining system, may require mineral prospecting studies. The difficulty of distinguishing between the two systems has led inevitably to persistent conflicts between the two main economic agents: mining companies and "garimpeiros".

Law No. 7805/89, which instituted the "Garimpo" Mining Permit, mainly aims to discipline "garimpo" activity. However, the concept of a simplified system was affected by the difficulty of establishing a homogeneous picture of the role of the "garimpo" activity in the mineral and even in the national scenario. The result of this situation was the regulation of dissimilar and even contradictory conceptions of the "garimpo" activity, which in practice brought about an overload of technical and bureaucratic requirements, in a attempt to regulate the Permit System, according to corporate reasoning, and ignoring that of the "garimpo".

All these incongruities and evident contradictions denote the difficulty of legally differentiating between the various mining systems. This deadlock has been adversely affecting the mining sector because of the increasing importance of the "garimpo" activity in recent years.

The priority given to the co-operative over other systems led the legislator to exclude the small and medium-size mining companies, meaning that a large part of the "garimpo" activity has evaded legal control. Such a rationale is much more business-oriented than cooperative or individual since "garimpo owners" are commonly known as "garimpo entrepreneurs". It seems necessary that the small and even medium-sized companies be recognized in the Brazilian mining scenario, not only because of the "garimpo", but principally for their own sake.

Reference to the small and medium-sized mining companies, means different rights and responsibilities from those of the so-called mining companies. This means a company with simplified legalization processes, taxed according to its size, although without losing its identity as a mining enterprise.

Equating the cooperative to the mining company is much more an enigma than a solution because, for logical and legal reasons, a cooperative is, and will always be, a cooperative and a company will always be a company. There are several types of companies, but they can never be mistaken for a cooperative.

Two points stand out in the current regulations: the priviledge of not having to carry out previous prospecting studies and doubts about the size of the area for the "garimpo".

Regarding the first point, both in the 1968 legislation and the current Law, one of the basic differences between the "garimpo"activity and mining companies was, and still is, precisely the non-existence or demand for mineral prospecting studies. This is not incidental, nor does this mean that the "garimpo" activity is being favored. The legislator's reasoning was to recognize the special nature of the "garimpo" activity due, essentially, to the type of deposit that can be mined. These are defined by law and are the alluvium, eluvium and colluvial deposits.

Regarding the second point, the size of the "garimpo" area, prevailling legislation determines that the "garimpeiros" are not allowed an area larger than 50 ha, and in spite of this restriction, this area is considered large, apparently without any plausible justification.

In short, this is a good reason, without knowing whether an area is larger than 50, 100 or 200 ha, for having large areas. "Garimpo" is an activity where prospecting studies are not required for the reasons mentioned above; therefore it does not have previously delimited mines or deposits: the limits and ore contents are uncertain and are defined as the work progresses.

It would make no sense, for example, if a cooperative requested a mining permit, which is presently a very complex process, and after one month's work has to abandon the area because the deposit is not in the requested area, or because it is not economically feasible.

Obviously, in large areas this may also occur but on a smaller scale, and as part of the risk involved in the mining activity in general. It seems that an exaggerated limitation (e.g. 50 to 100 ha) in the case of the "garimpo" and specially in the Amazon region, will make it an extremely high risk activity, making it impractical or leading to illegal practice, as currently occurs.

In the case of "garimpo", some concepts and beliefs must be clarified. One of them refers to the "garimpo" phenomenon itself. What is the reason for the existence of "garimpo" in Brazil? Generally, there is only one answer, whether from the progressive or conservative sectors: the reasons are exclusively social. The serious economic crisis in Brazil has brought to the "garimpo" a large number of unemployed people with no schooling or professional qualifications, who dedicate themselves to this activity as a last choice. Hence, if the social problem is solved, the "garimpo" problem would be solved.

Looking at the problem from another angle, there are geological reasons which motivated the appearance and increase of "garimpo"activity in Brazil. As long as these reasons persist, there will be "garimpo" activity in Brazil, regardless of the social reasons. These social reasons may aggravate the situation, but in themselves will never be determinant. This has to be proved not only empirically, but also technically and this means that if this is true, the solution is not outside the mineral sector and that the solution to the existing conflict between miners and "garimpeiros" must come from the mineral sector itself.

The attitude taken by current legislation shares the same idea: the creation of a new mining system is a clear example although, as explained above, this is still contradictory and incipient.

An aspect of the utmost importance in the solution to the "garimpo" problem is to know if is it possible to reconcile "garimpo" activity with environmental preservation? The answer to this question is crucial, since there is a progressive and inexorable movement in the direction of eliminating activities which are potentially and inevitably polluting. Certain activities can be considered as causes of greater environmental impacts than others and would be the "naturally polluting". To eliminate such impacts requires the development of technology and investment in the production processes so that these activities would become economically impractical.

In activities that are essential to mankind, the economic cost/environmental improvement ratio may be counterbalanced by subsidies, exemptions and other forms of economic and non-economic incentives. However, the tendency in the activities defined here as naturally polluting is their transformation, when possible, from a technological and economic point of view, or their elimination.

In this aspect, is "garimpo" a naturally polluting activity? The answer to this question is complex, because it involves a complete analysis of the work methods and relations, the technology used, the environmental impacts, among other relevant aspects of the "garimpo" activity in itself, meaning that the answer at this moment must come from reflections based on discussions of the matter, rather than from results of studies on it.

Politically speaking, the matter is addressed in another way, considering that the "garimpo" activity intrinsic nature could be described as disorganized, and consequently detrimental to the mineral sector (the ore would not be suitably mined), to the environment and to society. Nevertheless, in innumerable "garimpo" sites throughout Brazilian territory, including the States of Amazonas, Roraima, Pará, Goiás, Amapá, Acre, Tocantins and Mato Grosso, to mention only the most important, there are people working according to determined methods; the objectives and the social and professional status of each worker being clearly defined and structured. At a "garimpo" it is immediately apparent who is in command, and it is easy to discover which task each "garimpeiro" is responsible for. This is also the case for the methods and instruments which are used for extracting the ore, or even how and by whom a certain deposit was found, what classes of "garimpeiros" exist (much more will be revealed to those who are interested and ask properly).

It is often assumed that mining companies are characteristically organized, while the "garimpo" activity is characteristically disorganized. There are disorganized companies as well as organized "garimpos" and, of course, the opposite is also true.

If there is a disorganized characteristic in the "garimpo", this is due to the fact that the cooperative's legal nature, does not fit in with the "garimpo's" reality, nor that of the "garimpo" workers, because they are neither partners, nor individual workers, but someone else's employees. Any effort at regulating the "garimpo" must keep in mind the question of adapting the law to the "garimpo" reality. When the distortion of the work is mentioned, this refers to a "garimpo" system that has not existed since the sixties, although this is the concept of the Código de Mineração (Mining Code) and also, in part, of the recent law. This concept is perhaps mostly responsible for the current conflicts between "garimpeiros" and miners, which have led mining to become impractical in several regions of the country (*19,20*).

Disorganization is therefore not an intrinsic characteristic of the "garimpo". What are then the characteristics of the "garimpo"? What is referred to when talking about the "garimpo"? What are the differences between a mining company's activity and the "garimpo" activity? The answers to all these questions are found in the law; but are they satisfactory? These questions and answers could help understand the complex reality of the "garimpo" and the regulation of this activity.

If disorganization is not a "garimpo" attribute, and a conciliation of the "garimpo"activity with environmental preservation is possible, it remains to briefly present the environmental regulations that apply to "garimpo" activities. In the first place it should be noted that there are no substantial differences between the regulations applied to the Permit System and those applied to the other Mining Laws.

The previous Constitution (1967) on which the Mining Code was based, did not foresee environmental rules that would cover the activity of the different economic agents; hence, the Mining Code deals with this matter in a sporadic way, and only regarding one point or other (*21,22*). The eighties are particulary important for Brazilian environmental legislation. A set of rules and new concepts, like that of environmental preservation were introduced in the 1988 Constitution, as well as in subsequent common law (*23*). The 1988 Constitution puts a great emphasis on the environment and requires that States and Municipalities legislate and supervise environmental matters, and that class action may void any act harmful to the environment. The Amazon Forest, the Mata Atlântica, the Serra do Mar, the Mato Grosso wetlands (Pantanal) and the Coastal Zone were declared Protected National Properties (*24-26*).

This legislation applies to all economic activities, including mining, although some constitutional principles had been established for the mining activity (these were demands that were previously established by law). Among them are: all activities that may cause any environmental degradation must, before being established, be preceded by an environmental impact study; responsibility for recovering the degraded environment is required from the miners; and physical or corporate agents responsible for conduct and activities considered to be harmful to the environment are subject to penal and administrative sanctions, regardless of the obligation to repair the damage caused.

On one side, the 1988 Constitution defines the exclusive competence of the Federal Government to legislate on mineral deposits, mines and other natural resources, on the other, it establishes the competence of the Federal Government, of the States and of the Federal District (DC) to legislate on the preservation of nature, protection of the soil and mineral resources and protection of the environment and pollution control. Accordingly, federal control of prospecting and mining of mineral resources, must observe the federal environmental legislation and the Normas Suplementares Estaduais Específicas (Specific State Supplementary Rules) (*16*).

The "garimpeiro", as is the case of the miner, must request Environmental Licensing from the Órgão Estadual Ambiental (State Environmental Department) or from the Instituto Brasileiro do Meio Ambiente e dos Recursos Naturais Renováveis - IBAMA - (Brazilian Environment Institute) (Herrmann, H.; Fornasari Filho, N.; Loschl Filho, C., Universidade Estadual de Campinas, unpublished data).

Environmental licensing depends on an Environmental Impact Study - EIA. The Environmental Impact Report (RIMA) must reflect the conclusions of the Environmented Impact Study. The Environment Department holds a public hearing for presenting details about the project and its environmental impact as well as to discuss the RIMA.

Reducing the Problems

Adequate solutions to the many problems resulting from the "garimpo" in the Amazon region must be based on global proposals involving political, economic, social, technological and legal aspects. Before proposing measures to reduce pollution, it must be stressed that the use of mercury as part of any productive process, even by the "garimpo", should only be allowed where and when all technical resources available that guarantee that almost no loss to the environment will occur, are used. This means that the use of mercury should only be allowed in closed cyclic processes.

Amalgam burning is responsible for approximately 70% of the loss of mercury to the environment. The emission of mercury vapor is particulary harmful, because it reacts with atmospheric components (humidity, oxygen, radiation and suspended matter), oxidizes, and precipitates with rain in a highly reactive ionic form.

Clearly any measure that aims at decreasing the emission of vapor mercury by the "garimpo" must introduce burning procedures that avoid the emission of vapors. These already exist and are the well known retorts. The use of the retort, in spite of being relatively well known to the "garimpeiros", has met much resistance. One of the commonly cited reasons is based on the fact that the "garimpeiro" wants to see the moment of transformation of the amalgam and the appearance of gold, as if the act of observing it would make the quantity of gold increase. There are, however, much more practical explanations for this resistance. One would be the loss in gold through its incrustation on the retort; another would be the gold bleaching after distillation in the retort, possibly because of iron and arsenic compounds, decreasing the value of the gold obtained this way (*27*).

The "hardened" mercury could be purified through distillation in retorts with little heating, instead of being another source of mercury loss. Chemical purification should preferably use washing with strong detergents, followed by washing with water. Purification with strong acids should be avoided for technical reasons, cost, and the safety of the "garimpeiros". As for the amalgam tailings that still contain some quantity of gold besides mercury, it is important that technical support be present in the field, indicating ways of disposing of and treating these tailings.

In a "garimpo" that works primary gold deposits, it is still very common to recycle the amalgam tailings with new material. The mercury ends up being concentrated in the tailings pond and may eventually be carried away by leaching and revolving the particles, finally reaching drainage waters.

The situation is even worse in a "garimpo" that works alluvium deposits, where tailings are poured directly into the water. Tailings may contain up to 100 ppm of gold, according to tests, which justifies its further processing. The processing of this material highly contaminated by mercury could be done by cyaniding or by eletrochemical processes, such as:
- by technical staff hired by the "garimpo" organizations.
- be sold for processing to specialized companies.
- be processed in plants set up by environmental groups.

This way the "garimpo" would cease to be a dangerous source of contamination of the environment, at least regarding pollution caused by the loss of mercury. The familiarity of the technical staff with the problems of the "garimpo", as well as the mutual trust gained through the solution of the problems, would create the conditions to face the problems caused by the physical impacts, as already mentioned.

In areas already affected, mercury must be tracked and research on its possible methylation pathways be done. The recovery of the mercury dischaged into the environment must also be a constant concern and be permanently studied to make possible the use of processes that clean the affected areas.

Areas silted up by cassiterite mines or "garimpos", or by gold extraction, also deserve special attention to avoid permanent damage to the region's waterways. Recovery of silted up areas can be feasible, for example, if all the benefits which would come from the recovery of the local waterways system are contemplated.

The most important in all the "garimpo" restructuring process, adapting it to responsibility towards the environment, is the understanding that miraculous plans or military operations will not solve the problem satisfactorily.

Besides the merely technical aspects, a legal framework more suited to the nature of "garimpo" activity and its way of reasoning may contribute to better ordering of the "garimpo" activity, to prevent conflicts between "garimpeiros" and miners, as well as minimizing environmental impacts. For this it is necessary to immediately fill in the gaps in the law by enacting special laws until a new Código de Mineração (Mining Code) is approved.

The question of legal status seems to be essential for better legal classification of the "garimpo", since the cooperative did not manage to include all the "garimpo" world. In this respect, small and medium-sized mining companies could be a solution to bring the "garimpo" back to legality.

An analysis of Brazilian regulations, reveals that one of the causes of their not being efficiently applied, is the absence of technical norms, quantitative parameters and technical knowledge of each impact on the environment. An effort to classify the different impacts caused by mining seems appropriate. These technical parameters must guide the drafting of Impact Studies and Environmental Recovery Studies, considering the special features of the mining enterprises, and without which the lawmakers as well as the law enforcers will not be able to effectively evaluate environmental damage and determine the respective indemnity. The mining companies and the "garimpo" would in this way have automatic controls and safety mechanisms (28).

There is no specific legislation regulating just about any of the environmental impacts. If this process of making the law more specific does not occur, laws will continue inefficient (because of the difficulty of their enforcement) liable to the subjective interpretations of public agents whose absolute power may threaten the economic agents.

Conclusions

To forbid the "garimpo" is impossible at present, mainly because of the geological aspects of many gold deposits and because of the prevailing social and economic conditions in Brazil. The solution for by-passing the problems created by the "garimpo" must be based on making people become aware of the problems through systematic educational campaigns. Technical and institutional support, aimed at showing the "garimpeiro" that environmental preservation is not only absolutely necessary but may result in savings for himself, may be much more effective than mere police action. On the other hand the reality of the Amazon region - long distances, total absence of infrastructure, the severe climate and the forest itself - make the work of an individual "garimpeiro" impossible as described in the Mining Code and the subsquent common laws. Hence, in all the Amazon region, the "garimpo" will continue spreading through heterogenous communities, formed by groups of "garimpeiros" behaving like small mining companies, and emerging sudenly in the remote locations, whenever the news of a gold or cassiterite strike is anounced.

The problems created by the loss of mercury to the environment due to the "garimpo" activity are relatively well known to the technical and scientific community. Mercury is important in processing gold-bearing material, particulary that containing fine gold, however relatively simple procedures could minimize (or even eliminate) the problems of mercury pollution.

Literature Cited

1.   Pfeiffer, W. C.; Lacerda, L. D. *Environ. Technol. Lett.* **1988**, *.9*, 325-330.
2.   Ferreira, R. C. H.; Appel, L.E. *Fontes e Usos de Mercúrio no Brasil*; Série Estudos e Documentos n° 13; CETEM/CNPq: Rio de Janeiro, RJ, 1991.
3.   Silva, A.P. *Mercúrio em Áreas de Garimpo de Ouro*; Série Vigilância, n° 12; OPAS (Organização Panamericana de Saúde): Rio de Janeiro, RJ, 1993; pp 99-105.
4.   *Diagnóstico Preliminar dos Impactos Ambientais Gerados por Garimpos de Ouro em Alta Floresta/MT - Estudo de Caso*; Série Tecnologia Ambiental n° 2; CETEM/CNPq: Rio de Janeiro, RJ, 1992; 190 p.
5.   Marins, R. V.; Imbassay, J. A.; Pfeiffer,W. C.; Bastos, W. R. *Abstracts of Papers*, International Symposium on Environmental Studies of Tropical Forest. Manaus, AM, 1991.
6.   Malm, O. *Ph.D. Thesis*, Federal University of Rio de Janeiro, 1991.
7.   Pfeiffer, W. C.; Lacerda, L. D.; Malm, O.; Souza, C. M. M.; Silveira, E. G.; Bastos, W. R. *Sci. Total Environm.* **1989**, *87/88*, 233-240.
8.   Malm, O. In *Consequências da Garimpagem no Âmbito Social e Ambiental da Amazônia*; Belém, PA, 1992; pp 113-130.
9.   *Canga/Tajajós*. Programa de Controle Ambiental da Garimpagem no Rio Tapajós Canga - Tapajós; Governo do Pará, Relatório de Pré-Diagnóstico Ambiental, 1992; 185 p.

10.   Mitra S. In *Mercury in the Ecosystem*; Transtech Publs., Swtzerland, 1986; 327 p.
11.   Lacerda, L. D.; Pfeiffer, W. C.; Silveira, E. G.; Bastos, W. R.; Souza, C. M. M. *Anais Soc. Bras. Geoq.* **1987**, 295-9.
12.   Pfeiffer, W. C.; Mal, O.; Souza, C. M.; Bastos, W. R.; Torres, J. P. In *Conference Heavy Metals Environm., Proceedings*; vol.1, 1989.
13.   *Desenvolvimento de Tecnologia Ambiental*; Relatório Anual, CETEM/CNPq, 1989; 210 p.
14.   Salomons, W.; Forstner, U. In *Metals in the Hydrocycle*; Springer-Verlag: Berlim, 1984; 349 p.
15.   Rocha, L. L.; Lacerda, C. A. M. *Comentários ao Código de Mineração do Brasil: Revisto e Atualizado*; Editora Forense: Rio de Janeiro, RJ, 1983.
16.   Soares, O. *Comentários à Constituição Federativa do Brasil*; Editora Forense: Rio de Janeiro, RJ, 1990
17   Decreto n° 98812 de 9 de janeiro de 1990: Regulamenta a Lei n° 7805 de 18 de julho de 1989 e dá outras providências; *Diário Oficial da União*, Seção I: Brasília, DF, n° 7, 10 de janeiro 1990, pp 614-617
18.   Lei 7805, de 18 de julho de 1989: Altera o Decreto-Lei n. 227 de 28 de fevereiro de 1967; *Diário Oficial da União*, Seção I: Brasília, DF, n° 137, 20 de julho de 1989, pp 12027-12028.
19.   Salomão, E. P. *Rev. Bras. Tecnol.* **1982**, *13(2)*, pp 13-20.
20.   Rocha, G. A. (Org.) *Em Busca do Ouro: Garimpos e Garimpeiros no Brasil*; CoordenaçãoNacional dos Geólogos - CONAGE; Editora Morro Zero: Rio de Janeiro, RJ, 1984.
21.   Projeto de Lei n° 2277; do Poder Executivo, Mensagem n° 197, Brasil, 1989.
22.   Exposição de motivos n° 012/89 de 27 de abril de 1989, dos Senhores Ministros de Estado das Minas e Energia, do Interior e o Secretário-Geral da Secretaria de Assessoramento da Defesa Nacional: Brasília, DF.
23.   *Código de Mineração e Legislação Correlativa;* Ministério das Minas e Energia. Departamento Nacional da Produção Mineral: Brasília, DF, 1982.
24.   Lei n.6938 de 31 de agosto de 1981; *Diário Oficial da União*: Brasília, DF, 2 de setembro de 1981.
25.   *Legislação Básica do CONAMA*; Ministério da Habitação, Urbanismo e Meio Ambiente. Conselho Nacional do Meio Ambiente. Secretaria Especial do Meio Ambiente: Brasília, DF, 1988.
26.   *Resoluções do CONAMA: 1984/86*; Ministério do Desenvolvimento Urbano e Meio Ambiente. ConselhoNacional do Meio Ambiente. Secretaria Especial do Meio Ambiente, 2ª Ed.: Brasília, DF, 1988.
27.   *Small-Scale Gold Mining*; Priester, M.; Hentschell, T., Eds.;. Deutsches Zentrum für Entwicklungstechnologien - GKSS: Germany, 1992, pp 56-66.
28.   *Subsídios para Aperfeiçoamento da Legislação Relacionada à Mineração e Meio Ambiente: Cadastro da Legislação Ambiental*; IPT. Secretaria de Ciência e Tecnologia: São Paulo, SP, 1987.

RECEIVED October 28, 1994

Chapter 22

Mineral Extraction in the Amazon and the Environment

The Mercury Problem

Roberto C. Villas Bôas

Centro de Tecnologia Mineral, Rua 4, Quadro D, Cidade Universitária, Ilha do Fundão, CEP 21941–590 Rio de Janeiro, RJ, Brazil

Materials play a fundamental role in developing a nation. How significant such a role will be depends the overall economy, as expressed by its own domestic industrial production and innovation capacity, its consumer demands, and, last but not least, its worldwide bargaining power. In Brazil, particularly in the Amazon region, which accounts for the larger part of its territory, natural resources are an asset for providing the material bases for development. However, what are the environmental problems posed in developing mineral resources, specifically gold, and what is known of these ? This paper presents a summary of some of these problems, with emphasis on mercury pollution.

There are several the stages in a product life cycle and the way in which it interacts with the environment. From the extraction stage, the materials are processed, manufactured, utilized and wasted, being, in a greater or lesser fashion, reused, manufactured and recycled. Such an interaction with the environment is inherent to the production processes due to the very fact that they require inputs from the surroundings (energy, ores, fuel, etc.) as well as, like the Maxwell demon, their very presence interferes with the surroundings (1).

In this work attention is given to the mining stage, identified as extraction, as it relates to the Amazonian region. The main ore products extracted and commercialized in the Brazilian Amazon are iron ore, bauxite, alluvial gold, tin, kaolin and, to a lesser degree, manganese. Large kaolin project expansions are underway in the Rio Capim area (2).

The mineral potential in the Amazon region, however, seems to be much larger and has been reported in the literature (3-5). Figure 1 shows the mineral resources of the Brazilian Amazon.

0097–6156/95/0588–0295$12.00/0

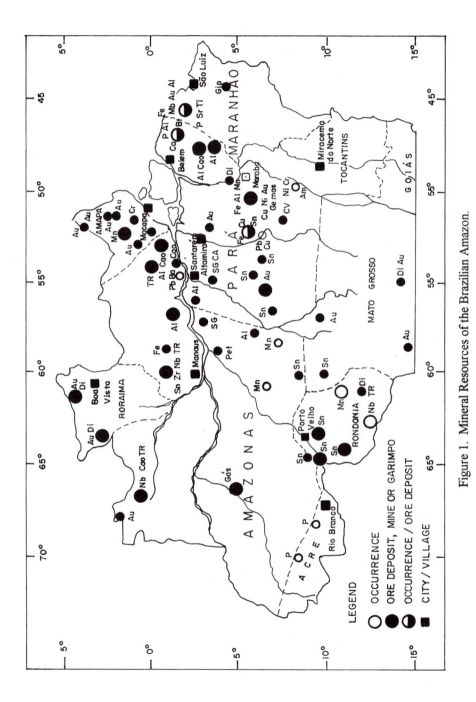

Figure 1. Mineral Resources of the Brazilian Amazon.

Environmental Impacts

A comprehensive study, under the Mining and Environmental Research Network, coordinated by the SPRU group at the University of Sussex, in England, and involving several organizations in Europe, United States, Africa, Asia and South America, dealing with the Brazilian aspects of mining/metallurgy and the environment was recently published (6). Of the interest to the Amazonic aspects of the discussion are the remarks on the Albras/Alunorte projects that resulted from an association of the Companhia Vale do Rio Doce (CVRD), a large mining concern that belongs to the Federal Government, with the Japanese, and the Alumar project of Alcoa in association with Shell/Billiton, dealing with the environmental concerns that were incorporated into the process design of the production stages. In this same study, coordinated in Brazil by the NAMA group at the University of the São Paulo, a review of the environmental aspects of tin extraction, largely represented by the Paranapanema group is given, as well as that of gold extraction in "garimpos" and mining companies. The iron ore project of CVRD, located in Carajás has been, subjected to an overall analysis of its environmental sustainability as well. Very interesting experiences were registered (7). A summary of some of the aspects linked to the production of the aforementioned commodities and the effects on the environment, is given in Table I.

Table I. Major Environmental Impacts of Selected Commodities of Interest to the Brazilian Amazon.

Al	Fe	Sn	Au
Red mud, HF, CO_2; Tar pitch; volatiles; spent pot linings; cyanide; earth movements; reclamation; dusts.	Waste water; solid wastes; heavy metals; CO_2; earth moving; reclamation; dusts.	Waste waters; heavy metals; CO_2; rivers and earth moving; reclamation; particulate matter.	Waste water; solid wastes; cyanide; mercury; particulate matter; rivers & earth moving; reclamation.

Sustainable exploitation methods are being sought, in a effort to introduce agrobusiness into the Amazonian rainforest area. However, timber productivity is low, 2-3 m^3 per year (8). The reasons for such a low productivity are the heavy leaching of nutrients from the soil and the fast decay of organic matter that follows. The soils of the Amazon are very poor in nutrients and lime is absent, showing a very low cation exchange capacity and high aluminium toxicity, so agriculture does not seem very appropriate.

Mining is a prospect for the sustainable development of the Amazon notwithstanding. Hoppe (9) has compared the total area of the Brazilian Amazon to that necessary to develop 350 mineral ore bodies and shown that they would require less than 0,1% of the total area!

Recently, Stigliani (10) advanced the concept of a "chemical time bomb", i.e., a waste deposit that does not offer any immediate danger, but which can eventually have disastrous environmental effects as geochemical conditions change. The effects of these time bombs are non-linear and delayed (e.g., toxic metals can "break through" once the specific buffering capacity of a sediment or soil system has been surpassed (1)). Consequently, these sites require "proactive assessment and management", as in the words of Förstner (11). Salomons and Lacerda applied this concept to the Amazon (12).

The "Index of the Relative Pollution Potential", as defined by Förstner and Müller (13), or "Technophility Index", as proposed by Nikiforova and Smirnova (14), may be used since the ratio of the annual mining activity to the mean concentration in the earth's crust, suggests that the highest degree of changes of the geochemical budget by man's activities, has occurred for metals such as Cd, Pb, and Hg (values in 5 x 10^7)

$$Mn = Fe < Ni < Cr < Zn < Cu = Ag < Hg = Pb < Au < Cd$$
$$1 \quad 1 \quad 2 \quad 4 \quad 10 \quad 20 \quad 20 \quad 30 \quad 30 \quad 60 \quad 140$$

This index does not imply in a constante value, changing in time reflecting the "societal demand" for that particular metal, or group of metals. In this respect, for instance, up to the end of the century, according to Nikiforova and Smirnova (14), the TP of mercury will increase 3 times, while that of lead by 4.5 times.

What About Mercury ?

In gold processing of alluvial ores in rainforest areas (15-19) mercury is widely utilized as a gold amalgam. Its utilization is not just a question of legislation, law enforcement and social pressures exerted upon the utilizers (20); it is a question of survival for the "garimpeiros"!

On the other hand, the environmental problems that may originate from the presence of mercury in the environment, being it through liquid, ionic form or vapor, are all well documented in the literature (21-23). However no alternative route to Hg amalgamation is in effect, and processes like cyanidization (a problem in itself), oil or wax agglomeration, halide extraction, etc. neither compete economically nor are suitable for the kind of operation usually employed in such tropical areas (24).

Fergusson (25) listed some of the alkyl derivatives of the heavy metals relevant to the environmental chemistry. Environmental interest in the compounds lies in their toxicity and the fact that methylation occurs for many of the heavy metal elements.

Methylation represents the transfer of a methyl group from one compound to another, the process occurring biologically or abiotically. Bacteria and fungi so far reported to methylate Hg, As, Se, Te, Pb, Cd, Tl and In, are usually aerobic (except *clostridium sp* and *methanobacterium* which are anaerobic). There is good evidence for the biomethylation of mercury, arsenic, selenium and tellurium; however there are doubts regarding that of the other heavy metals. The evidence of human poisoning led to the disclosure of environmental methylation: arsenic (nineteenth century); mercury (in the fifties: Minamata). In the sixties it was revealed that the main form of mercury in Minamata Bay fishes was CH_3Hg^+.

In 1964 the cobalt complex ion $[CH_3Co(CN)_5]^{3-}$ (model for vitamin B_{12}) was shown to methylate mercury. In 1968 Wood suggested that the methylating agent associated with methane-producing bacteria was methylcobalamin, i.e. the methyl derivative of vitamin B_{12} where the CN^- group is replaced by CH_3^- (*26*).

Methylation by non-enzymatic $MeCoB_{12}$ may be treated as abiotic, except that the reagent itself is produced biotically and may be re-methylated biotically. The two main abiotic methylation processes are transmethylation, and to a lesser extent photochemical.

Several features of the chemistry of mercury facilitate its existence in organo-species. Both Hg^{2+} and CH_3Hg^+ are soft acids and bond well to soft bases such as S^{2-} and SH^-. The cation is large and polarizable, and because of the dipositive charge, is itself a good polarizing cation and tends to form covalent bonds. The Hg-C bond (around 60-120 kJ/mole), though not that thermodynamically strong, is stronger than Hg-O bonds, therefore persisting in the environment. It is also non-polar.

Turning to the bacteria associated with methylation of mercury, they are located in the bottom sediments of rivers, estuaries and the oceans, in intestines and feces, in soils and yeast.

Several factors influence the formation of CH_3Hg^+, including temperature, mercury and bacteria concentrations, pH, type of soil, type of sediment, the sulphide concentration, and, of course, redox conditions. Seasonal variations in methylation in estuarine sediments, relate to bacterial activity, which may also apply to organic sediments, compared to sandy sediments.

Methylation of Hg occurs in both aerobic and anaerobic conditions, though the latter are the best; maximum methylation occurring in the Eh range +0,1 to -0,2V; the pH has a very significant effect as CH_3Hg^+ is more stable in neutral to acid conditions and $(CH_3)_2Hg$ in basic. The binuclear species $(CH_3Hg)_2$ OH seems to be of minor importance.

The influence of sulphide ion depends on the redox conditions: if anaerobic HgS (low solubility) will remove much of the Hg avoiding it from being methylated; if aerobic, the S^{2-} ion may be oxidized to SO_4^{2-} thus freeing Hg ion for methylation. The rate of production of CH_3Hg^+ depends on the exposed conditions and the reaction matrix, being generally greater in saline than fresh water.

Methylmercury accounts for circa 0,1 to 1,5% of the total mercury in sediments, and around 2% of the total in sea water, but in fish it accounts for over 80% of total mercury. It is not clear, however, if the CH_3Hg^+ is taken in by the fish from sea water or formed within the fish. Evidence suggests the former, though in rotting fish $(CH_3)_2Hg$ is formed. The changing chemistry of mercury, as pointed out by Renuka (27) is always a point of concern among those who study mercury.

Recently, the electroxidation method for removal of mercury was utilized to remove it from "garimpo"tailings, the chemistry of the process being:

$$2NaOCl + Hg^0 + 4HCl \rightarrow 2Na^+ + HgCl_4^{-2} + 2H_2O + Cl_2$$

the mercury being precipitated as stable tetrachloride (28).

And the Particulates ?

Besides mercury, the release of particulate material coming from earth-moving operations also contributes to the deleterious effects on the biota. Physical impacts on the environment coming from mining activities are related to the release of particulates into rivers, lakes, oceans and the air.

For the purpose of this article, particulate emissions in the region are related to dusts coming from the mining of bauxite, iron ore, manganese, and sediments coming from gold and tin extraction. These interact with the environment and may alter the biota. Little is known for the Brazilian Amazon in this regard, and not much research has been published, particularly for mercury (29). From the little that is known, mercury seems to be associated to the fines of the sediments, the "sands" being poorer in mercury than the "fines", due to the presence of iron-aluminate-hydroxides.

The partitioning of the contaminants may be determined via partition coefficients (Kd's) between the several compartments of interest, i.e. water, sediment, particulate matter and biota. As shown by Duursma (30) such partition coefficients might furnish useful empirical data in determining the percent distribution between dissolved and particulate matter, accumulation in organism, etc. A very interesting, and intriguing fact, along with the chemical-time bomb idea, is that a newly contaminated tropical estuary might be a sink for a long period, afterwards becoming a source as equilibrium is attained faster than in temperature climates, as discussed by Duursma (30).

A very extensive program, called CAMGA-TAPAJÓS, monitoring the garimpo activities at the Rio Tapajós area, is being conducted by the Government of the State of Pará (31). Table II, extracted from a report on this program, gives the relationship between the extraction and concentration techniques employed in the area and the impacts caused on the environment.

Table II. Environmental Impacts Derived from Extraction and Concentration Techniques for Gold Recovery in Rio Tapajós, as Related to Particulate Matter

Causes		Physical and/or Chemical	Biological	Anthropic
E	C	erosion/increasing suspended load		damaging fishing activities
X	O			
T	N	changes in color, turbidity and other organoleptic water properties	changes in ecological habitats	
R	C			
A	E			
C	N			increase in water treatment costs
T	T			
I	R	silting-out and changes in river courses		losses of natural resources
O	A			
N	T			
	I	water pollution (soaps and oils)	changes in ecological habitats	endemic diseases
	O			
	N			
				losses of natural resources

Literature Cited

1. Eyring,G. *Miner. Today*; U.S.B.M., Apr. 1993; pp 24-30.
2. Luz,A. B.; Damasceno,E. C. *Caulim: um Mineral Importante*; Série Tecnologia Mineral, no. 65; CETEM: Rio de Janeiro, RJ,1993; 32 pp.
3. Fernandes, F. R. C.; Portela, I. C. *Recursos Minerais da Amazônia - Alguns Dados Sobre Situação e Perspectivas*; Série Estudos e Documentos, no. 14; CETEM: Rio de Janeiro, RJ, 1991; pp 1-44.
4. Costa,M. L. *Potencial Metalogenético dos Minérios da Amazônia*; Monografia UFPA, 1991; 44 pp.
5. Amaral Jr.,A.; Capocchi, R. In: *Iron & Steel International Conference Proceedings*; A.B.M., Brazilian Metals and Material Association, S. Paulo, SP, 1990; pp 551-580.
6. Rattner, H.; Acero, L.; Rodrigues, T.; Hanai, M.; Barreto, M.L.; Coelho Neto, J.S. *Impactos Ambientais: Mineração e Metalurgia*; MERN/SPRU/NAMA/USP/CETEM, 1993; 281 pp.
7. Dall'Orto, Jr., V. C. *Abstract of Papers*, 1st International Symposium on Chemistry of the Amazon, Manaus, AM, Associação Brasileira de Química, 1993; 91 pp.
8. Hoppe, A. Annaherung, *Naturforsch. Ges* 1990, pp 1-264.

9.   Hoppe, A. *Nat. Resour. Forum* **1992,** 232-234.
10.   Stigliani,W. M. *Chemical Time Bombs: Definition, Concepts, and Examples*; E.R. 16, CTB Basic Document; IIASA: Luxembourg, Austria, 1991; pp 1-23.
11.   Förstner, U. In *Lectureship in Aquatic Sciences*; National Water Research Institute: Canada, 1992; pp 1-42.
12.   Lacerda,L. D.; Salomons,W. *Mercúrio na Amazônia: uma Bomba Relógio Química ?* Série Tecnologia Ambiental, no. 3; CETEM: Rio de Janeiro, RJ, 1992; 78 pp.
13.   Forstner, U.; Muller, G. *Heavy Metal Accumulation in River Sediments: a Response to Environmental Pollution*; Geoforum 14; 1973; pp 53-61.
14.   Nikiforova,E. M.; Smirnova, R. J. In: *Proceedings of the International Conference on Metals in the Environment*, Toronto, 1975; pp 94-96.
15.   Priester,M.; Hentscher,T. *Small-Scale Gold Mining*; GATE/GTZ; 1992; pp 1-96.
16.   Souza,V. P.; Lins,F. A. F. *Recuperação do Ouro por Amalgamação e Cianetação: Problemas Ambientais e Possíveis Alternativas;* Série Tecnologia Mineral, no. 44, CETEM: Rio de Janeiro, 1989; 27 pp.
17.   Lins, F. A. F.; Cotta, J. C.; Luz, A. B.; Veiga, M. M.; Farid, L. H.; Gonçalves, M. M.; Santos, R. L. C.; Barreto, M. L.; Portela, I.C.M.H.M. *Aspectos Diversos da Garimpagem de Ouro*; Série Tecnologia Mineral, no. 54; CETEM: Rio de Janeiro, 1992; 97 pp.
18.   Veiga, M. M.; Fernandes, F. R. C.; Farid, L. H.; Machado, J. E. B.; Silva, A. O.; Lacerda, L. D.; Silva, A. P.; Silva, E. C.; Oliveira, E. F.; Silva, G. D.; Pádua,H. B.; Pedroso,L. R. M.; Ferreira,N. L. S.; Ozaki,S. K.; Marins, R. V.; Imbassahy, J. A.; Pfeiffer, W. C.; Bastos, W. R.; Souza, V. P. *Poconé: um Campo de Estudos do Impacto Ambiental do Garimpo*; Série Tecnologia Ambiental, no. 1, CETEM: Rio de Janeiro, RJ, 1991; 113 pp.
19.   Farid, L. H.; Machado, J. E. B.; Gonzaga, M. P.; Pereira Filho, S. R.; Ferreira, A. E. F. C. N. S; Silva, G. D.; Tobar, C. R.; Câmara, V.; Hacon, S. S.; Lima,D.; Silva, V.; Pedroso, L. R. M.; Silva, E. C.; Menezes, L. A. *Diagnóstico Preliminar dos Impactos Ambientais Gerados por Garimpos de Ouro em Alta Floresta/MT: Estudo de Caso;* (versão Português/Inglês), Série Tecnologia Ambiental, no. 2, CETEM: Rio de Janeiro, RJ, 1992; 190 pp.
20.   Barreto, M. L. *Uma Abordagem Crítica da Legislação Garimpeira: 1967-1989*; Série Estudos e Documentos, no. 19, CETEM, Rio de Janeiro, RJ, 1993; 58 pp.
21.   D'Itri, F. M.. In: *Sediments: Chemistry and Toxicity of In - Place Pollutants;* Baudo, R.; Giesy, J. P.; Muntao, H., Eds.; Ann Arbor, Lewis, 1990; pp 163-214.
22.   Câmara, V. M. *Mercúrio em Areas de Garimpo de Ouro*; Série Vigilância, no. 12; OPAS/OMS: México, 1993.
23.   Pecora, W. T. *Mercury in the Environment*; Geological Survey Professional Paper 713; Washington, 1970; pp 1-67.

24. Cramer, S.W. *Problems Facing the Philippines*; Int. Min., July 1990; pp 29-31.

25. Fergusson, J. E. *The Heavy Elements: Chemistry, Environmental Impact and Health Effects*; Pergamon: Oxford, 1990; 614 pp.

26. Wood, J. M.; Rosen, C. G.; Kennedey, S. F. *Nature* **1968**, 173-174.

27. Rewuka, A. *J. Chem. Educ.* **1993**, *70*; 871-872.

28. Souza, V. P. In *Poconé: um Campo de Estudos do Impacto Ambiental do Garimpo*; Série Tecnologia Ambiental, no. 1, CETEM: Rio de Janeiro, RJ, 1991; pp 95-113.

29. Pessoa, A. *Mercúrio em Itaituba*; CETEM: Rio de Janeiro, RJ, 1994, (PT 009/94); 144pp.

30. Duursma, E. K. *The Tropical Estuaries Environmental Sinks or Sources?*; Monography, France, 1994; 20 pp.

31. *Estudos dos Impactos Ambientais Decorrentes do Extrativismo Mineral e Poluição Mercurial do Tapajós;* Monografia GEPA/SEICOM/PA, 1992; 24 pp.

RECEIVED November 10, 1994

Author Index

Affiliation Index

Subject Index

Production: Amie Jackowski
Indexing: Deborah H. Steiner
Acquisition: Barbara Pralle
Cover design: Bob Sargent

Printed and bound by Maple Press, York, PA

DATE DUE

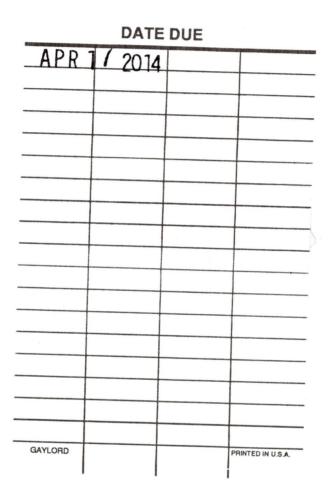

APR 1 / 2014			